Ecology and Evolution of the Grass-Endophyte Symbiosis

Ecology and Evolution of the Grass-Endophyte Symbiosis

Gregory P. Cheplick
Stanley H. Faeth

UNIVERSITY PRESS

2009

OXFORD
UNIVERSITY PRESS

Oxford University Press, Inc., publishes works that further
Oxford University's objective of excellence
in research, scholarship, and education.

Oxford   New York
Auckland   Cape Town   Dar es Salaam   Hong Kong   Karachi
Kuala Lumpur   Madrid   Melbourne   Mexico City   Nairobi
New Delhi   Shanghai   Taipei   Toronto

With offices in
Argentina   Austria   Brazil   Chile   Czech Republic   France   Greece
Guatemala   Hungary   Italy   Japan   Poland   Portugal   Singapore
South Korea   Switzerland   Thailand   Turkey   Ukraine   Vietnam

Published by Oxford University Press, Inc.
198 Madison Avenue, New York, New York 10016
www.oup.com

Oxford is a registered trademark of Oxford University Press

Library of Congress Cataloging-in-Publication Data

Cheplick, G. P. (Gregory Paul), 1957–
    Ecology and evolution of the grass-endophyte symbiosis /
    Gregory P. Cheplick and Stanley H. Faeth.
        p. cm.
    Includes bibliographical references and index.
    ISBN 978-0-19-530808-2
    1. Endophytic fungi.   2. Grasses—Ecophysiology.
    3. Plant-fungus relationships.   4. Endosymbiosis.
    I. Faeth, Stanley H., 1951–   II. Title.
    QK604.2.E53.C44 2009
    584'.91785–dc22   2008019162

9 8 7 6 5 4 3 2 1

Printed in the United States of America
on acid-free paper

Preface

Over the last several decades there has been tremendous growth in the literature and intensified research efforts on many aspects of the grass-endophyte symbiosis and its importance to the evolution and ecology of many ecosystems. While there have been edited volumes and book chapters devoted to investigations of the grass-endophyte symbiosis, especially for the agronomically important *Neotyphodium* endophytes and the forage grasses (e.g., *Festuca* and *Lolium*) they infect, this book is the first attempt to synthesize the growing literature within an ecological and evolutionary framework. We believe this is a timely publication that synthesizes the diversity of grass-endophyte research within modern ecological and evolutionary concepts.

The research examined spans the range from molecular to organismal to ecological and evolutionary. This includes investigations on the effects of endophyte infection on host growth, physiology, reproduction, abiotic and biotic stress tolerance, and competitive ability. The potential impact of endophyte infection of grasses on herbivores, multitrophic interactions, and community and ecosystem-level properties are also explored. By pulling together the diverse literature, we hope to place fungal endophytes firmly within the context of contemporary ecological and evolutionary theory. It is hoped that the reader will gain an appreciation of the importance of the symbiosis to the ecology and evolution of the interacting partners and the communities in which they occur.

The book should appeal to researchers and graduate students studying plant-microbial interactions within the broader context of physiological, population, community and ecosystem ecology. By summarizing much of the current information on the ecology and evolution of grass-endophyte systems, it should serve as a useful introduction for students and professional ecologists beginning the study of any aspect of plant-microbe symbioses which are ubiquitous in natural and agricultural ecosystems. Further, we hope researchers and students will discover through our book that fungal endophytes of grasses provide an ideal platform for testing a broad array of contemporary ecological and evolutionary concepts and theories.

We thank the many biologists who have contributed to the rapid growth in research devoted to understanding the many aspects of grass-endophyte symbiosis and those who have provided us with reprints and preprints of their latest work. In addition, A. Leuchtmann supplied comments and suggestions regarding tables 1.2, 4.1, and 4.2; D. S. Richmond kindly provided the original data used in figure 2.10; D. P. Belesky and J. M. Ruckle performed the original analysis of nonstructural carbohydrates reported in table 5.3; F. Bertoli supplied corrected data for figure 3.2; K. Morris provided the photograph for figure 3.3; and E. Chaneton kindly provided permission to reproduce figure 6.2. K. Saikkonen provided helpful advice, as always, and supplied data for figure 7.1.

G.P.C. and S.H.F. (July 2008)

Contents

Ecology and Evolution of the Grass-Endophyte Symbiosis

1

Introduction: The Grass-Endophyte Symbiosis

Symbiotic relationships between plants and fungi have a long evolutionary history. Plant fossil remains containing mycorrhizal fungi are at least 400 million years old (Brundrett 2002, Heckman et al. 2001) and fungal endophytes have been reported from a 400-million-year-old land plant (Krings et al. 2007). The widespread mutualistic association of photoautotrophs and fungi may have facilitated the evolutionary step from an aquatic to a terrestrial lifestyle (Selosse & Le Tacon 1998), and some land plant structures, tissues, and cells may be of fungal origin (Atsatt 1988, Barrow et al. 2007). Today, it is unlikely to find a plant species without any fungal associates.

The term symbiosis denotes a close association of two or more species with no implications of the ecological relationship involved (Douglas 1994, Lewis 1985, Paracer & Ahmadjian 2000). Symbiotic relationships between fungi and plants can range from highly parasitic to tightly mutualistic, with many gradations in between. Furthermore, the nature of the symbiosis is often highly dependent on environmental conditions and the genetics of the host plant and symbiont. For example, the well-known mycorrhizal symbiosis between plant roots and fungi is often described in ecology and biology texts as a classic example of mutualism (e.g., Begon et al. 2006, Solomon et al. 2006). The premise for this mutualism seems straightforward: mycorrhizal fungi benefit through the metabolic use of photosynthates supplied by the host, while the plant benefits through enhanced uptake of essential mineral nutrients, such as phosphorus, from the soil. Nevertheless, some plant species may not experience this benefit under conditions of reduced nutrient availability when photosynthate is limiting, or when infected with certain strains of mycorrhizae (Buwalda & Goh 1982, Parker 1995, Reynolds et al. 2005). In

Table 1.1 Range of possible fungal-plant interactions. Detrimental or beneficial effects are usually described based on the growth and/or reproductive capacity of host plants. The relative interaction intensity (RII) index is defined in section 1.3 (see also Armas et al. 2004).

Relationship	Symbol (fungus, plant)	Defining features	RII	Example
Parasitism	(+, −)	Beneficial to fungus, detrimental to host	−	Plant pathogens, *Epichloë* endophytes
Commensalism	(+, 0)	Beneficial to fungus, no effect on host	0	Some endophytes
Mutualism	(+, +)	Beneficial to fungus, beneficial to host	+	Mycorrhizae, *Neotyphodium* endophytes

situations where the net costs of the symbiosis exceed the net benefits to the host, mycorrhizal fungi can be considered parasitic on plants (Johnson et al. 1997). Furthermore, the effect on their host plants of some fungal symbionts that are generally considered pathogens, such as the rust *Coleosporium ipom-oeae*, can range from parasitism to commensalism, depending on microenvironmental conditions (Kniskern & Rausher 2006).

The categories of relationships for plant-fungal symbioses used in this book are a subset of the range of interspecific interactions studied by ecologists and follows Paracer and Ahmadjian (2000) (table 1.1). Note that all of the symbioses have in common a positive component for the fungal associate because the endophyte does not exist independently of its host. In contrast, the effect of the fungus on its plant host can be negative in a parasitic or pathogenic association, positive in a mutualistic association, or neutral in a commensalistic association (table 1.1). Thus the relationship between the two symbiotic partners is inherently asymmetrical (Bronstein 1994a, Law 1985). As will be seen in examples throughout this text, the grass-endophyte symbiosis spans this entire range. A general theme will emerge that much of the variation in the nature of the interaction is due to environmental and genetic contingency.

1.1 ENDOPHYTIC FUNGI AND GRASSES

Endophytes are any microorganism, typically bacteria or fungi that live within a plant (Clay & Schardl 2002, Stone et al. 2000, Wilson 1995). Often they are clandestine, unnoticed by the casual observer because infection does not always produce overt symptoms of plant disease (Wilson 1995). Endophytic fungi can be found in the stems and leaves of most, if not all, herbaceous and woody angiosperms, including grasses and trees (Arnold et al. 2000, Carroll 1988, Petrini et al. 1992). Their diversity is quite remarkable (Arnold & Lutzoni

2007). More than 50 fungal taxa were isolated from four cultivars of wheat (*Triticum aestivum*), (Crous et al. 1995) and 27 fungal taxa were isolated from three stands of sugar maple (*Acer saccharum*) (Vujanovic & Brisson 2002). In another study of the species richness of leaf-inhabiting fungi in four temperate forest tree species, 49 fungal taxa were identified, and with one exception, all were filamentous ascomycetes (Unterseher et al. 2007). An extensive and diverse endophytic community has also been documented from the perennial grass *Dactylis glomerata* sampled from 10 areas in Spain (Márquez et al. 2007). Foliar endophytes are clearly very common in many types of plants worldwide (Arnold & Lutzoni 2007).

This book will focus on grasses and their endophytic fungi, classified in the family Clavicipitaceae (phylum Ascomycota). Due to the clandestine nature of fungal endophytes, compilations of the numbers of grass species known to be infected are undoubtedly underestimates, as acknowledged by Clay (1989) for his list of 80 genera and 259 species of graminoids that can be hosts to clavicipitaceous endophytes. Leuchtmann (1992) stated that 290 grass species were reported to be endophyte infected and suggested 20–30% of grass species worldwide may be infected. Of 16 species of woodland grasses sampled for fungal endophytes, five previously unreported host species were discovered (Clay & Leuchtmann 1989). More recently, three more grass species were described as being previously unknown endophyte hosts following a survey in western Spain (Zabalgogeazcoa et al. 2003). Other *Neotyphodium* species have also been identified in endemic grasses from the Southern Hemisphere (Cabral et al. 1999, Moon et al. 2002) and from a native rangeland in Iran (Mirlohi et al. 2006). Undoubtedly, many more grass-endophyte symbioses await discovery. For example, in a recent survey of 41 grass species native to northern China, 25 were found to be infected by endophytes (Wei et al. 2006). Furthermore, two genera (*Cleistogenes* and *Koelaria*) as well as 20 grass species were not previously known to harbor fungal endophytes.

The fungal endophytes of grasses (family Clavicipitaceae, tribe Balansiae) are traditionally grouped into four genera: *Atkinsonella*, *Balansia*, *Epichloë*, and *Myriogenospora* (White 1997). Most of the 37 species in these genera can reproduce sexually via production of stromata and ascospores. Two other described genera have also been placed into the Balansiae: *Nigrocornus* and *Parepichloë* (table 1.2). An additional group of entirely asexual forms comprise the genus *Neotyphodium* (14 spp.) which apparently evolved from sexual *Epichloë* species (Schardl & Leuchtmann 2005, Wilkinson & Schardl 1997). Some of the grass genera that can be hosts to these endophytic genera are presented in table 1.2. As will become apparent in this book, much research effort has been devoted to the widespread genera *Epichloë* and *Neotyphodium* and the grasses they infect (Bacon & Hill 1997, Roberts et al. 2005, Schardl et al. 2004). This is partly because two genera (*Festuca* and *Lolium*) that include globally important forage crops (and turfgrasses) are commonly infected by *Neotyphodium* spp. Indeed, entire symposium volumes have been published devoted exclusively to *Neotyphodium*-grass symbioses (Bacon & Hill 1997, Roberts et al. 2005).

Table 1.2 The fungal genera (family Clavicipitaceae, tribe Balansiae) that are endophytes of grasses, the approximate number of fungal species, and examples of grass genera that can be infected.

Fungal genus	No. of fungal species	Examples of host grass genera
Atkinsonella	2	*Danthonia, Stipa*
Balansia	23	*Andropogon, Cenchrus, Panicum, Paspalum, Sporobolus, Tridens*
Epichloë	10	*Agrostis, Brachypodium, Brachyelytrum, Bromus, Dactylis, Elymus, Glyceria, Poa*
Myriogenospora	2	*Andropogon, Panicum, Paspalum*
Neotyphodium	14	*Festuca, Lolium, Triticum, Achnatherum, Bromus, Melica, Echinopogon, Poa*
Nigrocornus	1	*Bothryochloa, Cynodon, Sorghum*
Parepichloë	6	*Andropogon, Eragrostis, Sasa*

The related, well-known genus *Claviceps* (family Clavicipitaceae, tribe Clavicipitaceae) is parasitic on grass inflorescences, causing ergot disease. However, *Claviceps* spp. are not endophytic, although some alkaloids produced by them are also made by endophytic fungi (Bush et al. 1997). It should be noted that some of the Balansiae (e.g., *Atkinsonella* and *Balansia*) are not necessarily completely endophytic and can show epiphytic growth around meristems and developing inflorescences (Leuchtmann & Clay 1988a). Also, some *Neotyphodium* endophytes can show epiphyllous stages in which the hyphae grow onto leaf surfaces (Tadych & White 2007). In *Epichloë*, as fungal hyphae envelop the developing inflorescence, "choke disease" results (Bradshaw 1959) and the host plant can be rendered sterile. Endophyte effects on host reproductive systems will be considered further in chapter 2.

Endophytes are obligate biotrophs (Isaac 1992), obtaining nutrients from a living host plant. Their coiled hyphae are completely intercellular (figure 1.1) and cell penetration does not occur. Hyphae are typically concentrated in the leaf sheath and stems. Leaf and endophyte growth are synchronized (Christensen et al. 2002, Christensen & Voisey 2007, Tan et al. 2001) and the endophyte presumably obtains photosynthates while growing between its host's cells. Endophytic hyphae have also been shown to colonize vascular bundles in some grasses (Christensen et al. 1997, 2001). Asexual endophytes (*Neotyphodium*) grow into seeds matured by the infected host (White et al. 1993), thus ensuring vertical transmission of the symbiosis to the next generation of plants. For endophytes with a sexual cycle (*Epichloë*), wind or insects can act as vectors for horizontal transmission of spores (Bultman et al. 1995, Rao & Baumann 2004) (figure 1.2). Mature ascostromata of some endophytes such as *Epichloë* or *Atkinsonella* develop on the aborted host inflorescence, causing choke disease (figure 1.3). The stromata of other endophytes may emerge from host leaf surfaces (*Myriogenospora* and *Balansia*) (figure 1.4).

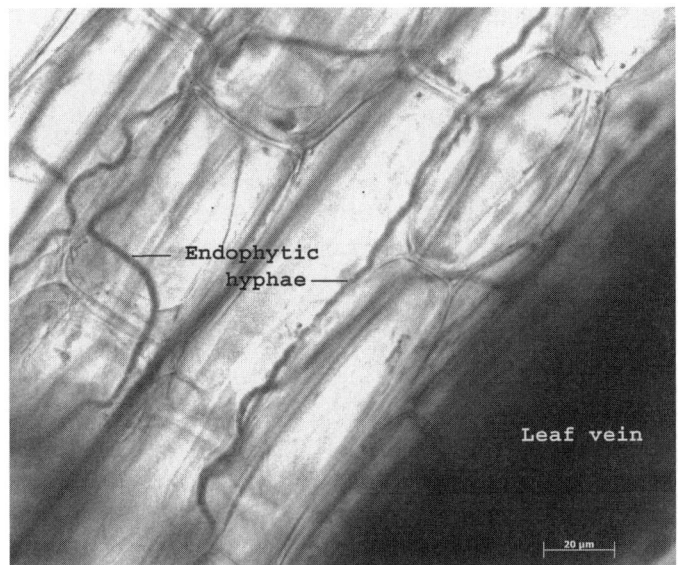

Figure 1.1 Endophytic hyphae of *Neotyphodium lolii* in the leaf sheath of perennial ryegrass (*Lolium perenne*) at 400×. Courtesy of W. L'Amoreaux, Advanced Imaging Facility, College of Staten Island.

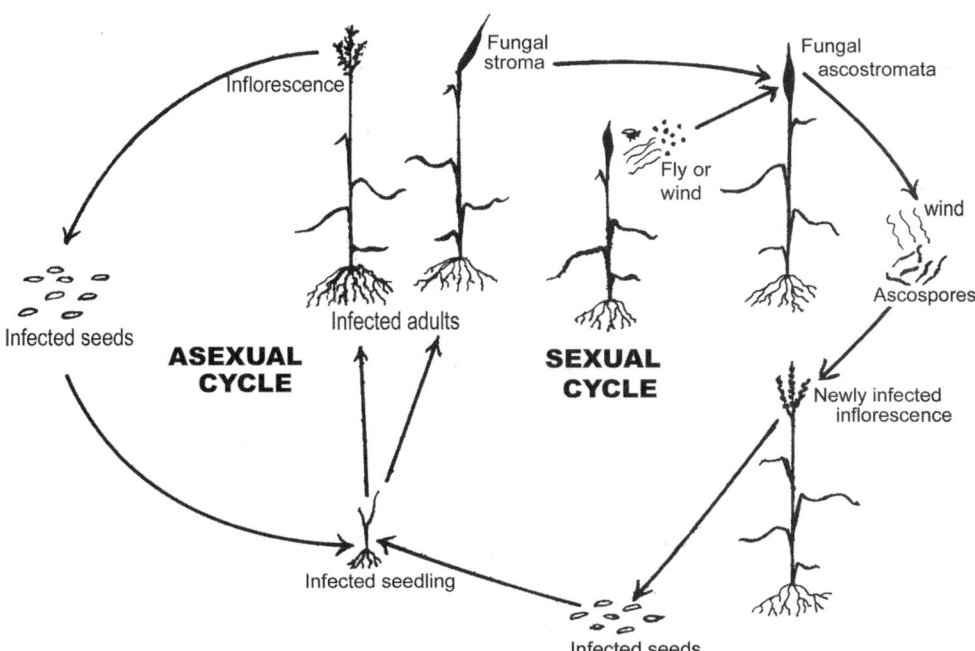

Figure 1.2 Asexual and sexual cycles possible in the grass-endophyte symbiosis. Some endophytes (e.g., *Neotyphodium*) are exclusively asexual and only vertically transmitted within host seeds. Other endophytes (e.g., *Epichloë*) may show asexual and sexual stages. Adapted from Schardl et al. (1997).

Figure 1.3 Ascostromata of *Epichloë elymi* (arrows) enveloping the aborted inflorescences of its host grass *Elymus virginicus* growing in Missouri. Photo courtesy of Thomas L. Bultman.

1.2 ECOLOGICAL, EVOLUTIONARY, AND ECONOMIC SIGNIFICANCE OF ENDOPHYTES

Due to the many ecological effects, and possible evolutionary impact, of endophytes on individual hosts, populations, and associated communities, there have been numerous reviews and several edited volumes published over the last few decades (table 1.3). Effects of endophytes on host growth, reproduction,

Figure 1.4 Dark ascostromata (arrow) of *Balansia epichloë* on leaves of *Tridens flavus* (purpletop grass) growing near Bloomington, Indiana. Photo by Keith Clay.

Table 1.3 A sampling of the many published review papers and edited volumes (italicized) covering various aspects of the grass-endophyte symbiosis from various perspectives. See reference list for complete citations.

Topic	Reference	Title
General	Siegel et al. (1987)	Fungal endophytes of grasses
	Clay (1988b)	Fungal endophytes of grasses: a defensive mutualism between plants and fungi
	Clay (1990b)	Fungal endophytes of grasses
	Redlin & Carris (1996)	*Endophytic Fungi in Grasses and Woody Plants*
	Clay (1993)	The ecology and evolution of endophytes
	Saikkonen et al. (1998)	Fungal endophytes: a continuum of interactions with host plants
	Clay & Schardl (2002)	Evolutionary origins and ecological consequences of endophyte symbiosis with grasses
	Saikkonen et al. (2004)	Evolution of endophyte-plant symbioses
	Schardl et al. (2004)	Symbiosis of grasses with seed-borne fungal endophytes
	Müller & Krauss (2005)	Symbiosis between grasses and asexual endophytes
	Popay & Thom (2007)	*Proceedings of the 6th International Symposium on Fungal Endophytes of Grasses*
Epichloë endophytes	Schardl (1996)	*Epichloë* species: fungal symbionts of grasses
	Scott (2001)	*Epichloë* endophytes: fungal symbionts of grasses
	Schardl & Leuchtmann (2005)	The *Epichloë* endophytes of grasses and the symbiotic continuum
Neotyphodium (= *Acremonium*) endophytes	White et al. (1993)	Taxonomy, life cycle, reproduction and detection of *Acremonium* endophytes
	Bacon & Hill (1997)	Neotyphodium/*Grass Interactions*
	Roberts et al. (2005)	Neotyphodium *in Cool-Season Grasses*
Host population dynamics	Clay (1998)	Fungal endophyte infection and the population dynamics of grasses
Host stress tolerance	Bacon (1993)	Abiotic stress tolerances (moisture, nutrients) and photosynthesis in endophyte-infected tall fescue
	Malinowski & Belesky (2000)	Adaptations of endophyte-infected cool-season grasses to environmental stresses: mechanisms of drought and mineral stress tolerance
	Malinowski et al. (2005)	Abiotic stresses in endophytic grasses
Herbivory	Clay (1987a)	The effect of fungi on the interaction between host plants and their herbivores
	Clay (1989)	Clavicipitaceous endophytes of grasses: their potential as biocontrol agents
	Latch (1993)	Physiological interactions of endophytic fungi and their hosts. Biotic stress tolerance imparted to grasses by endophytes

(continued)

Table 1.3 *Continued*

Topic	Reference	Title
	Breen (1994)	*Acremonium* endophyte interactions with enhanced plant resistance to insects
	Faeth (2002)	Are endophytic fungi defensive plant mutualists?
	Faeth & Bultman (2002)	Endophytic fungi and interactions among host plants, herbivores, and natural enemies
Communities and ecosystems	Hammon & Faeth (1992)	Ecology of plant-herbivore communities: a fungal component?
	Clay (1994b)	The potential role of endophytes in ecosystems
	Clay (2001)	Symbiosis and the regulation of communities
	Omacini et al. (2005)	A hierarchical framework for understanding the ecosystem consequences of endophyte-grass symbioses
	Rudgers & Clay (2005)	Fungal endophytes in terrestrial communities and ecosystems
	Malinowski & Belesky (2006)	Ecological importance of *Neotyphodium* spp. grass endophytes in agroecosystems

stress tolerance, and competitive ability are explored in chapter 2. There is now considerable evidence that endophytes can sometimes improve host vigor, and both sexual and asexual reproduction, especially in agronomically important grasses such as tall fescue (*Lolium arundinaceum*, formerly *Festuca arundinacea*, and now also reclassified into a genus closely allied with *Lolium*, designated *Schedonorus arundinaceus*) (Darbyshire 2007). Some of these same agronomic grasses exhibit an improved ability to tolerate abiotic stresses such as drought or low soil fertility when endophyte infected (Malinowski et al. 2005, West 1994). However, this may not necessarily occur in native grasses (Faeth & Sullivan 2003). Although the evidence is far from conclusive, under both intraspecific and interspecific competition, infected plants sometimes show better growth relative to their uninfected counterparts. However, for wild grasses, endophyte infection does not necessarily improve competitive ability and may instead decrease it (Brem & Leuchtmann 2002, Faeth et al. 2004).

It must be kept in mind that much of the evidence for the beneficial effects of endophyte infection comes from studies of just a few of the many grass species known to be potential hosts (Saikkonen et al. 2006). The two agronomic species investigated the most, *Lolium arundinaceum* (tall fescue) and *Lolium perenne* (perennial ryegrass), are not native to the places where they have been studied (e.g., New Zealand and the United States). Much additional work on native grasses in natural ecosystems remains to be done before realistic generalizations on the ecological impact of endophytes can be offered.

Another factor that can confound the interpretation of the ecological role of endophytes in grass populations is the dependence of many endophyte-mediated effects on host and endophyte genotype. Because in many studies the potentially confounding effects of different endophyte genotypes inhabiting different host genotypes are difficult to control, it is really particular host-endophyte genetic combinations that are being investigated. Thus reported interactions between host grass genotype and *Neotyphodium* endophyte infection (e.g., Cheplick & Cho 2003) for various plant growth traits in relation to environmental conditions actually reflect the response of a particular coevolved association of a vertically transmitted fungal genotype along with its host's genotype. The endophyte genotype can be important, and in one grass species infected by *Epichloë sylvatica*, the fungal genotype determines stroma formation and thus the extent to which the endophyte-grass symbiosis is pathogenic (Meijer & Leuchtmann 2001). Likewise, endophyte haplotype can dominate host physiological and growth responses in another grass species, Arizona fescue (Morse et al. 2007), and endophyte genotype likely determines alkaloid types and levels produced in infected plants (e.g., Faeth et al. 2002a, 2006). Evolutionary responses of host grasses to selection pressures in the natural environment may depend not only on plant genotype, but also on whether or not the genotype is infected. In addition, future microevolution of a population of infected and uninfected individuals should depend on specific host-endophyte combinations and how they respond together to abiotic and biotic factors (figure 1.5). Many additional aspects of the genotypic specificity of grass-endophyte interactions, including implications for evolutionary ecology and grass-endophyte coevolution, are considered in chapters 4 and 5.

The potential ramifications of endophyte-infected grass populations within ecological communities are likely to be very complex due to the many interactions that can occur among host plants and both primary and secondary consumers. Endophytic fungi may increase host defenses against invertebrate and vertebrate herbivores (Clay 1988b, Faeth & Bultman 2002, Koh & Hik 2007) due to their production of various alkaloids with potent biological effects (Bush et al. 1997, Powell & Petroski 1992, Schardl et al. 2007). The enhanced resistance of endophyte-infected grasses to invertebrate herbivores is one of the most thoroughly documented effects of endophytes and has been reviewed elsewhere (table 1.3). Again, much of the work has centered on *Neotyphodium* endophytes and the agronomically important *Festuca* and *Lolium* spp. they infect (Breen 1994, Latch 1993). However, the "defensive mutualism" concept (Clay 1988b) may not be broadly applicable to native grasses in natural settings where the frequency of infection typically shows wide variation (Faeth 2002, Saikkonen et al. 1999). The larger ecological and evolutionary context of endophyte-grass symbioses and associated herbivores is examined in chapter 3.

Community-wide effects on endophytes may be extended beyond herbivores to their associated parasites and predators in other trophic levels (Clay 2001, Faeth & Bultman 2002), as described further in chapter 6. For example, *Neotyphodium* endophytes of *Lolium multiflorum* lower plant quality, reduce the densities of aphid herbivores, and indirectly reduce the rate of parasitism on aphids by parasitoids (Omacini et al. 2001). The potential for alteration of the interactions of grasses with herbivores and their natural enemies will

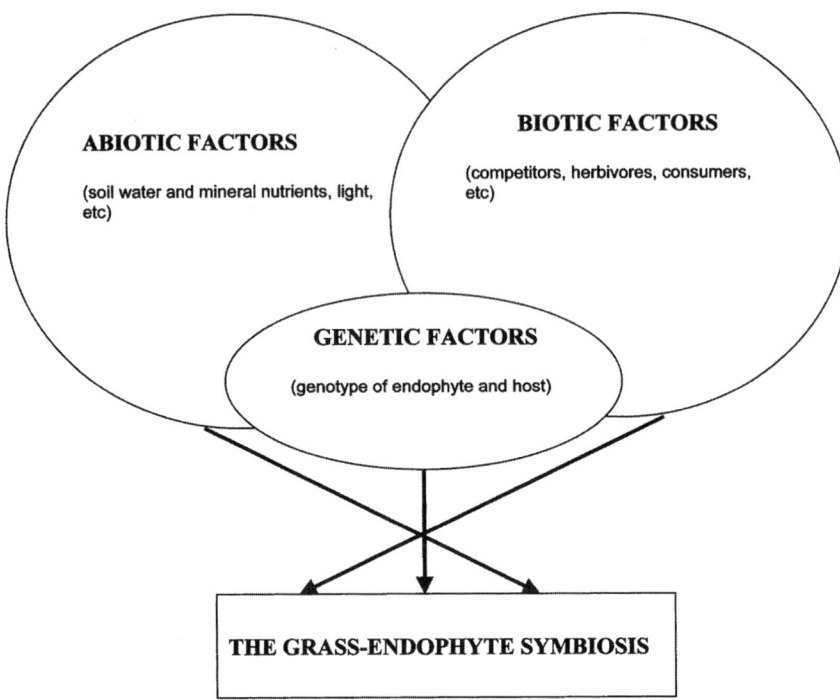

Figure 1.5 Schematic diagram depicting the interacting effects of abiotic, biotic, and genetic factors on the ecology and evolution of the grass-endophyte symbiosis. Based on Müller and Krauss (2005).

clearly depend on many factors, including the frequency of infected hosts within the community, the types and relative levels of bioprotective alkaloids, and the spatial distribution of grasses and herbivores. However, little is currently known about the impact of endophytes on food web dynamics in any natural ecosystem.

There is some evidence that endophyte-infected grasses can change the structure and composition of the associated plant community as it develops over time. Clay and Holah (1999) showed how infected *Lolium arundinaceum* (tall fescue) in successional fields increased in dominance over a 4-year period, resulting in a decline in plant species richness compared to plots with uninfected tall fescue. The mechanism for this effect probably involves improved competitive ability of infected plants coupled with reduced herbivory. In a later study (Clay et al. 2005), the frequency of endophyte infection in tall fescue was assessed over 4.5 years as the proportion of tillers with the endophyte. There was a significantly greater increase in endophyte frequency in the presence of herbivores compared to plots where herbivory was experimentally reduced. Additional evidence suggests that *Neotyphodium* endophytes with beneficial effects on their host grasses could make infected grass species such as tall fescue (Rudgers et al. 2005) and annual ryegrass (*L. multiflorum*) (Uchitel, A., Omacini, M., & Chaneton, E.J., unpublished data) better able to invade diverse communities. The possibility that other grass-endophyte symbioses in other systems act on plant communities in a similar manner has not been explored.

Finally, there is the potential for the grass-endophyte symbiosis to impact more than the associated animal and plant communities. The ecosystem-level consequences of endophytes have begun to be considered by a few researchers (Clay 1994b, Omacini et al. 2005, Rudgers et al. 2004). These consequences include carbon flow into soil food webs, such as the endophyte effects on litter decomposition and soil microbial activity (Franzluebbers et al. 1999, Lemons et al. 2005, Omacini et al. 2004), and changes in soil nutrient availability and secondary compounds that may feedback to the plant community (Matthews & Clay 2001). Long-term manipulative experiments will be necessary to better reveal whether or not fungal endophytes function as "keystone symbionts" (Omacini et al. 2005) in natural or managed ecosystems dominated by perennial grasses.

1.3 MUTUALISM, PARASITISM, OR COMMENSALISM?

Like many ecological interactions, the nature of the grass-endophyte relationship is variable and often dependent on the species involved and the environmental circumstances in which they are found. Because so many early studies focused on tall fescue (and a few other agronomically important species) that tended to show growth or reproductive benefits when infected by *Neotyphodium* endophytes, early researchers began to maintain that the grass-endophyte symbiosis was predominantly a mutualism. However, as noted in the previous section, there is presently enough additional information on other grass-endophyte associations, including some native grasses (e.g., Faeth & Sullivan 2003, Saikkonen et al. 2006), that most workers today would not consider these symbiotic interactions to necessarily be mutualistic. For the globally important forage and turfgrass *Lolium perenne* (perennial ryegrass), endophyte effects on host growth are notoriously inconsistent, ranging from parasitic to mutualistic and showing pronounced contingency on environmental conditions and host genetics (Cheplick 2004a, Cheplick et al. 1989, 2000, Hesse et al. 2004, Lewis 2004, Ravel et al. 1997a). Also, because *Neotyphodium* endophytes disproportionately infect perennial grass species, many of which likely have decade-long or greater life spans (Faeth & Hamilton 2006), assessing lifetime costs and benefits is challenging. Many more examples of the variable effects of endophytes on their hosts (and their herbivores) will be provided in the following chapters.

It is likely that for many grass-endophyte symbioses there are both benefits and costs to endophyte infection of the host. Whenever host benefits such as improved growth or reproductive fitness outweigh the potential costs of supplying photosynthetic products to the endophyte, the relationship is mutualistic, at least under the conditions and time frames examined. In contrast, if costs outweigh the potential benefits, perhaps under environmentally stressful conditions (e.g., Ahlholm et al. 2002, Cheplick 2007), then the relationship is parasitic.

One useful way to quantify any kind of ecological interaction in plants from competitive to symbiotic is to calculate an index of relative interaction intensity (RII) based on some measure of the performance of individuals growing with or without their interacting partner (Armas et al. 2004). If biomass is the response variable of interest, for grasses known to harbor endophytic fungi, the index is simply

$$\text{RII} = (B_W - B_{WO})/(B_W + B_{WO}),$$

where B_W is the mean biomass of the host with the endophyte and B_{WO} is the mean biomass of the host without the endophyte. Thus from the perspective of the host grass, the RII for a grass-endophyte symbiosis can range from highly parasitic (−1) to completely and obligatorily mutualistic (+1). A neutral value at or near zero, which means no effect of endophyte presence on host biomass, would be indicative of commensalism. This range of RII values nicely spans the range of possible fungal-plant interactions depicted in table 1.1.

It should be recognized that the RII index was developed for plants (Armas et al. 2004) and therefore only expresses the nature of a symbiotic association from the perspective of the host. For the endophytic fungus, hyphal concentration with host leaves, success at infecting viable seeds, and stromata production (in sexual species) will influence the extent of ecological and evolutionary success of the grass-endophyte relationship.

When one is working with infected and uninfected replicates of the same host genotypes (see section 4.1), RII can be calculated separately for each genotype (Cheplick 2007). To illustrate the utility of this index, RIIs were computed using the dry mass data for 10 genotypes of L. perenne cv. Yorktown III in a greenhouse drought experiment (Cheplick 2004a). For comparison, RIIs were also calculated for the same genotypes following 3 years of growth in a field plot in central New Jersey (Cheplick 2008). In the greenhouse experiment, total dry mass (shoot plus root) was obtained 7 weeks after three sequential drought periods (details in Cheplick 2004a). With few exceptions, most genotypes showed negative RII values (figure. 1.6), indicative of a detrimental effect of endophyte infection. However, the comparative performance of infected and uninfected replicates of these same genotypes in the 3-year field trial was markedly different: RII values were near zero or strongly positive (figure 1.6). Clearly for this ryegrass cultivar, endophyte effects on host growth were highly dependent on host genotype and environmental conditions.

For comparison, RII was also calculated from dry mass data for 13 endophyte-infected genotypes of L. perenne originally collected from native grassland habitats in Saxony-Anhalt, Germany, and grown in a field plot for 2 years in Halle (Hesse et al. 2004). In this trial, most RII values were negative (figure 1.6). It would appear that in this widespread, globally important forage crop the grass-endophyte symbiosis spans the range from parasitic to commensalistic to mutualistic. It remains to be determined whether or not this situation is typical for other endophyte-infected grasses.

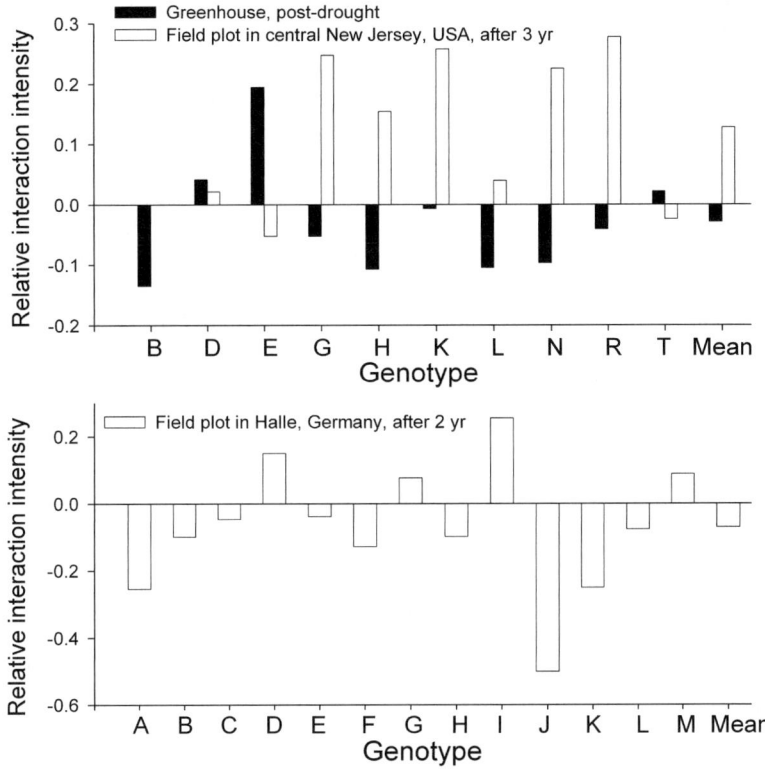

Figure 1.6 Relative interaction intensities (Armas et al. 2004) for the perennial ryegrass-endophyte symbiosis (see section 1.3). Top panel: Data were total dry mass of infected and uninfected replicates of 10 genotypes of *Lolium perenne* (cv. Yorktown III) exposed to severe drought in a greenhouse study and then allowed to recover for 7 weeks (Cheplick 2004a). Aboveground mass for nine of the same genotypes after three growing seasons in an outdoor plot in central New Jersey were used to calculate RIIs for comparative purposes (Cheplick 2008) (see section 5.1.2). Lower panel: RII based on dry mass data for 13 *L. perenne* genotypes from native grassland habitats in Saxony-Anhalt, Germany, and grown for 2 years in a plot in Halle (Hesse et al. 2004).

1.4 ASSESSMENT OF ENDOPHYTE INFECTIONS IN HOST POPULATIONS

Within grass species known to be capable of harboring fungal endophytes, the frequency of infection can vary among populations, individuals, and tillers within individuals (Bradshaw 1959, Saikkonen et al. 2000, Schulthess & Faeth 1998, Spyreas et al. 2001). In fact, explaining this extensive variation in infection frequency remains a challenge to researchers investigating the evolutionary ecology of grass-endophyte interactions (section 5.2). Clearly, in reporting the results of population surveys, it is important to note exactly how "infection frequency" was calculated. Most commonly, infection frequency denotes the proportion of the sampled individuals within a population that contained a fungal endophyte. It should be recognized that infection frequency is simply based on the presence or absence of the endophyte and does not provide information about the relative level or intensity of infection within the individual host. Because infection can vary among tillers (Meijer & Leuchtmann 2000) and the concentration of endophytic hyphae is typically unevenly distributed within host leaves (Christensen et al. 2002, Tan et al. 2001), different techniques may be necessary to delineate infection frequency from infection intensity.

The most widely employed technique to assess infection frequency within natural grass populations involves quick-staining of cleared leaf sheath sections followed by microscopic examination (Clark et al. 1983, Welty et al. 1986a). A 5 mm segment of leaf sheath is excised and placed into 70% ethanol for at least a day to clear. Alternatively, epidermal peels removed from the base of the leaf sheath where there is little or no chlorophyll can be stained directly without clearing (Clark et al. 1983, Keogh & Lawrence 1987). If placed into ethanol, the leaf segments can be stored indefinitely until they are stained (Hignight et al. 1993). The cleared tissue is then drenched in an aniline blue-lactic acid stain (Bacon & White 1994) for about 10 minutes. Note that other stains such as rose bengal or lactophenol cotton blue have also been employed to stain leaf sheath tissue and identify endophytic hyphae (Belanger 1996, Keogh & Lawrence 1987). Specimens are examined under a light microscope at 400×. Endophytic hyphae appear as convoluted tubes parallel to the orientation of the elongate leaf cells (figure 1.1).

Although histochemical techniques are somewhat time consuming, they are simple, effective, and economical. In addition, the number of hyphae observed across the 800 μm diameter field of view can be used as a quantitative estimate of infection intensity (Cheplick 1993a). For example, in *L. perenne*, infection intensities ranged from 2.7 to 8.3 hyphae per microscopic field of view (Cheplick et al. 2000). Unfortunately, if the endophyte is sparsely or unevenly distributed within the plant, histochemical techniques have the potential of false negatives unless many sheaths and many tissue samples are examined.

Endophyte infection in caryopses (hereafter referred to simply as seeds) can be assessed by similar methods. Seeds are first soaked overnight in a softening agent such as sodium hydroxide (NaOH) or nitric acid and then stained for several days with aniline blue (Clark et al. 1983, Wilson et al. 1991b). When squashed, infected seeds will reveal endophytic hyphae in the aleurone layer (Clark et al. 1983, White et al. 1993), a thin border between the seed coat and the underlying endosperm that is rich in proteins and lipids. Again, false negatives can arise if fungal transmission is imperfect (hyphae may occasionally fail to grow into seeds or new tillers of infected individuals) (do Valle Ribeiro 1993, Ravel et al. 1997b) and therefore multiple seeds should be examined. If earlier detection of endophytes is required before seeds have matured, a method is available for the detection of *Neotyphodium* spp. endophytes within developing ovaries in host grass flowers (Sugawara et al. 2004).

Other techniques for assessing endophyte infection are based on biochemistry or molecular genetics. Enzyme-linked immunosorbent assays (ELISA) are relatively expensive but can be used to process large numbers of samples infected by *Neotyphodium* (e.g., seeds) (Johnson et al. 1983, Welty et al. 1986b). Christensen et al. (1997) used ELISA to quantify the relative concentration of fungal hyphae in endophyte-infected forage grasses. Antibody-based immunoblot test kits are commercially available for use in the detection of *Neotyphodium* endophytes (Hiatt et al. 1999, Koh et al. 2006). Immunoblot kits are suitable for determining endophyte presence in seeds, seedlings, and adults, and apparently provide infection frequency estimates (e.g., Clay et al. 2005) comparable to those obtained with traditional microscopic examination. The intensity of color on the tissue prints positively correlates with hyphal density (i.e., infection intensity) and thus immunoblot kits should prove useful for large-scale studies (Koh et al. 2006).

Tissue print immunoassays (TPIAs) also allow reliable endophyte detection even when localized to specific tissues: microscopic examination of the tissue prints reveals red dots that indicate the presence of endophytic hyphae (Hahn et al. 2003). However, this method has not been useful in the quantification of infection intensity within individual hosts. Also, there has been some concern of false positives with TPIA if closely related fungi inhabit the plant.

Procedures based on polymerase chain reaction (PCR) using specific primers that amplify a specific part of the fungal genome (microsatellites) have been developed for *Neotyphodium* (Doss & Welty 1995) and *Epichloë* endophytes (Groppe & Boller 1997). Not only does PCR provide a method specific to the detection of endophytic fungi, even distinguishing between different *Epichloë* species (Groppe & Boller 1997), it also indirectly permits quantification of the concentration of fungi present within host tissues (Groppe et al. 1999, Rasmussen et al. 2007). Thus both infection frequency and intensity may be estimated using PCR methods.

Theoretically, measurement of DNA content or any biochemical unique to endophytic fungi, but not found in the host grass, might be used to quantify infection intensity. For example, Logendra and Richardson (1997) described a technique to quantify ergosterol produced by endophytes isolated from the leaf sheaths of four grass species. This sterol is highly specific to these fungi. They found that ergosterol content was tightly correlated with the mass of mycelia examined. Unfortunately this method may be time consuming, as fungal hyphae must first be isolated and then grown in culture to provide the mycelia. In another experiment, infected and uninfected seeds of *Festuca rubra* were mixed in variable proportions and ergosterol was extracted. The ergosterol content was strongly positively correlated with the proportion of endophyte-infected seed (Logendra & Richardson 1997). It remains to be seen whether or not ergosterol, specific endophyte-produced alkaloids whose concentration also positively correlates with fungal concentration (figure 3 in Rasmussen et al. 2007), or perhaps some other molecule common in all fungi (e.g., chitin) will prove to be the most effective way to assess the frequency and intensity of endophyte infection within grass populations and individuals, respectively.

2

Effects of Endophytes on Their Hosts

In geographically wide-ranging symbiotic interactions it is perhaps no surprise to observe great variation in ecological outcomes shaped by the underlying coevolutionary process (Thompson 2005). As will become evident, the grass-endophyte symbiosis is not unique in this regard. Conditional outcomes in interspecific interactions (Bronstein 1994a), driven by a combination of genetic variation and environmental effects, can be considered a major component of the raw material necessary for continued coevolutionary change (Thompson 1999). In this chapter, after exploring the diverse ways in which endophytes affect their hosts, the pronounced environmental contingency of ecological outcomes to the grass-endophyte symbiosis will become apparent as the impact of abiotic and biotic factors are examined.

2.1 HOST GROWTH AND REPRODUCTION

To determine the effects of endophytic fungi on host growth, physiology, and reproduction, one must first obtain infected (E+) and uninfected (E−) groups of the grass species of interest. Preferably the two groups should be as genetically homogeneous as possible so that the potentially confounding effects of host genetics are randomly distributed among the groups. Typically, possible genetic variation in endophytes is ignored, but this is usually not a problem unless multiple groups of E+ plants are being compared across experimental treatments. In the latter instance, it must be assumed that genetic variation in both endophyte and host is randomly distributed among the experimental E+ groups unless host (or endophyte) genotypes have been replicated among treatments. Genotypic specificity in grass-endophyte associations has been explored from the standpoint of both host and fungus (chapter 4) and has important implications for the evolutionary ecology of grass populations. In the present section, following a brief survey of methods used to eliminate the endophyte, the general effects of endophytic fungi on host growth, physiology, reproduction, and competitive ability will be described.

2.1.1 Eliminating the endophyte

Endophyte researchers have obtained E+ and E− plants of their study species in one of three ways: (1) direct collection (or measurement) of plants or seeds that are naturally infected or not in the field, (2) treatment of infected seeds with heat or fungicide, or (3) treatment of entire E+ plants or excised tillers with a systemic fungicide. Artificial inoculation of uninfected individuals with endophytic hyphae has also proven to be an effective and powerful technique for investigating potential effects of specific endophytes on their hosts (Brem & Leuchtmann 2003, Christensen et al. 2000, Johnson-Cicalese et al. 2000, Leuchtmann & Clay 1988b, Tintjer & Rudgers 2006). The utility of inoculation techniques for assessment of endophyte and host genotypic compatibility in studies of grass-endophyte interactions will later become apparent (chapter 4).

An example of the first technique of distinguishing the endophyte effect by direct field collection of E+ and E− plants is provided by an investigation of the impact of the endophyte *Balansia henningsiana* on the perennial caespitose grass *Panicum agrostoides* (Clay et al. 1989). Clumps of the host were removed from a field site along the shore of Lake Monroe, Indiana, in February. Infected clumps were readily identified by fungal stromata remaining on dried leaves from the previous growing season. Putative uninfected clumps lacking stromata were also collected. In the greenhouse, ramets were planted individually and grown for 15 weeks. Reclassification of individuals as to their infection status was necessary, as some of the clumps (and ramets) originally identified as uninfected produced fungal stromata. It was possible to identify E+ and E− ramets from the same clump because infected individuals can produce tillers of both types. This study underscores the importance of assessing endophyte infection for all plants used to determine the effect of endophytes on host growth and reproduction. Because infections can sometimes be lost from individual tillers of infected plants or seeds (e.g., Kover & Clay 1998), even during the course of a single experiment it may be prudent to assess endophyte infection several times.

One of the major drawbacks with direct collection of infected and uninfected plants from a field population is that it unearths the age-old question regarding correlation versus causation in scientific research. Endophyte presence can be correlated with a specific host response (e.g., increased growth), but is not necessarily the direct cause of that response. This is because other factors such as genotypic differences among potential hosts in growth rate, physiology, etc., are likely to be widespread in natural populations. Therefore caution must be exercised when making conclusions based on these types of endophyte studies. Although experimental methods are preferred and probably most effective in grass-endophyte studies purported to examine endophyte-mediated effects on their hosts, correlational studies can be very useful in ecological research, as elegantly discussed in detail by Shipley (2000). Preliminary investigation of a previously unstudied grass-endophyte symbiosis such as the system examined by Clay et al. (1989) may dictate simple comparisons of field-collected E+ and E− plants as a necessary first step.

Experimental treatment with heat or fungicides of seeds infected by endophytes has been used to generate E− plants, but care must be taken to ensure that such treatments do not influence subsequent host growth (Leyronas et al. 2006, Siegel et al. 1987, Williams et al. 1984). For example, Latch and Christensen (1982) reported complete eradication of the endophyte from infected *Lolium perenne* seeds following 15–30 minutes of immersion in hot water at 57°C or 10 minutes at 59°C. However, both seed germination and subsequent seedling size were significantly reduced relative to control seeds. For three fungicides tested, phytotoxicity was apparent at application concentrations that efficiently killed the endophyte (Latch & Christensen 1982). Hot water treatment of E+ seeds of *Lolium arundinaceum* (= tall fescue, *Festuca arundinacea*) for 20 minutes at 60°C effectively killed the endophyte, but reduced germination (Williams et al. 1984). A more recent study of chemical control of *Neotyphodium* endophytes in *L. perenne* and *L. arundinaceum* employed five fungicides tested at different doses (Leyronas et al. 2006). The fungicide prochloraz provided the best balance between high eradication of endophytes and low phytotoxicity in both hosts.

An effective way to eliminate the endophyte from adult plants with multiple ramets is to treat excised ramets (tillers) with a systemic fungicide that shows minimal phytotoxicity. Although a variety of fungicides have been tested (Bacon & White 1994), benomyl (methyl 1-[butylcarbamoyl]-2-benzimidazolecarbamate) has proven to be quite effective at eliminating *Neotyphodium* endophytes without showing phytotoxic effects (Latch & Christensen 1982). Propiconazole has also been used for the extermination of endophytic fungi in hydroponically grown host ramets (Faeth & Sullivan 2003).

The following technique has been successfully used by one of us (G.P.C.) to eradicate the endophyte from *L. perenne* and is a synopsis of methods described by Latch & Christensen (1982), Latch (personal communication), and Cheplick et al. (2000). Using a sharp razor blade, a single ramet with a few adventitious roots is first excised from the infected parent plant and both shoot and roots are trimmed to about 3 cm. Although ramets without attached roots are sometimes able to initiate new adventitious roots following replanting, greater survival is achieved using ramets with a few attached roots (G. P. Cheplick, personal observations). The ramet is immediately placed into a disposable plastic container filled with fine sand or vermiculite. The sand is saturated with a benomyl solution at a concentration of 1–2 g/L, although lower concentrations are sometimes used (Latch & Christensen 1982). To encourage tillering, a weak solution of mineral fertilizer can be added, although this is not necessary.

After a month or two of growth in a sunny environment, one or two new tillers will usually be available to sample. During the growth period, more benomyl solution can be added if the sand becomes dry. A leaf sheath from a new tiller (marked with a permanent marker) is collected, stained, and microscopically examined to determine if it is free of endophyte. If the new tiller appears to be endophyte-free, it is carefully excised from the original, treated ramet and replanted into whatever soil will be used in a future experiment. The original, treated ramet should not be used because it may still retain some viable endophytic hyphae. After the replanted E− ramet has grown and produced additional tillers, the endophyte status of the new plant should be verified before its use in an experiment. Other ramets sampled from the original parent plant are used to retain the E+ condition and should be planted into water-saturated sand. To ensure that there are no residual effects of the fungicide on future growth, plants should be allowed to grow for as long as possible prior to their use in an experiment. Alternatively, one can grow fungicide-treated plants to maturity, collect their E− seeds, and germinate the seeds to provide new E− plants (Faeth & Hamilton 2006, Morse et al. 2007, Vila-Aiub et al. 2005). This protocol should help greatly to ameliorate potential side effects of fungicide treatment. In addition, comparison of the size and early growth rates of fungicide-treated and untreated plants that are all endophyte-free can be used to determine whether or not fungicide treatment per se directly affects the host being studied. Sometimes some fungicide-treated ramets will remain infected; measured variables of these E+ plants can be compared to those of E+ plants that were not treated with fungicide. When no differences in the measured variables are detected, one can be reasonably confident that the fungicide treatment is not having an extraneous effect on the host (Faeth 2008, Faeth & Hamilton 2006, Faeth & Sullivan 2003).

Note that fungicide treatment is useful for controlling the effect of host genotype because multiple E− and E+ replicates of the same genotypes can be propagated indefinitely. When coupled with artificial inoculation procedures, endophyte genotype by host genotype interactions may be explored. Replicated host genotypes may then be used in experiments designed to investigate the genotypic specificity of host responses to endophyte infection (section 4.1).

2.1.2 Growth

Many features have been considered when examining plant growth (Körner 1991). For grasses, the most commonly used quantitative traits are the number of tillers, leaf area, and the dry mass of shoots and roots. It should be recognized that often these variables are recorded at one point in time and represent the results of growth rather than growth rates. Bradshaw (1959) reported one of the earliest indications that endophytic fungi could alter host growth. He showed that the choke fungus *Epichloë typhina*, which completely suppressed panicle production, resulted in a significant increase in tiller density in three populations of *Agrostis tenuis*.

Following a lag of more than 20 years after Bradshaw's (1959) seminal work, there are now numerous studies showing the impact of endophytes on host growth. In a field population of *Danthonia spicata*, plants infected by *Atkinsonella hypoxylon* had 57.1 + 5.3 (mean + SE) tillers, compared to 22.1 + 1.7 tillers in uninfected plants (Clay 1984). In addition, survival and growth by tiller production were significantly greater for E+ (relative to E−) plants when competing with *Anthoxanthum odoratum* in a field experiment. However, a later study of this same system showed that infected *D. spicata* plants could have a performance disadvantage compared to uninfected plants under some environmental conditions such as low nutrients and low soil moisture (McCormick et al. 2001).

There is little doubt that under controlled conditions *Neotyphodium* endophytes can sometimes be beneficial to growth of their host grasses, or at least be positively correlated with host growth, although many opposite results have also been reported. In a controlled-environment experiment in which E− clones of *L. perenne* were obtained by benomyl treatment of plants infected by *Neotyphodium lolii*, following 8 weeks of growth, E+ plants showed significantly greater leaf area, tiller numbers, and dry mass of shoots and roots compared to E− plants (Latch et al. 1985). *Neotyphodium* endophytes also increased dry matter accumulation in *L. perenne* and *L. arundinaceum* (tall fescue, KY-31) over a 14-week growth period in a greenhouse (figure 2.1) (Clay 1987b). However, separate cultivars (highly infected cv. Repell and uninfected cv. Yorktown) of *L. perenne* were used for the E+ versus E− comparison and thus differences ascribed to infection might be due to other differences between the two cultivars. In contrast, in another early study (Keogh & Lawrence 1987), tiller numbers and shoot and root mass were unaffected by endophytes in seedlings of a ryegrass hybrid (*Lolium* × *hybridum*) harvested 40 days after emergence. In that study, endophytic infection status was determined at the end of the experiment by microscopic examination of leaf sheaths and, as only very young plants were harvested, it is possible that other endophyte-mediated effects on host growth had not yet become manifest. More recently, in actively growing *L. perenne* in a controlled environment, endophyte infection decreased the rates of tillering and leaf extension compared to E− plants previously generated by systemic fungicide treatment (Spiering et al. 2006a).

Additional associations between endophyte presence and greater growth in tall fescue have been described. For example, Cheplick et al. (1989) reported that greenhouse-grown tall fescue infected by *Neotyphodium coenophialum* had significantly more tillers and dry mass than uninfected plants. Regrowth rates of E+ tall fescue following clipping are reportedly greater than E− plants for some genotypes in environment-controlled growth experiments (Belesky & Fedders 1996, Rahman & Saiga 2005). In addition, tall fescue (Kentucky 31) grown in a controlled environment showed greater tiller production, but not aboveground mass, when infected relative to E− plants previously generated by systemic fungicide treatment (Belesky et al. 1987).

In another growth chamber experiment, uninfected plants of meadow fescue (*Lolium pratense*, formerly *Festuca pratensis*) were compared to those

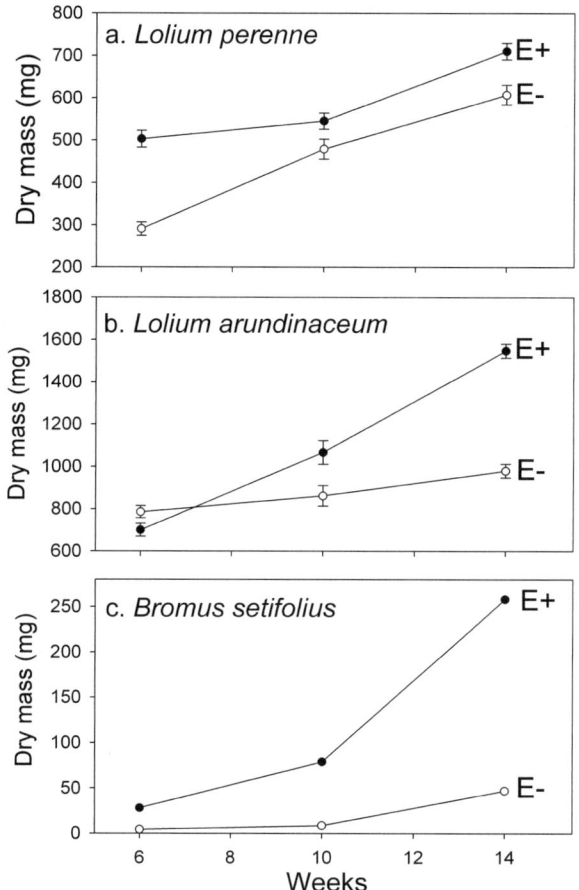

Figure 2.1 Accumulation of dry mass (mean ± SE) in endophyte-infected (E+) and uninfected (E−) plants of (a) *Lolium perenne* (perennial ryegrass), (b) *Lolium arundinaceum* (tall fescue), and (c) *Bromus setifolius* over a 14-week growth period. (a) and (b) adapted from Clay (1987b); (c) adapted from Novas et al. (2003). SEs were not available for the *B. setifolius* study.

naturally infected by *Neotyphodium uncinatum* (Malinowski et al. 1997a). Infected plants showed greater root and shoot dry mass compared to E− plants. More recently, Vila-Aiub et al. (2005) reported that in the annual grass *L. multiflorum*, E+ plants produced 18% more tillers and 7.5% more root mass than E− plants. In that study, half of the E+ seeds collected from natural communities were treated with fungicide to eliminate the endophyte and a second generation of seeds collected from E+ and E− plants were used for the experiment. Therefore the genetic backgrounds of the host plants used to compare E+ and E− conditions were likely to have been similar. Biomass accumulation over time in the perennial *Bromus setifolius* also was significantly greater in a population of plants infected by *Neotyphodium* endophytes compared to an uninfected population in a controlled environment (figure 2.1c) (Novas et al. 2003). However, it should be recognized that endophytes were not experimentally exterminated; naturally infected or uninfected populations were examined (Novas et al. 2003). Thus there is the potentially confounding effect of endophyte presence with the source population.

Examples of other endophyte species whose presence is correlated with enhanced host growth include the following:

- Tubers from E+ and E– individuals of *Cyperus rotundus* (a sedge) were collected from four sites in Baton Rouge, Louisiana, and variables were recorded on sprouts grown under greenhouse conditions. Tiller numbers, dry mass, and tuber production were greater for plants infected by *Balansia cyperi* (Stovall & Clay 1988).
- E+ and E– individuals of the caespitose grass *Panicum agrostoides* were collected from field sites along the shore of Lake Monroe, Indiana. When reared in a greenhouse, tiller numbers, dry shoot mass, and root mass were greater for plants infected by *B. henningsiana* (Clay et al. 1989).
- E+ and E– individuals of the caespitose grass *Tridens flavus* (purpletop) were collected from three field sites in Monroe County, Indiana. When reared in a controlled environmental chamber, leaf area was greater for plants infected by *Balansia epichloë* (Marks & Clay 1990).
- E+ and E– individuals of the perennial grass *Bromus erectus* were collected from a calcareous grassland in the Swiss Jura Mountains. These were clonally propagated and planted into an experimental plot. Tiller numbers and aboveground dry mass were greater for plants infected by *Epichloë bromicola* (Groppe et al. 1999).
- E+ and E– individuals of the stoloniferous grass *Glyceria striata* were collected from three populations in southern Indiana. The dry mass of stolons and daughter ramets were greater for plants infected by *Epichloë glyceriae* (Pan & Clay 2002).

The study by Groppe et al. (1999) is notable in that it is one of the few examples of a dosage-dependent effect of an endophyte (*E. bromicola*) on the growth of its host (*Bromus erectus*). By using a quantitative polymerase chain reaction (PCR) method (Groppe & Boller 1997) they were able to estimate the amount of fungal DNA, and by extension, the concentration of the endophyte expressed as the concentration of fungal PCR product per gram of host DNA (Groppe et al. 1999). When fungal DNA content was grouped into size classes ranging from 0 to 5 fmol/g host DNA, there was a highly significant linear increase in both the number of tillers ($r = 0.88$, df = 5, $P < 0.01$) and vegetative dry mass ($r = 0.89$, df = 5, $P < 0.01$) with increasing fungal concentration (figure 2.2). In contrast, endophyte-mediated effects on the growth of perennial ryegrass appeared to be independent of in planta endophyte concentration (Spiering et al. 2006a).

Given this brief summary of a set of studies showing host growth benefits when endophyte-infected, it is little wonder that so many researchers in the past have maintained that the grass-endophyte symbiosis is a mutualism. However, many of these investigations have been conducted in controlled environments (e.g., growth chambers, greenhouses) where conditions

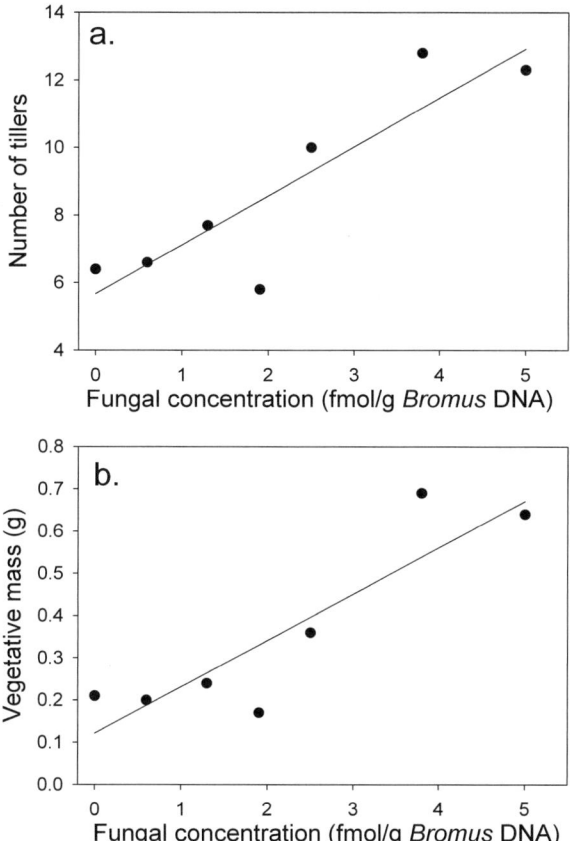

Figure 2.2 (a) Tiller production and (b) vegetative mass of *Bromus erectus* infected with variable concentrations of the endophyte *Epichloë bromicola*. Data from Groppe et al. (1999).

were benign and highly favorable to plant growth. Light, water, and soil nutrients typically were not limiting. Mineral fertilizer, for example, is often supplied in controlled-environment studies that compare E+ and E− hosts and plants are watered as needed. In experiments where environmental conditions have been varied as a subject of the investigation, potential costs of infection can sometimes outweigh the benefits (Ahlholm et al. 2002, Cheplick 2007, Cheplick et al. 1989, Faeth & Sullivan 2003, Morse et al. 2002). Also, particular host genotype-endophyte combinations can show reduced growth relative to endophyte-free replicates of the same host genotypes (e.g., Cheplick 2004a, Hesse et al. 2004). As will become apparent from the research results presented in upcoming sections (2.2, 2.3, and chapter 4) in which abiotic stress, competitive ability, and host genotypes are examined in greater detail, it is now recognized that grass-endophyte relationships can and do span the entire mutualism to parasitism continuum.

2.1.3 Photosynthetic physiology

Host physiological parameters have been measured for only a few grass-endophyte symbioses. Considering that host grasses can inhabit environments that range from sunny, open fields to shady, closed forest understories, it is remarkable how few investigators have examined the impact of light intensity on physiological properties in endophyte-infected grasses (Malinowski et al. 2005, Marks & Clay 2007). Nevertheless, several researchers have examined physiological variables in relation to endophyte infection and environmental conditions. Greater photosynthetic rates in infected grasses could benefit both endophyte and host. If photosynthesis can be maintained at a relatively high level in infected hosts growing under abiotically stressful conditions, the host may show better stress tolerance and subsequent growth following alleviation of the stress. This could lead to greater sexual reproduction by the host and eventual dissemination of vertically transmitted endophytes.

An early, often-cited study by Belesky et al. (1987) showed that tall fescue plants infected by N. coenophialum had significantly lower net photosynthetic rates at irradiances of 280–1700 μmol/m^2/sec compared to uninfected plants. Slightly elevated photosynthetic rates and stomatal conductances were reported for E+ plants of two tall fescue genotypes (Richardson et al. 1993). Marks and Clay (1996) reported similar photosynthetic rates for E+ and E– tall fescue at leaf temperatures from 24°C to 33°C, but greater photosynthesis in E+ plants at higher leaf temperatures (37°C–42°C). Photosynthetic rates in E+ tall fescue was 16% greater than in E– plants under high (but not low) soil nitrogen conditions (Newman et al. 2003).

Net photosynthetic rates of E+ and E– plants of Arizona fescue (*Festuca arizonica*), an understory perennial common in semiarid Ponderosa pine communities in the American southwest, were highly dependent on water availability (Morse et al. 2002). Infected and uninfected plants were grown outdoors and randomly placed into two water treatments. Plants were watered to field capacity three times a week in the "high water" treatment, while plants were initially watered once a week in the "low water" treatment and then once every other week to impose a more severe water stress. When water was readily available, E– plants showed greater net photosynthesis than E+ plants (Morse et al. 2002). It was only after the more severe water stress had been imposed that E+ plants showed significantly greater photosynthesis. Growth variables tended to mirror these patterns. In a later study (Morse et al. 2007), E– plants of Arizona fescue showed higher photosynthetic rates and stomatal conductance than E+ plants over a 49-day growth period. Additional studies of drought and the grass-endophyte symbiosis are considered later (section 2.2.1).

Photosynthetic rates (and other physiological parameters) have also been examined in *L. perenne* infected by *N. lolii*, typically in relation to environmental conditions. Whether under water stress or not, E+ plants of one perennial ryegrass cultivar showed greater net photosynthetic rate, stomatal conductance, transpiration rate, and tiller numbers than E– plants (Amalric et al. 1999). As with most growth variables, the photosynthetic activity of perennial ryegrass in relation to endophyte infection depends on environmental conditions. Monnet et al. (2005) were interested in physiological variables of E+ and E− plants reared under zinc stress. Measurements of leaf gas exchange were made on plants growing in nutrient solutions with five zinc concentrations at 400 or 2000 µmol carbon dioxide (CO_2) and irradiances of 400 or 1000 µmol/m²/sec. There were no differences between E+ and E− plants in stomatal conductances or mineral and chlorophyll concentrations within leaves. However, under some conditions there were detectable differences in net photosynthetic rate: at the highest irradiance and CO_2 levels, E+ plants exhibited greater photosynthesis than E− plants (figure 2.3). In two genotypes of a different perennial ryegrass cultivar (Grasslands Nui), Spiering et al. (2006a) reported significantly lower rates of photosynthesis for E+ plants relative to E− plants, but dark respiration rates were not significantly different (figure 2.4).

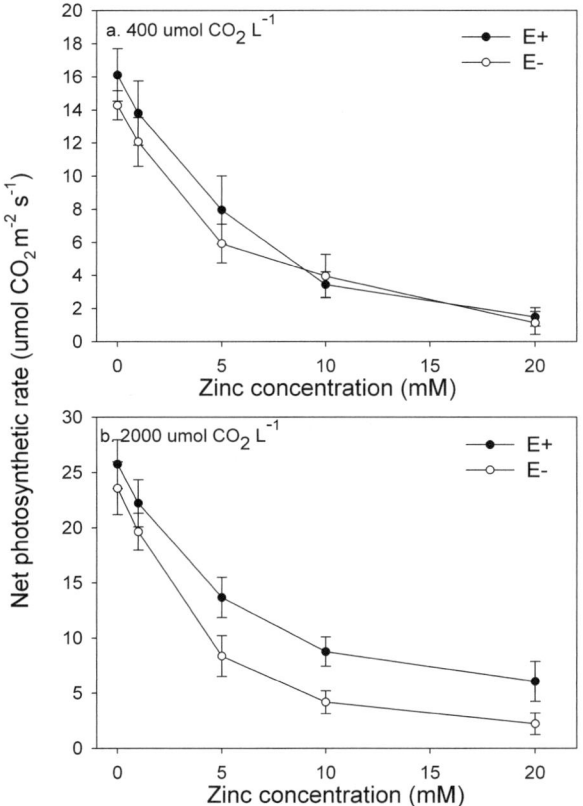

Figure 2.3 Net photosynthetic rates (mean ± SE) in endophyte-infected (E+) and uninfected (E−) plants of *Lolium perenne* at variable concentrations of zinc under (a) low and (b) high levels of CO_2. Data from Monnet et al. (2005).

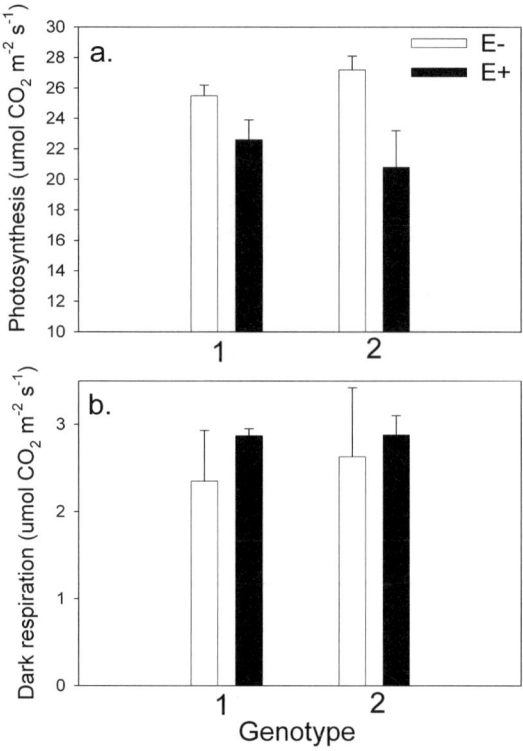

Figure 2.4 Mean (± SE) rates of (a) photosynthesis and (b) dark respiration in endophyte-infected (E+) and uninfected (E–) plants of two *Lolium perenne* genotypes. Data from Table 1 of Spiering et al. (2006a).

While there appear to be differences between infected and uninfected plants in photosynthetic rates (and related physiological parameters), at least under some conditions for the few grass species examined, the important question arises as to how relevant these differences are to ecological and evolutionary success in nature. Körner (1991) maintained that the contribution of gas exchange studies toward a deeper understanding of plant growth has been "greatly overestimated." This is because developmental processes (such as those that stimulate tiller production) and growth variables (such as leaf area) often determine whole-plant photosynthesis (Nelson 1988). Although genetic variation for photosynthetic rates in forage crops such as tall fescue has been reported (Asay et al. 1974), in "most cases" there is "no association, and in some cases even a negative association, between photosynthesis per unit area of single leaves and yield of field and forage crops" (Nelson 1988). There have been so few ecophysiological studies of the grass-endophyte symbiosis in natural environments, especially with native grasses, that it is presently unknown to what extent, if any, photosynthetic rates correlate with survival, growth, and reproductive fitness, or with general ecological success as assessed by competitive ability or tolerance of abiotic stresses.

2.1.4 Reproduction

The coevolutionary dynamics of endophyte and host populations are likely to depend on the extent of genetic variation in both partners, as will be detailed in chapter 4. However, genetic variation within and among populations is strongly influenced by the reproductive system of both host and fungus. Vertically transmitted endophytes such as *Neotyphodium* spp. have not retained the ability for sexual reproduction and consequently may exhibit reduced genetic variability. The breeding system of host grasses may be predominantly sexual when infected with such endophytes, but can be changed dramatically when infected by parasitic endophytes that can reproduce sexually. In this section, after a survey of the diversity of endophyte-mediated effects on host reproduction, changes in the reproductive systems of both symbionts will be related to the dynamics of host-endophyte coevolution along the parasitism-mutualism continuum.

In general, the effects of endophytic fungi on host reproduction are extremely variable, depending on the particular grass-endophyte symbiosis under consideration. Effects range from parasitic castration (i.e., no sexual reproduction) (Clay 1991b) in an *Epichloë*-infected host (Groppe et al. 1999) to improved sexual reproduction (i.e., increased seed production) in *Neotyphodium*-infected tall fescue (Rice et al. 1990). In addition to documented effects on host sexual reproduction, increased asexual reproduction via greater production of clonal offshoots has been reported in *Glyceria striata* infected by *E. glyceriae* (Pan & Clay 2002).

The most severe manifestation of the effects of endophyte infection on host sexual reproduction occurs in grass-endophyte symbioses that are parasitic by definition. Complete abortion of some host inflorescences typically occurs as *Epichloë* spp. endophytes produce choke-inducing stromata during their own sexual cycle (Groppe et al. 1999, Meijer & Leuchtmann 2001, Schardl 1996). Although choke disease can be relatively widespread and frequent in some grasses (Pfender & Alderman 1999, Wennström 1996), the incidence of disease symptoms is a poor indicator of infection levels (Clay & Brown 1997) because asymptomatic infected plants are frequently found in host species (Bucheli & Leuchtmann 1996, Meijer & Leuchtmann 2000, Tintjer et al. 2008). Furthermore, an individual infected by *Epichloë* may produce a combination of both choked and normal flowering tillers and so sexual reproduction is not necessarily completely occluded in infected populations. Nonetheless, infection-induced sterility is likely to reduce the potential reproductive fitness of individual hosts. When host genotype has a pronounced influence on the expression of choke disease, as in *Brachypodium sylvaticum* infected by *Epichloë sylvatica* (Meijer & Leuchtmann 2000), there is the potential for natural selection to impact the coevolutionary relationship between host and endophyte within populations.

In their review of ten *Epichloë* species, Schardl & Leuchtmann (2005) noted that vertical transmission through seeds only occurs in some host species. For endophytes that transmit spores via the sexual cycle, ejected ascospores from the fungal stromata can infect the flowers of neighboring grasses. This form of horizontal transmission results in infected seeds that produce infected seedlings upon germination rather than simply infected adult plants (Schardl & Leuchtmann 2005). When infected individuals show a growth advantage in nature, perhaps showing greater competitive ability or abiotic stress tolerance than E– individuals, very high infection frequencies are possible within host populations (e.g., *B. sylvaticum* infected by *E. sylvatica*).

From an evolutionary standpoint, parasitism like that exhibited by some *Epichloë* species may favor sexual reproduction and outcrossing among hosts whenever there exists a selective advantage to rare host genotypes (Busch et al. 2004, Clay & Kover 1996). This can occur if rare host genotypes have a reduced chance of becoming infected by parasites that are locally adapted to more common, widespread host genotypes (Clay & Kover 1996, Thompson 2005). Sexual reproduction via outcrossing by the host can generate novel genetic combinations in the offspring that are better able to resist infection, or at least reduce the severity of effects caused by parasite infection. For the parasite, suppression of host sexual reproduction would thereby be advantageous and could represent a critical fungal adaptation essential for the parasitic lifestyle (Clay 1991b). When higher virulence results in a reduced rate of vertical transmission of the parasite, selection may favor greater ability of the parasite to castrate its host, as this would improve horizontal transmission (Kover & Clay 1998). The ensuing coevolutionary battle between host and fungus is likely to be driven by contrasting selection pressures on the symbiotic partners: increased defense against parasitic castration for the host versus increased virulence to minimize host sexual reproduction for the fungus. The symbiotic continuum from pathogenic to mutualistic associations represented by *Epichloë* endophytes and their asexual *Neotyphodium* derivatives (Schardl & Leuchtmann 2005, Schardl et al. 1997) provides a superb opportunity for further investigation into the evolution and ecology of fungal-plant interactions.

The asexual *Neotyphodium* spp. endophytes, which evolved from *Epichloë*, can show a positive effect on sexual reproduction in a few grass hosts. Rice et al. (1990) showed the clearest indication of this in tall fescue (cv. Kentucky-31) infected by *N. coenophialum*. In field trials, E+ and E– replicates of the same host genotypes were grown for 1–2 years. Infected plants consistently produced greater numbers of panicles and seeds than E– plants. In another tall fescue cultivar (Kenhy), however, seed production was not affected by endophytes (Siegel et al. 1984a). For *L. perenne* genotypes collected from native grasslands in Germany, effects of *N. lolii* endophytes on seed production were variable, but seed yield was greater in some, but not all, genotypes when infected (Hesse et al. 2003, 2004). Seed germination was also higher for at least some genotypes when infected. Similarly, Clay (1987b) reported a higher final percent germination of E+ seeds in perennial ryegrass and tall fescue. Seed lots of *L. multiflorum* with high levels of endophyte infection (>85%) showed a reduced germination rate and a lower proportion of germination under increasing levels of water stress compared to seed lots with very low infection levels (<10%) (Gundel et al. 2006). Also, Faeth and Sullivan (2003), Hamilton and Faeth (2005), and Neil et al. (2003) reported that infection by *Neotyphodium* did not improve seed germination of Arizona fescue seeds at varying osmotic potentials.

Although seedling establishment might be enhanced by increasing levels of endophyte infection in species such as perennial ryegrass (Eerens et al. 1998b, van Heeswijck & McDonald 1992), it is by no means certain as to whether or not endophyte-enhanced seed production and germination would be of any ecological significance in mature grasslands where seedling establishment is rare. However, in situations where seedling establishment does occur, selection should favor improved seed production and germination from hosts with seed-borne endophytes, as this is clearly beneficial to both the grass and symbiont. The establishment of new populations, even if this involves only a relatively low number of seedlings, could still have coevolutionary significance when E+ seedlings show improved survival and growth relative to E− seedlings (perhaps under abiotically stressful conditions). This is because such a scenario will increase the frequency of endophyte infection in the ensuing generation of plants.

In addition to the range of possible effects in the *Epichloë*/*Neotyphodium*-grass symbioses, from complete or partial castration to enhancement of host sexual reproduction, some intriguing effects on host breeding systems have been reported for other Balansieae fungi. One system studied by Clay (1984, 1994a) and Kover (Kover 2000, Kover & Clay 1998) involves the epiphytic fungus *Atkinsonella hypoxylon* (Leuchtmann & Clay 1988a) which infects North American grasses in the genus *Danthonia*. In uninfected *D. spicata*, individuals produce open-pollinated, chasmogamous (CH) spikelets on emergent terminal panicles and additional self-fertile cleistogamous (CL) spikelets completely enclosed by leaf sheaths near the tiller base. Infected individuals show abortion of the terminal panicles and no seeds in CH florets can be matured. However, basal CL spikelets can mature seeds and the resulting seedlings are infected by *Atkinsonella*, indicating vertical transmission of the fungus (Clay 1994a). Thus the fungus is effectively changing a mixed breeding system that normally entails some combination of both outcrossing and selfing into one of self-fertilization only. Clay (1994a) maintained that selection would favor *Atkinsonella* genotypes that did not abort CL spikelets because vertical transmission within the resulting seeds could be a fungal adaptation that ensures dispersal and host colonization. Among *Atkinsonella* isolates, genetic variation in their ability to cause castration has been documented (Kover & Clay 1998). As in *Epichloë* spp., by preventing the formation of seeds on outcrossed panicles, *Atkinsonella* may prevent the generation of potentially resistant novel host genotypes. The enforced production of infected seeds on self-fertilizing CL spikelets by *Atkinsonella* has the added benefit to the fungus of maintaining susceptible and compatible host genotypes (Clay 1991b).

An even more severe change in reproduction occurs in the sedge *Cyperus virens* when infected by *Balansia cyperi*. Following envelopment of the developing inflorescences by this epiphytic parasite (Leuchtmann & Clay 1988a), viviparous plantlets can emerge from some of the aborted inflorescences (Clay 1986). For five host populations in southwestern Louisiana, the frequency of infection typically exceeded 50% and about 30% of the infected plants, which are larger than uninfected plants, showed evidence of vivipary. Viviparous plants produced two to six plantlets and microscopic examination revealed that plantlets contained abundant hyphae of *B. cyperi* (Clay 1986). Thus vertical transmission of this fungus is possible via the induction of asexual reproduction in the host. On the aborted inflorescences of two *Andropogon* spp. infected by *Myriogenospora atramentosa*, Clay (1986) also observed viviparous plantlet production. Although the ecological and evolutionary ramifications of endophyte-induced vivipary have yet to be explored, the genetic structure and evolutionary potential of host populations are likely to be heavily impacted by the change in reproductive mode from sexual (when uninfected) to asexual (when infected).

Increases in the ability to reproduce asexually when infected also occur in two other plant species. In the widespread weed *Cyperus rotundus* (purple nutsedge), subterranean tubers are the predominant mode of propagation as few viable seeds are set in inflorescences. For E+ and E– plants emerging from tubers sampled in Louisiana and then grown in a greenhouse, infected plants showed inflorescence abortion but were larger and produced almost twice as many tubers as uninfected plants (Stovall & Clay 1988). In the clonal grass *Glyceria striata*, plants infected by *E. glyceriae* allocated more biomass to clonal growth (figure 2.5), showing increased numbers and lengths of stolons and greater mass of daughter offshoots, than infected plants (Pan & Clay 2002, 2003). In this way the symbiotic fungus could improve genet spread and persistence within the community (Cheplick 2004b) as the infected asexual offshoots root and spread themselves.

The continuum of effects that clavicipitaceous fungi impart on the reproductive capacity of a diversity of grasses and sedges provide useful models for examination of the evolutionary development of symbiotic interactions that range from parasitism to mutualism (Clay 1988a, Saikkonen et al. 2004, White 1988, Wilkinson & Schardl 1997). The reciprocal coevolutionary dynamics between plants and pathogens in natural ecosystems has often been recognized (Frank 1993, Gilbert 2002, Thompson 2005) and several authors (e.g., Clay 1988a, Doebeli & Knowlton 1998, Ewald 1987, Schulz & Boyle 2005) have described evolution along the parasitism-mutualism continuum. It is typically assumed that the evolutionary pathway to mutualism begins from a parasitic symbiosis and then proceeds via reciprocal genetic interactions between parasite and host populations over time, perhaps passing through a commensalistic stage along the way (Boucher et al. 1982, Clay 1988a, Ewald 1987, White 1988). It has been shown with an evolutionarily stable strategy model that vertical transmission (as occurs in asexual endophytes) can be a significant factor that promotes the evolution of reduced virulence in parasites and

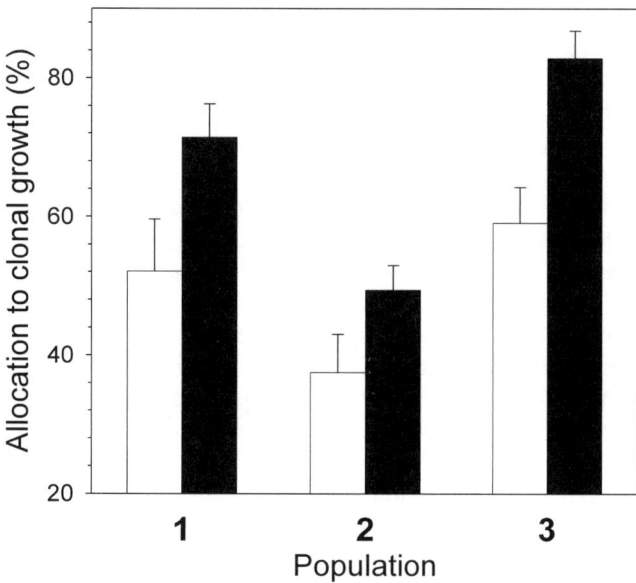

Figure 2.5 Mean (± SE) percentage of the total genet mass allocated to clonal growth (i.e., stolons and daughter ramets) in three populations of *Glyceria striata* infected by *Epichloë glyceriae* (shaded bars) or uninfected (open bars). Data from Pan and Clay (2003) and reprinted from Cheplick (2004) with permission of the *New Phytologist* Trust.

favors mutualism (Yamamura 1993). Another model showed how evolutionary processes could promote the differentiation of avirulent, vertically transmitted symbionts from virulent, horizontally transmitted parasites (Ferdy & Godelle 2005). However, it must be emphasized that the parasitism-mutualism continuum is remarkably labile (see section 5.3.2.3). Over evolutionary time, mutualisms can break down and shifts from mutualism to parasitism have been documented in some lineages of various symbiotic associations (Sachs & Simms 2006, Thompson 2005). Furthermore, there is phylogenetic evidence of the "abandonment of mutualism" in some types of symbioses (Sachs & Simms 2006). Such a process could occur in grass-endophyte symbioses where the interaction is obligatory for one partner (the fungus), but not the other (the host). If the cost of harboring an endophytic fungus exceeds the benefits to the host under some conditions, selection could favor abandonment of the endophyte by the host population.

Underlying any coevolutionary scenario that describes the dynamics of the grass-endophyte interaction is the considerable evidence for genetic variation in both the responses of hosts to the endosymbiont and the ability of the endosymbiont to affect its host (chapter 4). For example, the expression of choke disease in *B. sylvaticum* infected by *E. sylvatica* was highly dependent on host genotype (Meijer & Leuchtmann 2000). Consistent differences between maternal families of *Elymus hystrix* in the production of stromata by *E. elymi* have been demonstrated (Tintjer et al. 2008). Variation in the propensity of endophyte isolates (*Epichloë* spp.) to form stromata has also been demonstrated (Meijer & Leuchtmann 2001, Tintjer & Rudgers 2006). Indeed, Meijer and Leuchtmann (2001) suggested that the fungal genotype may "be responsible for whether an endophytic association is beneficial or pathogenic." The genetic changes needed to convert a plant pathogen to an endophytic mutualist are not necessarily complex (Redman et al. 2001) and may involve only a single gene mutation (e.g., Freeman & Rodriguez 1993, Tanaka et al. 2006).

Closely tied to the coevolutionary dynamics of host and endophyte along the parasitism to mutualism transition are changes in the reproductive systems of both symbiotic partners. White (1988) proposed the choke disease strategy ("type 1," now commonly designated with a roman numeral as type I) of endophytes such as some *Epichloë* spp. as the original parasitic stage that eventually evolved into the asymptomatic, asexual strategy ("type 3" [type III]) of endophytes such as *Neotyphodium* that represents a predominantly mutualistic stage. The partial castration that occurs in some grass-endophyte symbioses whereby fungal stromata are made on only a few of the infected plants or on only some fraction of infected tillers ("type 2" [type II]) represents an intermediate stage. Clay (1988a) observed how the transition from parasitism to mutualism in grass-endophyte symbioses involved a change from mostly sexual to completely asexual reproduction in the endophyte with a concomitant change from mostly asexual to completely sexual reproduction in the host (figure 2.6). This simple model includes intermediate conditions along the parasitism-mutualism continuum (e.g., commensalism) whereby the endophyte would employ a moderate level of sexuality and the host would show a mix of asexual and sexual reproduction. The scenario that mutualism originated

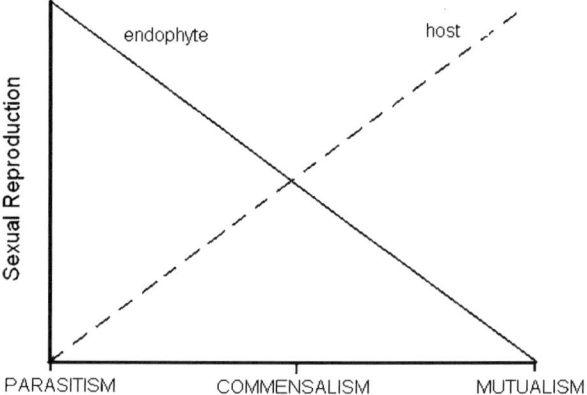

Figure 2.6 The transition from parasitism to mutualism in grass-endophyte symbioses in relation to the reproductive system of host and endophyte. Based on Clay (1988).

from parasitism by way of coevolutionary changes in the reproductive systems of both endophytes and hosts (Clay 1988a) has provided a useful theoretical framework for the study of grass-endophyte symbioses. Further empirical research should focus on how ecological and genetic factors interact to cause evolutionary shifts between the extremes of sexuality and asexuality, and parasitism and mutualism (Saikkonen et al. 2004).

2.2 ABIOTIC STRESS TOLERANCE

The introductions to many papers on the grass-endophyte symbiosis refer to the improved tolerance or growth of infected hosts (relative to their uninfected counterparts) when exposed to abiotic stresses such as low water or soil nutrient availability. However, the reported cases for such tolerance mainly come from two agronomic grass species: tall fescue and perennial ryegrass infected by *Neotyphodium* spp. (reviews in Bacon 1993, Belesky & Malinowski 2000, Malinowski & Belesky 2000, Malinowski et al. 2005, West 1994). Even for these well-studied grass-endophyte associations, the evidence for abiotic stress tolerance has varied greatly from study to study. Note that host tolerance to abiotic stress, which can involve the ability to survive and maintain physiological activity under adverse conditions such as drought or low soil fertility, differs somewhat from tolerance to biotic stresses such as herbivory. As will be further discussed in section 3.8, tolerance to herbivory typically depends on a specific ability to compensate for tissue lost to herbivores by regrowth, without a concomitant loss of fitness.

Perhaps nowhere else is the environmental contingency of the grass-endophyte relationship better illustrated than in studies that have investigated the symbiosis in relation to abiotic stress. In this section, endophyte-mediated effects on host tolerance of and responses to drought, low soil fertility, and other potential stresses will be described. An abiotic stress is considered here to be any environmental factor that severely limits plant growth when present in low quantity. Well-designed experiments provide evidence of stress following the imposition of specific treatments, measurement of host responses, and comparison to unstressed controls.

2.2.1 Drought

Drought stress is the most thoroughly studied of the abiotic stresses in relation to the grass-endophyte symbiosis (Malinowski & Belesky 2000, West 1994). Given the direct effect drought has on the productivity of forage grass crops (Frank et al. 1996), it is no surprise that much of the investigation of drought and its impact on endophyte-infected grasses has focused on tall fescue and perennial ryegrass. The intensive research effort devoted to tall fescue and *N. coenophialum* has likely been spurred on by tantalizing field data suggesting better survival and persistence of E+ plants during drought (Read & Camp 1986, West et al. 1993). In addition, the frequency of infection in tall

Table 2.1 Tiller variables and yield of uninfected (E−) and endophyte-infected (E+) tall fescue (cv. Kentucky 31) grown in a field plot in Arkansas for 2 years with or without supplemental irrigation.

| Variable | Not irrigated | | Irrigated | | ANOVA results | |
	E−	E+	E−	E+	Endophyte	Endophyte × treatment
Tiller density (per m²)	2117	2396	2704	3400	**	ns
Leaf area (cm²/tiller)	0.62	0.75	6.74	9.77	*	**
Dry mass (mg/tiller)	23	26	55	66	*	**
Yield (g/m²)	48	60	149	227	**	ns

Adapted from West et al. (1993).
The significance of endophyte and water treatment in an analysis of variance is indicated (ns = not significant; $*P < 0.05$; $**P < 0.01$).

fescue, meadow fescue, and perennial ryegrass in Europe appears to be greatest in the warmer and drier southern regions relative to the cooler and moister northern regions (see figure 1 in Malinowski & Belesky 2006).

In one influential early study (Arachevaleta et al. 1989), E+ plants of tall fescue (cv. Kentucky 31) showed greater survival and biomass production than E− plants under soil moisture stress. Leaf rolling during drought and regrowth during the recovery period was also greater for E+ plants. A field study of the same tall fescue cultivar along a water supply gradient supported these results: tiller density and size were significantly greater in E+ plants (table 2.1) (West et al. 1993). Overall dry matter yield per square meter was also greatest in E+ monocultures (table 2.1). These and several other studies (reviewed by Bacon 1993, Belesky & Malinowski 2000, West 1994) strongly suggest that endophyte-enhanced drought tolerance might be an ecologically important component of the perceived success of tall fescue and its fungal endophyte in natural ecosystems.

The putative physiological mechanisms for drought stress tolerance of E+ plants are diverse and include decreased stomatal conductance resulting in reduced transpiration, greater water-use efficiency, and enhanced osmotic adjustment (Belesky & Malinowski 2000, Elmi & West 1995, Malinowski & Belesky 2000, West 1994). The latter would entail an increase in the ability of E+ plants to accumulate cell solutes and thereby maintain turgor during the drought period (Frank et al. 1996). The solutes that might be important to osmotic adjustment include sugars and amino acids made by the host or sugar alcohols and alkaloids made by the endophyte (Malinowski & Belesky 2000, Richardson et al. 1992).

Despite the widely cited evidence for improved drought tolerance of tall fescue when endophyte infected, counterexamples exist even for this well-studied symbiosis:

- In three tall fescue collections from Mississippi and Georgia subjected to severe water stress under greenhouse conditions, endophyte infection did not affect osmotic potential, relative water content, or osmotic adjustment (White et al. 1992). Tiller production, survival and recovery from drought did not differ between E+ and E− plants. The authors concluded "no evidence for endophyte-mediated drought tolerance was observed in this study" (White et al. 1992).

- In a greenhouse study of photosynthesis for two genotypes of tall fescue (cv. Kentucky 31) subjected to varying degrees of water stress (Ψ_w from −0.15 MPa to −2.5 MPa), stomatal conductances were higher and transpiration losses greater for E+ plants of both genotypes (Richardson et al. 1993).

- In a controlled growth chamber study of three genotypes of tall fescue (cv. Grasslands Roa), E+ and E− plants in a dry treatment received 25% of the water supplied in a wet treatment (Maclean et al. 1993). Greater tiller production occurred in E− plants, although dry mass was greater in E+ plants. However, photosynthetic physiological measurements were not significantly affected by endophytes in wet or dry conditions and the authors concluded "results...do not support the view that endophyte-infected plants are likely to be more resistant to drought stress" (Maclean et al. 1993).

- In a field study of growth and water relations in three genotypes of tall fescue (two of Kentucky 31 and one from Russia) planted into plots in Arkansas, there were no consistent effects of endophytes on leaf elongation or the density and mass of tillers (Elbersen & West 1996). Stomatal conductances measured on seven dates did not differ between E+ and E− plants on five days, but were greater for E+ plants on two days. The authors concluded "*Neotyphodium coenophialum* had no consistent effect on leaf growth in tall fescue under field conditions regardless of soil water status" (Elbersen & West 1996).

- In a greenhouse study of drought stress effects on a common genotype of tall fescue infected by three different endophyte isolates, two isolates provided "no apparent benefit [to the host] when long-term drought stress was imposed," although the other endophyte isolate was capable of stimulating host growth (Hill et al. 1996).

- In a greenhouse study of two cultivars of tall fescue (Maris Kasba and El Palenque), E+ plants had lower dry mass and tiller numbers during water deficit relative to E− plants (Assuero et al. 2000).

- In tall fescue (cv. Grasslands Flecha), no endophyte effects on the rate of development of drought stress measures in a greenhouse experiment or post-drought stand recovery (assessed by tiller counts) in a field trial could be detected (West et al. 2007).

This ambiguity in the ability of endophytes to consistently mediate drought stress tolerance in their hosts becomes even more apparent when examining other grass-endophyte associations. In the water availability experiment with *F. arizonica* described earlier (section 2.1.3), following an 87-day growth period E+ plants (infected by *Neotyphodium starrii*) produced significantly more aboveground (but not total) mass than E− plants under low water availability (Morse et al. 2002). However, root mass, which could be important when water is scarce, did not differ between E+ and E− plants at low or high water levels, and E+ plants had significantly lower aboveground and total mass when water availability was high. In *F. rubra* (red fescue) infected by *Epichloë festucae* and grown for two seasons in a greenhouse, increased vegetative growth of E+ plants was only observed under well-watered and fertilized conditions (Ahlholm et al. 2002). In the same study, E– plants of *L. pratense* (formerly *F. pratensis*) produced more biomass than E+ plants infected by *N. uncinatum* under both low and high water availability. For a single clone of *F. elatior* (meadow fescue) also infected by *N. uncinatum* and studied in a growth chamber experiment, neither cumulative dry mass or tiller numbers differed between E+ and E– plants following 4 weeks of regrowth after a 26-day water stress treatment (Malinowski et al. 1997a). Both leaf water potentials and stomatal conductances were lower in E+ plants during the imposed water stress. In general, effects of endophyte infection on water-use efficiency and other physiological features important to drought tolerance have not been clearly demonstrated in field populations of host grasses (Malinowski & Belesky 2000, 2006, West 1994).

The perennial ryegrass-*N. lolii* system has been closely examined in relation to drought tolerance and recovery from drought in both field and greenhouse experiments. Lewis et al. (1997) quantified the percentage of infected seeds in 57 populations of *L. perenne* throughout France and performed a multiple regression with six climatic variables. Cumulative evapotranspiration from March to October was positively correlated and cumulative water supply deficit from July to August was negatively correlated with infection frequency. These analyses suggested that "summer drought conditions may impart a selection pressure in favour of infection" (Lewis et al. 1997). However, despite these intriguing field patterns, experimental studies of drought tolerance and recovery in *L. perenne* have mostly yielded mixed results.

In field trials in the United Kingdom conducted over 2 years during which severe drought occurred, endophyte infection of perennial ryegrass did not increase the survival of seedlings or plant growth (Lewis 1992). Leaf extension growth, a parameter that is sensitive to drought and important to recovery from drought, was greater in uninfected ryegrass (cv. Grasslands Nui) under conditions of variable moisture stress (Eerens et al. 1993). For three perennial ryegrass genotypes grown in a controlled environment and subjected to a 6-day drought stress, E+ plants had lower osmotic potential and more tillers than E– plants, suggesting greater drought tolerance in infected plants (Ravel et al. 1997a). A 3-year field experiment of three perennial ryegrass cultivars grown at five locations in France showed that E+ plants had greater productivity and persistence in a variety of environments, including those characterized by high evapotranspiration (Ravel et al. 1999). In another study, leaves of E+ plants maintained significantly higher relative water content under severe drought stress (Anzhi et al. 2006). In contrast, a greenhouse study of one cultivar during which a water stress treatment was imposed revealed greater root and shoot mass and water-use efficiency in E– plants (Eerens et al. 1998a). Although genotypic variation in the ability to recover from drought stress is readily demonstrable in perennial ryegrass, the relative effect of endophytic fungi depends greatly on host genotype (Cheplick 2004a, Cheplick et al. 2000, Hesse et al. 2003, 2005).

2.2.2 Minerals

The performance of endophyte-infected plants has been examined in tall fescue, perennial ryegrass, and a few other hosts in relation to mineral stresses, predominantly low levels of soil nitrogen (N) or phosphorus (P). Much of the key literature on the agronomic grasses has been reviewed (Belesky & Malinowski 2000, Malinowski & Belesky 2000, Malinowski et al. 2005). Here, the focus will be on the general patterns that have emerged from mineral stress experiments and the potential relevance of the results to natural populations and communities.

Simple growth studies in which E+ and E− plants were exposed to variable levels of added fertilizer suggest that the beneficial effects of endophyte infection are most likely to be manifested under high soil fertility. In one study, tall fescue (cv. Kentucky 31) was exposed to three N supply rates: 11, 73, or 220 mg N/L (Arachevaleta et al. 1989). Dry mass of E+ plants was greater than E− plants at medium and high N; at low N supply there was no difference. In another study of tall fescue (cv. Kentucky 31) seedling growth, plants were grown in pure vermiculite and then exposed to 1.5 g/L of Peter's 20–20–20 (N-P-K) fertilizer, a 0.10 dilution of the same fertilizer, or no fertilizer (Cheplick et al. 1989). Dry mass of E+ seedlings after 56 days was significantly greater at high nutrient levels, not different at intermediate nutrients, but significantly lower at the lowest nutrient level. The authors suggested that under strongly resource-limited conditions there might be a metabolic cost to harboring fungal endophytes due to the limited availability of photosynthates (Cheplick et al. 1989).

Endophyte infection of tall fescue can result in higher concentrations of some forms of N within host tissues (Belesky & Malinowski 2000, Lyons et al. 1990). Some genotypes may also be better able to take up or accumulate minerals such as P (and others) when infected, but this ability typically depends on the host genotype used and soil nutrient conditions (Malinowski et al. 1998, 2000, Rahman & Saiga 2005). For example, in a low P soil, E+ plants showed greater uptake and transport of P, potassium (K), calcium (Ca), and magnesium (Mg), and produced more dry mass, than E− plants; however, in a high P soil this advantage was not apparent and E+ plants produced 31%–47% less dry mass (depending on "ecotype") than E− plants (Rahman & Saiga 2005). Malinowski et al. (1999b) showed that E+ tall fescue increased root hair length in response to P deficiency more than E− plants and suggested that the resulting increase in root surface area could provide greater tolerance of mineral stress in infected tall fescue.

For perennial ryegrass, effects of endophyte infection in relation to mineral nutrient stress have not been consistent. In the study of tall fescue described earlier (Cheplick et al. 1989), other experiments were performed to investigate the effects of nutrient limitation on *L. perenne* seedlings and adults. More biomass was produced by E+ seedlings at high and medium, but not low, nutrient levels. For adults, there was no significant effect of infection on total dry mass (Cheplick et al. 1989). In another study of *L. perenne* (cv. Yorktown

III) (Cheplick 1997a), plasticity in the phenotypic responses (tiller number, leaf area, and leaf mass) to increasing levels of nutrients supplied as N-P-K fertilizer depended greatly on host genotype, but did not change predictably with endophyte infection. Lewis et al. (1996) did not find any effect of endophyte infection on dry mass or uptake of ammonium (NH_4+) and nitrate (NO_3^-) for one genotype of perennial ryegrass grown in solutions at two concentrations of N. However, in another study Lewis (2004) found greater growth of E+ plants of three genotypes at low N. In addition, Ravel et al. (1997a) reported greater performance of E+ plants under conditions of N deficiency for three *L. perenne* genotypes. Recent research showing a 40% reduction in endophyte concentration of perennial ryegrass under high N supply (Rasmussen et al. 2007) leads to the intriguing speculation that the reduced benefits of endophyte infection that become manifest under some conditions may be the result of low infection intensity within individual host plants.

In an effort to determine if there might be a growth cost associated with endophyte infection for *L. perenne* under nutrient stress, Cheplick (2007) grew four to five genotypes each of four accessions originally collected from Eurasia and North Africa, the probable geographic area of origin for this species. Both E+ and E− ramets of each genotype were grown for 5 months in an extremely nutrient-poor substrate (vermiculite) in temperature- and light-controlled growth chambers. Because the objective was to explore the putative cost of endophyte infection under stressful conditions only, and not to compare the performance of E+ and E− hosts under stressed versus unstressed conditions, soil nutrient availability was not explicitly varied in the experiment. At harvest, root and shoot dry mass was recorded and the proportion of the shoot mass that consisted of living, green, and photosynthetically active tissue was

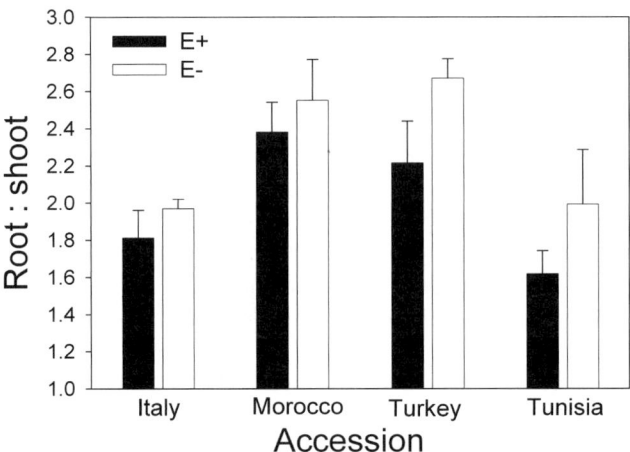

Figure 2.7 Mean (± SE) root:shoot ratios of endophyte-infected (E+) and uninfected (E−) *Lolium perenne* accessions from Italy, Morocco, Turkey, and Tunisia after 5 months growth in an extremely nutrient-poor substrate. Data compiled from Cheplick (2007).

calculated. This proportion was significantly lower for plants that were endo-phyte infected ($F_{1,\ 150} = 7.25$, $P = 0.02$). Furthermore, the root:shoot ratio was significantly lower for E+ plants ($F_{1,\ 150} = 5.78$, $P = 0.03$) (figure 2.7). It was suggested that the reduction in root:shoot ratio and the proportion of the shoot comprised of photosynthetically active tissue found in E+ plants (relative to E− plants) indicates that the costs of endophyte infection may outweigh the benefits under extreme mineral nutrient stress (Cheplick 2007). It remains to be seen whether or not these differences would translate into fitness advantages for uninfected plants growing in impoverished soils in the field.

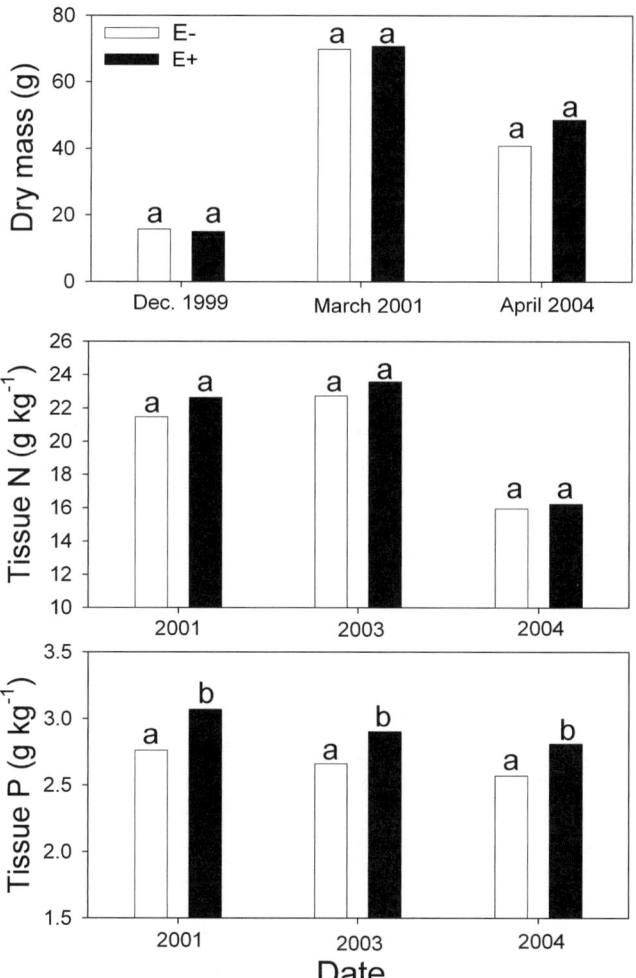

Figure 2.8 Dry mass, tissue nitrogen, and tissue phosphorus content of endophyte-infected (E+) and uninfected (E−) *Festuca rubra* plants growing in a nutrient-poor field site in Spain over 3 years. Data from Zabalgogeazcoa et al. (2006a). Bars with different letters are significantly different at $P < 0.05$.

The impact of low-nutrient substrates on grass-endophyte interactions have also been investigated in *L. pratense* (= *F. pratensis*) infected by *N. uncinatum* (Ahlholm et al. 2002) and in *Festuca rubra* infected by *Epichloë festucae* (Ahlholm et al. 2002, Zabalgogeazcoa et al. 2006a). Under low nutrients, E+ plants of *L. pratense* produced fewer tillers and lower mass relative to E− plants, suggesting a possible growth cost to the host when endophyte infected (Ahlholm et al. 2002). However, this cost was not evident in *F. rubra* and E+ plants showed greater growth in fertilized and watered soils. In the study by Zabalgogeazcoa et al. (2006a), E+ and E− plants of *F. rubra* were grown in a field site for 5 years under the relatively nutrient-poor conditions typical of semiarid grasslands in Spain. Dry mass recorded in 3 years did not differ between E+ and E− plants (figure 2.8). Tissue N content also did not differ between E+ and E− plants, but tissue P content was significantly greater in the former (figure 2.8). This result is interesting in that it parallels that described earlier for tall fescue (Malinowski et al. 1998, 2000, Rahman & Saiga 2005).

In short, there is some evidence that endophytes can improve host growth under some types of mineral stress, for example, under low P conditions (e.g., Ren et al. 2007). Unfortunately, there are too few long-term field studies like that of Zabalgogeazcoa et al. (2006a) and too many inconsistencies among other studies to permit any general conclusion regarding the ecological impact of endophytes to plants, populations, and communities in nutrient-poor soils. However, host benefits to endophyte infection do appear more likely in high-nutrient environments, as they have most often been detected under such conditions (Saikkonen et al. 2006). Although the grass-endophyte interaction may represent an "adaptive symbiosis" in relation to some environmental stresses, as noted in other plant-fungal systems (Rodriguez et al. 2004, 2005), it is presently not wise to assume that endophytes impart greater abiotic stress tolerance to host grasses in natural ecosystems without additional experimental evidence.

2.2.3 Other stresses

There have been a limited number of studies on the impact of other environmental factors on the grass-endophyte symbiosis. Not all of these factors necessarily act as abiotic stresses, however. They include soils contaminated by aluminum (Malinowski & Belesky 2000) or zinc (Monnet et al. 2005), soil pH (Belesky & Malinowski 2000, Cheplick 1993a, Malinowski & Belesky 2000), elevated temperature (Eerens et al. 1998a, Faeth et al. 2002b), and shading (Lewis 2004). Most of the studies cited used either tall fescue or perennial ryegrass as the host grass. In general, experimental results have again been highly variable with regard to the ability of endophytes to ameliorate such stresses, and additional research is warranted.

Documentation of environmental parameters related to global climate change and how they could impact symbiotic interactions among microorganisms and plants is a potentially important avenue for ongoing research (see section 7.2.1). Some of the factors that have been explored in relation to endophyte-infected tall fescue and perennial ryegrass are elevated atmospheric CO_2 (Hunt et al. 2005, Marks & Clay 1990, Newman et al. 2003), elevated ultraviolet (UV) radiation (Newsham et al. 1998), and simulated acid rain (Cheplick 1993a).

Endophyte infection of tall fescue did not appear to alter physiological or morphological variables with changes in ambient CO_2 (Newman et al. 2003); however, some indicators of plant quality for consumers, such as the concentration of crude protein, were greater in E+ plants at elevated CO_2 levels. Infected plants of perennial ryegrass showed greater levels of carbohydrates compared to E– plants, but only in ambient CO_2 conditions, while E– plants had 40% lower concentration of soluble protein under elevated CO_2 (Hunt et al. 2005). The authors suggested that E– plants would be better able to accumulate carbohydrates under elevated CO_2. In another study, both *L. perenne* and *T. flavus* failed to show any interactions between endophyte infection and physiological responses to CO_2 enrichment (Marks & Clay 1990). Also, the expression of choke disease caused by *E. sylvatica* in the perennial grass *B. sylvaticum* was unaffected by elevated CO_2 (Meijer & Leuchtmann 2000). In contrast, Groppe et al. (1999) reported that a CO_2-enriched environment enhanced the production of stromata by *E. bromicola* on infected *Bromus erectus.*

At elevated levels of UV radiation, the number of inflorescences and seeds made by E+ plants (but not E– plants) of perennial ryegrass was significantly reduced, suggesting that the dynamics of the host may be impacted in infected populations as ozone depletion continues (Newsham et al. 1998). For tall fescue exposed to simulated acid rain supplied as foliar sprays at pH 3, 4.5, and 6, E+ plants had reduced dry mass relative to E– plants after 4 weeks at the two lowest pH levels; however, at later harvests (up to 23 weeks) infected plants performed as well or better than E– plants regardless of acid rain exposure (Cheplick 1993a).

One abiotic stress that is prominent in many grasslands is fire. In a unique set of studies designed to investigate the hypothesis that endophytes would increase host resistance to fire, Faeth et al. (2002b) determined that infection frequency in natural populations of Arizona fescue in the southwestern United States was not related to fire frequency or intensity. Following dry-heat treatments at 95°C, 105°C, 110°C, and 115°C, and at room temperature, E+ seeds typically showed lower germination than E– seeds. Furthermore, survival of E+ and E– plants after a prescribed fire in an experimental field plot did not differ. Thus in Arizona fescue, the high levels of *Neotyphodium* infection found in most populations appeared to be unrelated to the potential stress imposed by fire.

As this brief overview shows, results of experiments designed to investigate the possible effects of variables linked to climate change or fire are clearly mixed for the few grass-endophyte symbioses studied to date. Nevertheless, the population and community-level consequences of global climate change and widespread factors such as fire, disturbance, and habitat fragmentation should continue to be examined in relation to the potentially subtle alterations that may be mediated by microbial symbionts of plants. Both ecological and coevolutionary changes in grass-endophyte interactions should be anticipated.

2.3 COMPETITIVE ABILITY

Many endophyte-infected grass species occur in successional fields, pastures, or other densely populated habitats where competitors are a major component of the biotic environment (Bazzaz 1996, Fales et al. 1996, Turkington & Mehrhoff 1990). Assuming that there is a broad overlap in niches between grasses and other herbaceous species due to their requirements for similar resources (light, water, minerals), both intraspecific and interspecific competitive pressures are expected in many ecosystems (Harper 1977). Competitive ability depends upon a combination of plant characteristics, including growth rate, size, and reproductive capacity (Aarssen 2005, Grime 1979). Thus, by expressing the extent to which resources may be denied to competitors via increased growth rate and size of a competitive individual, and the extent to which low levels of resources (preempted by competitors) may be tolerated (Aarssen 2005), competitive ability integrates many of the plant traits considered earlier in this chapter.

Regardless of whether a specific grass-endophyte association is antagonistic or mutualistic, competitive interactions among plants in general are likely to be profoundly impacted by fungal symbionts (Clay 1990c). If endophyte-mediated effects on host competitive ability occur, then this may provide a bridge to additional community consequences, including effects on other trophic levels such as herbivores (Faeth & Bultman 2002), plant community composition (Clay & Holah 1999, Clay et al. 2005, Francis & Baird 1989), and invasiveness potential of infected species (Rudgers et al. 2005) (chapter 6).

Some of the indirect evidence that infected grasses can be more competitive than their uninfected counterparts comes from reports of decreased productivity of legumes such as white clover or alfalfa when growing in the field with E+ tall fescue or perennial ryegrass (Eerens et al. 1998b, Hoveland et al. 1999, Stevens & Hickey 1990, Sutherland & Hoglund 1990, Watson et al. 1993). For example, after 3-years' growth in a field site in Georgia, the yield of alfalfa (*Medicago sativa*) in a mixture with E+ tall fescue was only 71% of its yield when growing with E− tall fescue (Hoveland et al. 1999). Endophyte-infected perennial ryegrass suppressed the growth of white clover (*Trifolium repens*) during summer drought at a field site in Southland, New Zealand (Eerens et al. 1998b). These studies suggest that infected agronomic grasses may be better interspecific competitors in pastures (Francis & Baird 1989), but in other studies E+ hosts have not evinced greater suppression of associated legumes or endophyte effects have varied with environmental conditions (Hoveland et al. 1997, Lewis 1992, Malinowski et al. 1999a). Furthermore, controlled greenhouse experiments of interspecific competition using E+

and E– plants have sometimes found that endophyte infection has a negative effect on host competitive ability (Brem & Leuchtmann 2002, Marks et al. 1991, Richmond et al. 2003).

For field populations and communities, probably the best evidence for enhanced competitive ability of infected hosts comes from studies of tall fescue and its endophyte *N. coenophialum*. Hill et al. (1991) performed two field experiments designed to measure the competitive ability of E+ and E– plants of two tall fescue genotypes in both pure and mixed stands at variable densities. At all densities, E+ plants of both genotypes consistently had greater total mass, tiller production, and seed yield when competing with E– plants. A follow-up report of the same experiment after 5 years revealed that dry mass yield of E+ plants was typically greater than E– plants, regardless of whether they were in pure or mixed stands (Hill et al. 1998). In separate experimental plots in Indiana, E+ or E– tall fescue (cv. Kentucky 31) seeds were sown into a plowed field and biomass data were taken each spring and fall over the next 4 years (Clay & Holah 1999). Other species also established within these plots and the result was a diverse successional community in which the proportion of biomass represented by infected tall fescue increased over time (compared to E– plants). In other plots followed for 4.5 years in the presence or absence of herbivores, E+ tall fescue became more dominant (in terms of relative biomass within the community) over time under ambient mammalian herbivory (Clay et al. 2005). Finally, endophyte infection appears to improve the ability of tall fescue to invade experimentally established plant communities within a controlled greenhouse environment (Rudgers et al. 2005).

Given the increasing evidence that endophyte infection may have major ecological implications for competitive ability in tall fescue, impacting plant diversity and composition, as well as other trophic levels, it is surprising that relatively few investigators have examined host competitive ability in other grass-endophyte symbioses. The possible outcomes of both intraspecific and interspecific competition for E+ and E– hosts will influence conceptions related to the nature of the grass-endophyte relationship (table 2.2). The range

Table 2.2 Possible outcomes of intraspecific and interspecific competition for endophyte infected (E+) and uninfected (E−) grasses.

Competition	Competitors	Outcome	Relationship
Intraspecific	E+ × E−	E+ wins	Mutualistic
Intraspecific	E+ × E−	Equivalent	Commensalistic
Intraspecific	E+ × E−	E− wins	Parasitic
Interspecific	E+ × 2nd species	E+ wins	Mutualistic
	E− × 2nd species	2nd species wins	
Interspecific	E+ × 2nd species	Equivalent	Commensalistic
	E− × 2nd species	Equivalent	
Interspecific	E+ × 2nd species	2nd species wins	Parasitic
	E− × 2nd species	E− wins	

of possible outcomes is likely to span the symbiotic continuum from parasitic to mutualistic, depending again on particular host genotype-endophyte combinations and environmental conditions. The competitive ability of host grasses may be depressed (parasitism), unaffected (commensalism), or enhanced (mutualism) by endophytes (table 2.2). Indeed, all three possible outcomes have been reported in different grass-endophyte systems.

In one of the earliest field experiments of endophyte effects on host competitive ability, ramets of genotypes of *Danthonia spicata* that were uninfected or infected by *Atkinsonella hypoxylon* were forced to compete with ramets of eight genotypes of the co-occurring grass *Anthoxanthum odoratum* (Clay 1984). Survival and flowering proportion were significantly greater for E+ ramets, as were growth (based on tiller production) and reproductive effort (based on inflorescence production). Ramets of *D. spicata* were also grown in intraspecific competition in which two ramets from the same host genotype competed. Infected ramets showed competitive superiority to uninfected ramets in "intragenotypic competition" (Clay 1984). In a subsequent competition study with a similar design, using *D. spicata* and *A. odoratum* grown in the field for 2 years, increases in the number of vegetative and reproductive tillers were significantly greater for E+ compared to E– host genotypes (figure 2.9) (Kelley & Clay 1987). Also, reproductive tiller production was greatest for E+ plants of *D. spicata* in intraspecific competition (figure 2.9).

Because infected individuals of *D. spicata* show abortion of the terminal panicles (see section 2.1.4), it is possible that the resources that would have been used for sexual reproduction were diverted to vegetative growth, thereby increasing competitive ability. However, Kover (2000) could not find evidence to support this speculation, instead noting that infected *D. spicata* reallocated resources that would have been used for host reproduction to the endophyte rather than to vegetative growth. Additional work on allocation patterns of hosts infected by castrating fungi such as *Atkinsonella* and *Epichloë* spp. is needed to better address the hypothesis of resource reallocation to vegetative growth as a mechanism for greater competitive ability in E+ plants.

A few additional studies have provided supporting evidence that the grass-endophyte symbiosis can be mutualistic under competitive conditions. Whether growing in intraspecific E+/E– mixtures or with competing orchardgrass (*Dactylis glomerata*), plants of a single clone of meadow fescue (*L. pratense*) showed greater root and shoot dry mass when infected by *N. uncinatum* (Malinowski et al. 1997b). In the caespitose grass *Bromus benekenii*, which is often infected by asexual strains of *E. bromicola*, E+ target plants had greater tiller and dry matter production, independent of the infection status of competitors (Brem & Leuchtmann 2002). Note that neither of these fungal endophytes causes inflorescence abortion of its host and thus the results do not support the reallocation hypothesis for enhanced competitive ability.

In several other grass-endophyte symbioses, endophyte infection has no effect on, or is detrimental to, host competitive ability. Faeth et al. (2004) investigated *F. arizonica* (Arizona fescue), a native caespitose perennial grass

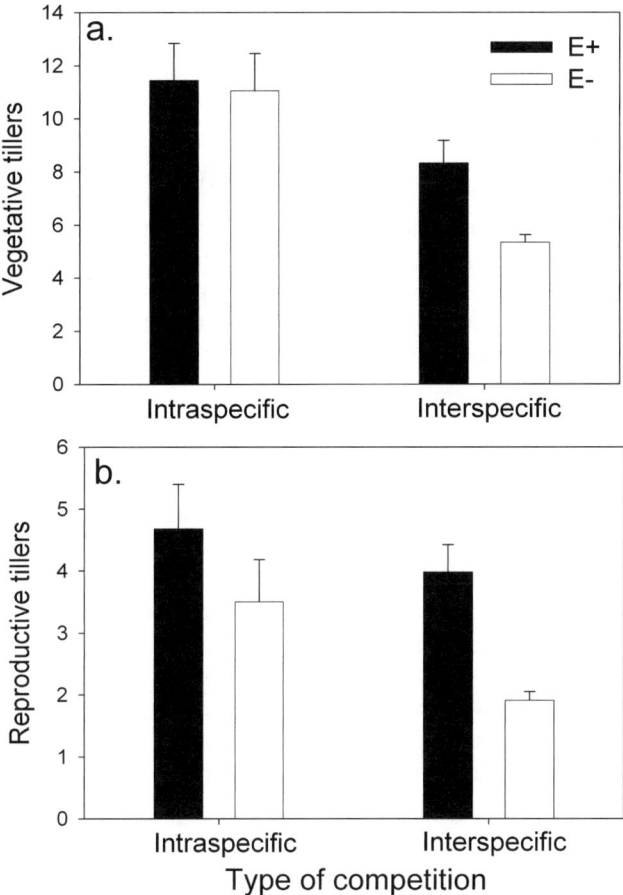

Figure 2.9 Mean (± SE) vegetative and reproductive tiller production of endophyte-infected (E+) and uninfected (E−) *Danthonia spicata* in intraspecific or interspecific competition. Data from Kelley and Clay (1987).

found in Ponderosa pine woodlands in the western United States. Depending on population, 60–100% of the plants are infected by *Neotyphodium starii* (Schulthess & Faeth 1998). Intraspecific competition experiments were established in the field and greenhouse. Plants were paired in three combinations (E+, E+; E+, E−; E−, E−) when planted into the field in spring and subjected to two levels of water and soil nutrients. Growth measurements were recorded at the beginning of the second growing season. Regardless of the infection status of the paired partner, E− plants had significantly greater aboveground dry mass and volume than E+ plants. In addition, the volume of E− plants was greater than that of E+ plants under low water and low nutrients, and under high water and low nutrients (i.e., there was a significant infection by treatment interaction). In the high water–high nutrient treatment, there was no difference between the volume of E+ and E− plants. In the greenhouse experiment, two E+ and two E– seedlings were planted per pot and again subjected to two levels of water and soil nutrients. There was no significant infection by treatment interaction, but E– plants consistently showed significantly greater root, shoot, and total dry mass than E+ plants. The authors concluded "*Neotyphodium* infections in Arizona fescue appear to act parasitically by reducing competitive abilities" (Faeth et al. 2004).

A similar parasitic effect of endophyte infection on intraspecific competitive ability was reported for *B. sylvaticum*, another caespitose woodland grass, infected by *E. sylvatica* (Brem & Leuchtmann 2002). One target plant (E+ or E–) was surrounded by four competing plants (all E+ or all E–) and placed into an experimental garden. Uninfected target plants produced significantly greater aboveground dry mass regardless of the infection status of the competitors.

Competition experiments with perennial ryegrass have revealed a range from negative to positive effects of infection by *N. lolii* under various conditions. Both intraspecific and interspecific experiments were conducted in a greenhouse by Marks et al. (1991) using *L. perenne* cv. Repell. E+ and E– plants were grown together and also with E+ and E– plants of tall fescue in three trials (densities equivalent to 4762/m² or 8163/m²). In both intraspecific and interspecific combinations, E– target plants produced more dry mass than E+ plants, regardless of the infection status of competitors. The authors concluded that endophyte infection "may be detrimental" to competitive ability in this particular cultivar (Marks et al. 1991). In subsequent greenhouse experiments, E+ and E– perennial ryegrass of the same cultivar produced similar dry mass in the absence of herbivory, but E+ plants produced almost twice

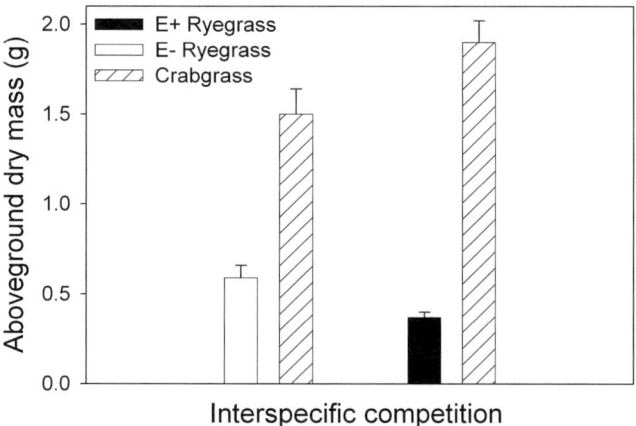

Figure 2.10 Mean (± SE) aboveground dry mass of endophyte-infected (E+) and uninfected (E–) *Lolium perenne* (perennial ryegrass) in interspecific competition with *Digitaria sanguinalis* (crabgrass). The dry mass of crabgrass is also shown for comparison. Redrawn from Richmond et al. (2003).

the dry mass of E− plants when subjected to herbivory by the fall armyworm (*Spodoptera frugiperda*) (Clay et al. 1993). This result, which also occurred for infected tall fescue in interspecific competition, is related to the preference of this herbivore for uninfected plants that do not contain the alkaloids produced by the fungus.

A more recent investigation of interspecific competition between E+ and E− *L. perenne* (cv. Palmer II) and crabgrass (*Digitaria sanguinalis*) revealed a significant detrimental effect of infection on the competitive ability of *L. perenne* (Richmond et al. 2003). In this greenhouse study, the production of tillers and aboveground dry mass was reduced for E+ plants (relative to E− plants) (figure 2.10) when growing with crabgrass at a constant density (1111/m²). Crabgrass showed greater mass when competing with E+ perennial ryegrass (figure 2.10), and also had a greater seed yield (0.46 ± 0.05 g versus 0.35 ± 0.05 g when competing with E− perennial ryegrass). It remains to be determined whether or not endophyte infection is important to the competitive ability of perennial ryegrass under natural field conditions where other weedy plant species and herbivores would also be present (e.g., Richmond et al. 2004a, 2006).

As with many studies of plant competitive ability (Aarssen & Keogh 2002), competition studies with endophyte-infected grasses have typically been conducted in controlled environments over a single growing season, with growth parameters as the primary indicators of competitive ability. A recent meta-analysis of the published studies did not support the contention that endophytes significantly improve the competitive ability of their hosts (Saikkonen et al. 2006). The effects of endophytes on grasses are variable and can range from parasitic to mutualistic, even within a single grass-endophyte system. Thus it is still too early to make any sweeping generalizations regarding the possible impact of endophytes on competitive interactions within grass populations and the communities they inhabit.

3

Endophytes and Host-Plant Herbivore Relationships

3.1 ENDOPHYTES AS DEFENSIVE MUTUALISTS

Historically, resistance to herbivores has been the raison d'être for concluding that systemic endophytes are mutualistic symbionts of grasses. For more than a century, toxicity of some cool-season pooid grasses to livestock had been noted anecdotally. For example, 130 years ago, Hance (1876) reported consumption of a grass in Mongolia caused intoxication in cattle. Toxicity in this grass, commonly called drunken horse grass, is now known to be caused by the recently described *Neotyphodium ganuense* (Li et al. 2004). *Neotyphodium* in this grass produces extraordinarily high levels of ergot alkaloids, up to 2500 ppm (Miles et al. 1998). Levels of about 40 ppm are known to cause severe toxic effects in livestock (Siegel & Bush 1996). Likewise, Freeman (1904) described a fungus in the seeds of *Lolium temulentum* (darnel) and speculated that the fungus was responsible for the toxic effects of darnel seeds when eaten by livestock. Nearly 100 years later, Moon et al. (2000) isolated and described the endophytic culprit, *Neotyphodium occultans*.

The causative agent of toxicoses in cool-season grasses was first directly linked to systemic endophytes and their alkaloids in the 1970s. Bacon et al. (1977) determined that fescue toxicosis in cattle resulted from alkaloids produced by *Neotyphodium coenophalium*, an asexual, seed-borne fungus that inhabits agronomic cultivars of tall fescue (*Lolium arundinaceum*). This particular endophyte has been largely responsible for widespread livestock production losses in North America due to widespread planting of the infected cultivar Kentucky 31, released in the 1940s as a panacea for degraded and drought-stricken pasturelands (Hoveland 1993). Likewise, the alkaloids produced by *Neotyphodium lolii* in another widely cultivated pasture grass, perennial ryegrass (*Lolium perenne*), caused staggers (a neurological disorder), weight loss, and tetanic muscle spasms in cattle and sheep in New Zealand and Australia (Eerens et al. 1998c, Prestidge & Gallagher 1988). Currently, intensive agronomic research is attempting to manipulate the endophyte in these two grasses such that desirable agronomic properties remain (e.g., increased drought tolerance and resistance to invertebrate pests) without the negative effects on livestock (e.g., Hunt & Newman 2005, Yue et al. 2000a).

Although systemic endophytes in pooid grasses have been long known to produce alkaloids that may deter herbivores and grazing, this phenomenon is not limited to the grasses. Recent evidence (Braun et al. 2003) showed that the long-known toxicity of white locoweed (*Oxytropis sericea*) to livestock is due to production of the alkaloid swainsonine by the fungal endophyte *Embellisia*. Interestingly, the level of swainsonine is dependent upon another plant symbiotic microbe (*Rhizobium*) that fixes nitrogen in root nodules of locoweed (Valdez Barillas et al. 2007). Another recent discovery indicates that the unusual production of loline alkaloids by some dicotyledons in the family Convolvulaceae is due to the presence of a clavicipitaceous fungus (Steiner et al. 2006). This fungus is epibiotic rather than endophytic, but is apparently vertically transmitted on seeds of species of *Ipomoea*. These recent studies suggest that endophytic fungi may be responsible for other cases of plant toxicity to herbivores, especially those in which characteristic fungal alkaloids are found.

Past and ongoing studies of infected agronomic grasses have provided a rich empirical basis for concluding that systemic endophytes in grass provide protection to their host against herbivores. Undoubtedly endophytes and their associated alkaloids can have negative effects on invertebrate (e.g., Breen 1994) and vertebrate (e.g., Siegel et al. 1987) herbivores. As a consequence, endophytes in grasses (e.g., Cheplick & Clay 1988, Clay 1988b, Schardl & Clay 1997, Schardl et al. 2004, Vicari & Bazely 1993) and in plants in general (e.g., Carroll 1988, 1991) have long been viewed as defensive mutualists via alkaloids that reduce herbivory. However, more recent studies have questioned the generality of endophytes as protective symbionts against herbivores (Faeth 2002, 2008, Müller & Krauss 2005, Saikkonen et al. 1998, 1999, 2006, Tibbets & Faeth 1999). Much of the earlier work on endophytes in grasses focused on tall fescue and perennial ryegrass where, at least in the agronomic cultivars, resistance to invertebrate and vertebrate herbivores due to infection is often emphatic. However, as studies of endemic infected grasses in their native communities have accumulated, defense against herbivores appears much more erratic, and spans the spectrum from reduced resistance to enhanced resistance (Faeth 2002, Saikkonen et al. 1998, 2006). In this chapter we explore the reasons for this variability.

3.1.1 Alkaloids—the key to herbivore resistance?

It is increasingly clear that endophyte-mediated resistance to herbivores is largely dictated by the levels and types of alkaloids produced by systemic endophytes in host grasses (Clay & Schardl 2002, Faeth 2002, Saikkonen et al. 1998, Schardl et al. 2004, 2007). The four major groups of alkaloids (table 3.1)

Table 3.1 The four major types of alkaloids (and common alkaloids found in each class) in *Neotyphodium* or *Epichloë* endophytes of grasses and their anti-biological effects on vertebrate and invertebrate herbivores, other microbes, and non-host plant competitors. Types of alkaloids may be inferred from infected host grass species used in the studies.

Alkaloid	Vertebrate herbivores?	Invertebrate herbivores?	Microbes?	Plants (allelopathy)?
Pyrrolizidine alkaloids (lolines)	Yes – Small mammals[10,17] – Large mammals[5,41]	Yes – Insects[3,4,7,44,45] No – Insects[45,47]	Yes[11,22,23,24] No[21,25]	Yes[19,27,33] No[18]
Ergot alkaloids (clavines, ergovaline, lysergic acid, ergopeptines)	Yes – Small mammals[10,17,37,38,39,49] – Large mammals[5,41,48] – Avian seed predators[39]	Yes – Insects[8,15,16,35,36,40,43] – Nematodes[12,29] No – Insects[29,47], – Nematodes[32]	Yes[11,22,23,24] No[21,25]	Yes[19,27,33] No[18]
Pyrrolopyrazine alkaloid (peramine)	No[7,11,26]	Yes – Insects[1,7,11,36,42,43,46] No – Insects[30,31,42,47] – Nematodes[32]	No[20,25]	Yes[19,27,33] No[18]
Indole diterpene alkaloids (lolitrems)	Yes – Small mammals[17] – Large mammals[1,2,5,13]	Yes – Insects[2,6,9,15,33,34] – Nematodes[28] No – Insects[47]	Yes[24]	Yes[4]

[1]Prestidge et al. 1982, [2]Prestidge & Gallagher 1988, [3]Yates et al. 1989, [4]Eichenseer et al. 1991, [5]Prestidge 1993, [6]Popay & Wyatt 1995, [7]Siegel et al. 1990, [8]Breen 1994, [9]Rowan & Latch 1994, [10]Coley et al. 1995, [11]Siegel & Bush 1996, [12]Ball et al. 1997, [13]Eerens et al. 1998c, [14]Sutherland et al. 1999, [15]Leuchtmann et al. 2000, [16]Yue et al. 2000a, [17]Conover 1998, 2003, [18]Renne et al. 2004, [19]Orr et al. 2005, [20]Hamilton & Faeth 2005, [21]Burpee & Bouton 1993, [22]Gwinn & Gavin 1992, [23]White & Cole 1985, [24]Yue et al. 2000b, [25]Trevathan 1996, [26]Saikkonen et al. 1999, [27]Matthews & Clay 2001, [28]Eerens et al. 1998d, [29]Kunkel et al. 2004, [30]Lopez et al. 1995, [31]Tibbets et al. 1999, [32]Timper et al. 2005, [33]Peters & Zam 1981, [34]Carrière et al. 1998, [35]Clay & Cheplick 1989, [36]Rowan et al. 1990, [37]Bhursari et al. 2006, [38]Fortier et al. 2000, [39]Wolock-Madej & Clay 1991, [40]Knoch et al. 1993, [41]Thompson & Stuedemann 1993, [42]Krauss et al. 2006, [43]White et al. 2001a, [44]Bultman et al. 2004, [45]Bultman & Bell 2003, [46]Meister et al. 2006, [47]Ball et al. 2006, [48]Bazely et al. 1997, [49]Koh & Hik 2007

found in infected grasses include (1) peramine, the only pyrrolopyrazine alkaloid found in infected grasses (Siegel & Bush 1996), active mainly against invertebrate herbivores (Leuchtmann et al. 2000, Rowan & Latch 1994); (2) loline or pyrrolizidine alkaloids, which deter insects and small mammals, but not livestock (Leuchtmann et al. 2000); (3) ergot alkaloids, which provide protection against vertebrate and invertebrate herbivores; and (4) lolitrems or indole diterpene alkaloids, which are active against mammalian and insect herbivores (Leuchtmann et al. 2000, Prestidge 1993, Siegel & Bush 1996).

The alkaloids—lolitrems in agronomic perennial ryegrass, lolines, ergot alkaloids, and peramine in tall fescue cultivars—are thought to be largely responsible for the well-known toxic and neurological effects on livestock and the resistance to generalist pest insect herbivores (Clay & Schardl 2002, Schardl et al. 2004, 2007, Siegel & Bush 1996). Similarly the ergot alkaloids in some wild grasses infected with *Neotyphodium* are well known for causing toxicity in domesticated livestock in Australia (e.g., Miles et al. 1996, 1998), North America (e.g., Jones et al. 2000), and Asia (Li et al. 1997a,b). However, the most common alkaloid found in grasses infected with *Neotyphodium* or its sexual counterpart, *Epichloë*, in native grasses appears to be peramine (Leuchtmann et al. 2000). With a few exceptions, infected wild grasses in their native habitats do not appear to harbor either the variety or high levels of alkaloids that are found in the two agronomic grasses, tall fescue and perennial ryegrass (Faeth 2002). In keeping with alkaloids as the major route of herbivore resistance, it also appears that infected native grasses exhibit less toxicity to vertebrates or invertebrates than agronomic tall fescue and perennial ryegrass (Faeth 2002), including wild varieties of these two grasses (e.g., Bony et al. 2001, Saikkonen et al. 2002). However, relatively few infected native grasses have been thoroughly examined for alkaloid content and for herbivore resistance due to systemic endophyte infections, especially for resistance to invertebrate herbivores.

3.2 EFFECTS ON VERTEBRATES

3.2.1 Agronomic grasses

By far, the best known effects of endophyte infection on herbivores involve livestock, undoubtedly because of the economic repercussions of reduction in production due to infected tall fescue and perennial ryegrass (e.g., Hoveland 1993). In tall fescue, loline and especially ergot alkaloids contribute to fescue toxicoses, including elevated temperature, heat stress, increased vasoconstriction, and reduced reproduction and milk production, and to fescue foot or hoof gangrene (Siegel & Bush 1996, Thompson & Stuedemann 1993). In perennial ryegrass, ergot alkaloids and lolitrems result in livestock staggers, a neurological disorder associated with reduced weight gain, decreased fertility, and impaired neuromuscular function (Gallagher et al. 1984, Prestidge 1993, Siegel & Bush 1996).

Infected tall fescue and perennial ryegrass cultivars may also have harmful effects on native small mammals and migrating birds. Proliferation of infected tall fescue variety Kentucky 31 since the 1940s has likely resulted in reduced abundance of multiple insectivorous and herbivorous small mammal species in North America (Coley et al. 1995). Consumption of infected tall fescue alters sex ratios and delays female sexual maturation in prairie voles (Fortier et al. 2000), but not female vole home range (Fortier et al. 2001). Likewise, consumption of infected perennial ryegrass cultivars reduced weight gain in male, but not female, meadow voles (Conover 2003).

Infected tall fescue also alters feeding preferences and weight gain of granivorous native birds and herbivorous Canadian geese. Wolock-Madej and Clay (1991) demonstrated that five species of passerine birds preferred E− to E+ tall fescue seeds (cv. Kentucky 31). One species, the dark-eyed junco, fed mixed diets of E− and E+ tall fescue seeds (cv. Kentucky 31) and millet had reduced weight gain on diets higher in E+ seeds. Conover and Messmer (1996) likewise showed that Canadian geese avoided agronomic cultivars of E+ tall fescue (cv. Kentucky 31) and showed reduced weight gain relative to E− tall fescue. It is not clear which alkaloid or alkaloids mediated these effects on native birds and small mammals, but ergot alkaloids, lolines, and lolitrems are likely suspects. It is also not known whether the consumption of endophyte-infected grass seeds by these birds in nature is great enough to have any significant effect on their growth, survival, and reproduction.

In a comprehensive study of the effect of systemic endophyte infections in *Festuca rubra* on vertebrate and invertebrate herbivory in several Scottish Islands, Bazely et al. (1997) found that grazing by sheep was positively correlated with infection frequency on one island (Hirta) but not two others. The systemic endophyte, later identified as *Epichloë festucae* (Bazely et al. 2007), can be horizontally transmitted, but is often seed-borne. Although *F. rubra* is a wild grass in Europe, is not clear from the study whether the pastures used in the study were natural or planted. Nonetheless, they found that infected *F. rubra* from Hirta also reduced the growth and survival of locusts in bioassays, and higher levels of ergovaline, the suspected antiherbivore compound, could also be induced by clipping.

3.2.2 Wild grasses

About a dozen wild grasses are known for strong toxicity to vertebrates because of seed-borne endophyte infections (Faeth 2002). Most of these vertebrates are domesticated livestock (e.g., horses, cattle, and sheep) and we know little about how endophyte-infected wild grasses affect native vertebrates. These cool-season grasses are endemic forage grasses, and livestock losses to toxic grasses result in attention by pastoralists and ranchers, and subsequently researchers. High levels of ergot alkaloids, lolitrems, and lolines in these infected grasses appear responsible for the toxicity (Jones et al. 2000, Miles et al. 1996, 1998, Siegel & Bush 1996). Alternatively, nontoxic but infected grasses garner little attention, and only recently have these nontoxic endophyte-grass associations been studied (e.g., Faeth 2002, Faeth & Sullivan 2003, Lewis & Clement 1986, Mirlohi et al. 2006, Nan & Li 2001, Saikkonen 2000, Saikkonen et al. 1999, Vinton et al. 2001, Wei et al. 2006).

Within the infected toxic grasses, variability in deterrency and toxicity is the hallmark of these grasses. Even the poisonous agronomic cultivars of tall fescue (Piano et al. 2005, Saikkonen 2000) and perennial ryegrass (Bony et al. 2001) show a wide range of alkaloid levels in their native habitats. In their range of origin, infected tall fescue (Leuchtmann et al. 2000, Piano et al. 2005, Saikkonen 2000) and perennial ryegrass (Bony et al. 2001) range from no alkaloids to moderate and high levels, unlike their uniformly toxic agronomic counterparts. Bony et al. (2001) surveyed 83 wild European populations of perennial ryegrass and found the different morphological strains of *N. lolii* exhibited remarkable variation in levels of lolitrem B, ergovaline, and peramine (figure 3.1). Some infected plants produced no lolitrem B, ergovaline, or peramine. Of particular note is that, on average, infected plants produced lower peramine levels (median 6.3 ppm) than agronomic perennial ryegrass cultivars (range 10–30 ppm), but infected plants from two wild populations had extremely high levels, greater than 40 ppm.

For native tall fescue from three European populations, Leuchtmann et al. (2000) found that loline concentrations ranged from 0 to 2286 ppm, ergovaline from 0 to 2 ppm, and peramine from 0 to 3 ppm. The enormous range in loline alkaloids that are largely responsible for toxicity to livestock in agronomic cultivars of tall fescue, of more than four orders of magnitude, is particularly impressive in these native populations. Similarly Piano et al. (2005)

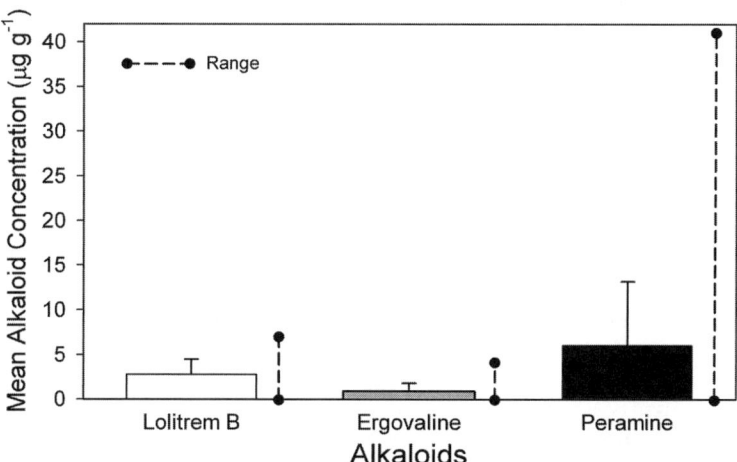

Figure 3.1 Mean (± SD) alkaloid variation and the range in alkaloid levels (lolitrem B, ergovaline, and peramine) in 83 wild populations of perennial ryegrass (*Lolium perenne*). Means were calculated from Bony et al. (2001, figure 4a,b,c) and are conservative in estimating variation because values were reported as frequency classes. Actual variation is likely greater.

found that alkaloid levels varied greatly among 60 populations of wild tall fescue populations on Sardinia (figure 3.2), and were related to morphological variants of the endophyte *Neotyphodium coenophialum*. One morphological ..rain (with long conidia) produced loline and ergovaline levels comparable to that of three infected tall fescue cultivars. However, another strain (with short conidia) had alkaloid levels much lower than the agronomic cultivars. These two tall fescue types on Sardinia may be different hexaploid and hybrid species of tall fescue with different endophytes, one as *N. coenophialum* and the other identified as FaTG-2 by Christensen et al. (1993) and detailed in Tsai et al. (1994) (C. Schardl, personal communication). Nonetheless, the examples in Piano et al. (2005) demonstrate the wide variability in grass and endophyte types and their alkaloids in natural communities.

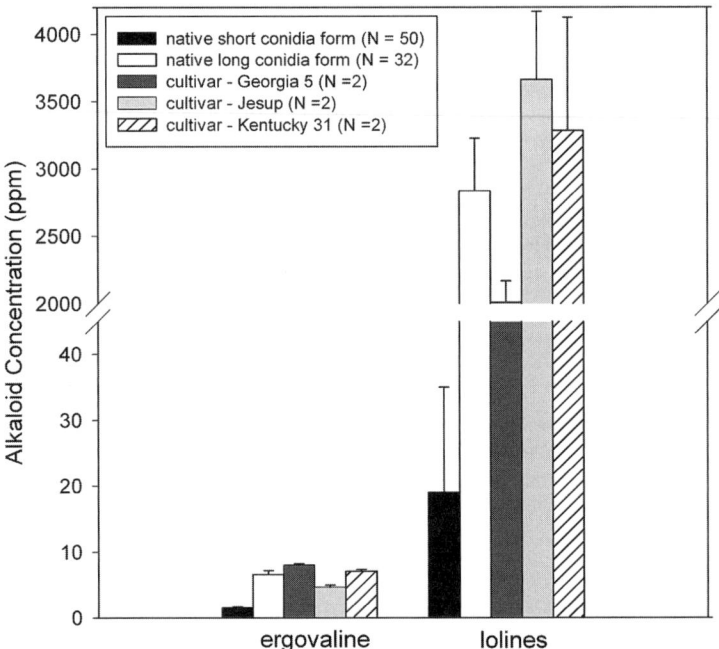

Figure 3.2 Variation in loline and ergovaline alkaloid levels from 58 native tall fescue populations in Sardinia, Italy (Piano et al. 2005). Populations were distinguished by two morphological forms of the *Neotyphodium* endophyte, a short-conidia ($N = 50$) and a long-conidia ($N = 8$) form. Generally, tall fescue with the more common short-conidia form produced far less of both loline and ergovaline alkaloids. Within each form of *Neotyphodium*, considerable variation in alkaloid levels also exists, as evidenced by the standard error bars. Note that ergovaline levels were incorrectly reported in Piano et al. (2005, table 3) as µg/g instead of µg/kg (F. Bertoli, personal communication); corrected data are shown here. Data courtesy of Federico Bertoli. Figure reprinted from Faeth and Saikkonen (2007) with permission of the New Zealand Grassland Association.

In a recent study of sleepygrass (*Achnatherum robustum*, formerly *Stipa robusta*), a North American native grass long known for its toxicity to livestock (e.g., Marsh & Clausen 1929, Petroski et al. 1992), Faeth et al. (2006) showed that the notorious toxicity of this grass exists in only a tiny fraction of its range. Other populations, although they are highly infected, exhibit very low or no alkaloids at all. Thus it is dangerous to conclude that any of the 10 or so well-known cases of endophyte-mediated toxicity to vertebrates are uniformly noxious, even if infected. Indeed, the variability within an endophyte-grass association is often equal or even greater than that among infected host grasses species. Therefore attempts to characterize any given endophyte-grass symbiosis by some static level of resistance or toxicity to vertebrates or livestock is a fruitless endeavor.

This is not to say that endophytes and their alkaloids in native grasses do not have negative effects on native vertebrates or that grazing by vertebrates, in turn, does not alter infection frequencies or alkaloid distribution in wild grass populations. For example, in a recent study, Koh and Hik (2007) found a positive correlation between grazing intensity by native marmots and pikas and *Neotyphodium* infection frequency in the subalpine native grass, *Festuca altaica* near and away from boulder fields. Although causation for this pattern was not established in their study, at least one hypothesis is that intense grazing selects for infected grasses with herbivore-deterring alkaloids, similar to livestock grazing by livestock that increases infection in agronomic grasses (e.g., Clay 1998). In some cases of native grasses and vertebrate herbivores, we should expect that vertebrate grazing, endophyte frequency, and alkaloid levels may be tightly linked, where in others, little relationship may exist for the reasons explored below in section 3.7.

3.3 INVERTEBRATE HERBIVORY

Like resistance to vertebrates, deterrence of herbivory by invertebrates in infected grasses is largely linked to alkaloid varieties and levels produced by *Neotyphodium* or *Epichloë* endophytes. For example, resistance to insect pests in agronomic tall fescue seems linked to levels of peramine and loline alkaloids (Clay 1989, 1991a, Siegel & Bush 1996, 1997) and in agronomic perennial ryegrass to levels of lolitrems (Breen 1994, Carrière et al. 1998, Popay & Wyatt 1995, Prestidge & Gallagher 1988). In perennial ryegrass, resistance to insect pests, such as the chinch bug, in field plots of E+ plants is often visually dramatic, as shown in figure 3.3. Clay (1989, 1991a: table 17.1) lists 14 infected pooid grasses with known resistance to invertebrate herbivores. Because alkaloids are often concentrated in seeds (Leuchtmann et al. 2000), resistance to invertebrate seed predators is also predicted in agronomic grasses. There have been few experimental tests of infected seed toxicity to invertebrate seed predators. Cheplick and Clay (1988) found that infected tall fescue seed was highly toxic to *Tribolium* flour beetles. Knoch et al. (1993) experimentally showed that one species of seed-harvesting ant preferred E– tall

Figure 3.3 Endophyte-mediated resistance of perennial ryegrass (*Lolium perenne*) to the chinch bug. The plot labeled Entry 20 is E+ perennial ryegrass; Entry 25 consists of E– perennial ryegrass. Photograph courtesy of Kevin Morris, National Turfgrass Evaluation Program, Beltsville, MD.

fescue to E+ seed, while another species showed no preference. Various alkaloids appear to act as deterrents (e.g., peramine), while others decrease growth and survival (e.g., lolitrems) (Schardl et al. 2004, 2007).

Most of the research to date on invertebrate resistance has been conducted with agricultural invertebrate pests feeding on agronomic cultivars of tall fescue (*L. arundinaceum*), perennial ryegrass (*L. perenne*), or Italian ryegrass (*Lolium multiflorum*). However, even in infected agronomic grasses, resistance to invertebrate herbivores, most of which are generalist pest species, is not a certainty. In a review, Saikkonen et al. (1998) found that only 66% and 71% of bioassays involving tall fescue and perennial ryegrass, respectively, showed increased resistance to insect herbivores. Far less is known about invertebrate resistance of infected grasses in their native habitats. Some wild grasses infected with *Epichloe* or *Neotyphodium* appear more resistant to invertebrate herbivores (e.g., Brem & Leuchtmann 2001, Clement et al. 1992, 1997, 2005, Tintjer and Rudgers 2006), while others are not different (e.g., Clement et al. 1997, 2005, Lopez et al. 1995) or less resistant (e.g., Saikkonen et al. 1999, Tibbets & Faeth 1999) than their uninfected host grasses. Saikkonen et al. (2006) compiled studies of endophyte effects on herbivory in a meta-analysis and found that 53% of all studies of the effects of grass endophytes on herbivore resistance showed variable outcomes, ranging from positive to neutral to negative.

Thus accumulating studies suggest enormous variability in invertebrate resistance to endophyte-infected native grasses (Clement et al. 1997, Faeth 2002, Faeth & Bultman 2002, Lopez et al. 1995, Meister et al. 2006, Saikkonen et al. 1999, Tibbets and Faeth 1999, Tintjer & Rudgers 2006), much like resistance to vertebrate herbivores. Adding to this variability is that some herbivore-damaged grasses may show increased (induced) levels of alkaloids which increase subsequent resistance to herbivores, at least in tall fescue (Boning & Bultman 1996, Bultman et al. 2004, Bultman & Ganey 1995) and in *F. rubra* (Bazeley et al. 1997). In general, the variability in resistance to invertebrates is much greater in wild infected grasses than that found in agronomic grasses. For example, Tintjer and Rudgers (2006) found that resistance to native invertebrate herbivores in the wild grass *Elymus hystrix* depends upon the strain of *Epichloë elymi* that was inoculated into different plant genotypes. Although they found no significant reduction in herbivory between E+ and E− plants, resistance to more generalized leaf scrapers and chewers varied among E+ plants with the 12 different endophyte strains, and non-stroma-forming tillers experienced greater damage than stroma-forming tillers within plants. Overall Saikkonen et al. (2006), in a meta-analysis of systemic endophyte studies, showed that results of herbivory studies in native grasses were significantly more variable in their outcomes relative to herbivory studies involving agronomic tall fescue and perennial ryegrass.

3.4 CAUSES FOR VARIABILITY IN HERBIVORE RESISTANCE

What drives this variability in herbivore resistance? The obvious answer would seem to be variations in alkaloid types and levels in host grasses are directly and positively related to greater herbivore resistance (e.g., Clay & Schardl 2002). But alkaloid variation may only be a partial answer. In some cases, alkaloid levels and types may even be inversely related to herbivore resistance, indicating alkaloid types and levels alone may not be good predictors of herbivore resistance.

If herbivore resistance is largely linked to alkaloid types and levels, then genetic and environmental factors that mediate alkaloid types and levels should also be linked to herbivore resistance. The same generally holds for constitutive and induced plant allelochemicals that influence herbivory (e.g., Fritz & Simms 1992, Karban & Baldwin 1997). Plant allelochemical types and levels vary by plant species and genetic variability within species, as well as environmental factors, such as soil nutrients and light availability (e.g., Dudt & Shure 1994). For infected grasses, however, genetic variation in the endophyte also comes into play (e.g., Clement et al. 1992, 2001; Tintjer & Rudgers 2006), and in some cases overrides other factors.

Genetic variation in grass endophytes includes among- and within-species variation. The latter includes the strain and haplotype of the endophyte. Genetic variation in the endophyte at either the among- or within-species level often dictates alkaloid types and concentrations (Bony et al. 2001, Bush et al. 1997, Christensen et al. 1993, Leuchtmann et al. 2000, Piano et al. 2005, Rasmussen et al. 2007, Siegel et al. 1990, Timper et al. 2005), although host genotype and environmental factors also play a role (e.g., Easton et al. 2002). Indeed, specific genes and gene clusters for some alkaloids, such as lolines (Spiering et al. 2006b, Wilkinson et al. 2000) and lolitrems (Young et al. 2005), have been identified and manipulated in some grasses. Schardl and Craven (2003) suggested that hybridization in some *Neotyphodium* endophytes was a mechanism by which asexual endophytes could acquire genes for production of novel alkaloids, and thus resistance to herbivores. However, to date there is no convincing evidence that hybridization increases the variety or levels of alkaloids. In fact, in two native grasses, Arizona fescue (*Festuca arizonica*) and sleepygrass (*A. robustum*), hybrid endophytes are associated with either equal or even lower levels of alkaloids (Faeth et al. 2006, Sullivan & Faeth 2004). Although the jury is still out on hybrid endophytes and alkaloid production, clearly the haplotype or strain of the endophyte influences alkaloid varieties and amounts. Within natural grass populations, the haplotype of even asexual endophytes varies greatly (see chapter 4), presumably because of variation in *Epichloë* ancestors or from hybridization between co-occurring *Neotyphodium* strains or between asexual *Neotyphodium* and sexual *Epichloë* endophytes (Schardl & Craven 2003). For example, Sullivan and Faeth (2004) examined multiple populations of Arizona fescue and found three nonhybrid and five hybrid *Neotyphodium* haplotypes that were highly spatially structured among populations.

But endophyte haplotype is not the only factor influencing alkaloid production. In agronomic infected grasses, greater soil nitrogen may increase alkaloid levels (Agee & Hill 1994, Bacon 1993), particularly when host plants are under moderate water stress (Arachevaleta et al. 1992, Belesky et al 1989, Lane et al. 1997). However, in some other cases, increased nitrogen appears to decrease alkaloid concentrations. In agronomic perennial ryegrass, Hunt et al. (2005) found that alkaloid concentrations were reduced when nitrogen was increased. Likewise, Rasmussen et al. (2007) recently found that nitrogen supplements reduced peramine and lolitrem B concentrations in two perennial ryegrass cultivars, Aberdove and Fennema, infected with three strains of the *Neotyphodium* endophyte (AR1, AR37, and the common or wild strain). In addition, the high carbohydrate cultivar, Aberdove, showed even further reductions in alkaloids under nitrogen supplementation. Alkaloid levels in tall fescue can also be induced by previous herbivore damage and thus affect subsequent herbivores (e.g., Bultman et al. 2004, Sullivan et al. 2007).

Less is know about native grasses, but Faeth et al. (2002a) found no effects of soil moisture or nutrient levels on levels of peramine in infected Arizona fescue. Because alkaloids are nitrogen-rich compounds, restricted nitrogen may limit alkaloid production (Faeth 2002). Alkaloids also vary seasonally for agronomic (Ball et al. 1993, Belesky et al. 1987, Roylance et al. 1994, Spiering et al. 2005, 2006b) and some native grasses (Faeth et al. 2002a, 2006, Leuchtmann et al. 2000), with alkaloid levels generally increasing through the growing season (Leuchtmann et al. 2000, Siegel & Bush 1996, Spiering et al. 2006b). Host genotypic background also influences alkaloid production (Easton et al. 2002). Faeth et al. (2002a) found that the same endophyte haplotype in different maternal hosts of Arizona fescue produces significantly different levels of peramine. In herbivory assays of an endophyte that was artificially inoculated into a novel host for agronomic reasons, Hunt and Newman (2005) found that the novel combination afforded less herbivore resistance than the wild-type infected hosts harboring the same endophyte, suggesting that host background influences herbivore resistance in agronomic grasses as well. Similarly Bultman et al. (2006) showed resistance to the aphid *Rhopalosiphum padi* differed in two cultivars of agronomic tall fescue (Georgia 5 and Jesup) containing different isolates of *N. coenophialum*, although both infected grasses were more resistant than their uninfected counterparts.

Alkaloid levels may also vary within plants. Of interest, the distribution and density of the endophyte within perennial ryegrass plants accounts for little of the within-plant variability in alkaloids, and specific alkaloid levels appear to be regulated differently in different plant tissues (Spiering et al. 2005). Yet, in a recent paper, Rasmussen et al. (2007) found that endophyte density within infected perennial ryegrass cultivars is linearly related to concentrations of peramine, ergovaline, and lolitrem B. This study suggests that hyphal density, which is a function of endophyte strain and environmental factors, may be directly related to alkaloid concentrations. Certainly, additional research, especially on infected wild grasses, is needed to understand the relationship between host and endophyte genotype, hyphal density, and environmental factors and alkaloid concentration within and among infected grass hosts.

It should also be noted that systemic endophyte infection alters other aspects of host morphology, physiology, and chemistry in addition to alkaloids, which may alter host plant susceptibility to herbivores. The well-known changes in root morphology, leaf rolling, water-use efficiency, leaf water potential, photosynthesis, plant hormonal changes, and nutrient uptake (e.g., Bacon & Hill 1997, Malinowski et al. 2000, Morse et al. 2007) all potentially influence choice, growth, and reproduction of herbivores. Furthermore, infection may produce nonalkaloid allelochemicals in grasses that may have biological activity with herbivores. For example, Powell et al. (1994) isolated resveratrol from infected *Festuca versuta*, although it was also isolated in lower concentrations from uninfected plants. Resveratrol is a phytoalexin that may have biological activity against herbivores and pathogens. Rasmussen et al. (2008) found that endophyte infection in agronomic varieties of perennial ryegrass interacted with nitrogen availability to alter nitrate, several amino acids, water soluble carbohydrate, magnesium, lipids, organic acid, chlorogenic acid, and carbon/nitrogen levels. Some or all of these potentially affect herbivores and thus the action of endophyte infection on herbivory is likely much broader and more complex than the production of alkaloids by fungal endophytes.

Finally, adaptation and evolution of the herbivores themselves may lead to variation in susceptibility to endophytic alkaloids. Many invertebrate herbivores evolve the ability to tolerate or detoxify plant allelochemicals (e.g., Fritz & Simms 1992). Some specialized insect herbivores not only detoxify plant allelochemicals but may also require them for oviposition, development, and their own defenses against their natural enemies (e.g., Barbosa & Letourneau 1988, Bowers 1990). We should expect likewise for alkaloid-based defenses conferred by endophytes. For example, the bird cherry-oat aphid *R. padi*, commonly used in bioassays to test the effect of endophytic alkaloids (e.g., Leuchtmann et al. 2000, Siegel and Bush 1996, Seigel et al. 1990), appears highly sensitive to low levels of alkaloids in perennial ryegrass and tall fescue (Hunt & Newman 2005, Krauss et al. 2007, Meister et al. 2006). In contrast, *Metopolophium festucae*, the fescue aphid, is relatively insensitive to endophytic alkaloids in perennial ryegrass (Krauss et al. 2007).

3.5 ANTIMICROBIAL EFFECTS OF ENDOPHYTES AND THEIR ALKALOIDS

Host plant pathogens, including fungi, yeasts, bacteria, and viruses can be viewed simply as microherbivores and parasites that consume plant tissues, usually from within (Faeth & Wilson 1996). Systemic endophytes of grasses, as well as endophytes in general (e.g., Arnold et al. 2003, Wiyakrutta et al. 2004), have been hypothesized to inhibit plant pathogens and thus benefit their hosts (Clay 1988b, Siegel & Bush 1996). Again, the mode of resistance is generally viewed as endophyte-associated alkaloids (Siegel & Bush 1996). Some endophyte-produced alkaloids show antimicrobial activity against fungal pathogens in culture (Siegel & Latch 1991, White & Cole 1985, Yue et al. 2000b) and in plants (Gwinn & Gavin 1992, Stovall & Clay 1991). Field tests of endophyte resistance to plant pathogens are relatively few. For *Epichloë*-infected grasses, Clarke et al. (2006) demonstrated that infection by *E. festucae* increased resistance to *Sclerotinia homoeocarpa* (dollar spot disease) in fine fescues, and Koshino et al. (1987) demonstrated increased resistance of *Epichloë*-infected timothy grass (*Phleum pratense*) to *Cerospora* leaf spot disease. For *Neotyphodium*-infected grasses, Trevathan (1996) could not detect differences in growth and survival between E+ and E− tall fescue when exposed to the soil pathogen *Cochliobolus sativus*.

There have been few tests of resistance of infected adult wild grasses to pathogens. Clay et al. (1989) showed that infected wild *Panicum agrostoides* was more resistant to fungal pathogen attack than uninfected hosts. Nan and Li (2001) reported that *Neotyphodium* in adult plants is negatively correlated with abundance of other and potentially pathogenic fungi in some endemic Asian grasses. However, Schulthess and Faeth (1998) found the opposite in Arizona fescue: adult plants infected with *Neotyphodium* had increased abundance of other endophytic and possibly pathogenic fungi.

In recent in vitro greenhouse and field tests of susceptibility of E− and E+ meadow ryegrass (*Lolium pratense*) cultivars to strains of the pathogenic grass-snow mold *Typhula ishikariensis*, Wäli et al. (2006) found highly variable outcomes depending on the type of test, pathogen strain, and grass cultivar. In vitro cultures showed that the *Neotyphodium uncinatum* endophyte clearly inhibited growth of the snow mold. But in the greenhouse and field experiments, E+ meadow ryegrasses were more susceptible to the snow mold, and E+ grasses suffered more winter damage. However, E+ grasses recovered better than E− grasses, indicating greater tolerance to snow mold damage. Furthermore, there were significant interactions between the pathogen strain and infected cultivar type, suggesting genetic effects of both host grass and pathogen in the outcome of the interaction. These types of genetic interactions that alter host resistance to both macro- (e.g., Popay & Bonos 2005) and microherbivores (Hamilton & Faeth 2005, Trevathan 1996) are likely common in infected grasses, and complicate any general conclusions about the effects of endophytic infection on host resistance.

The fact the some grass diseases are vectored by insect herbivores further renders predictions of endophyte-mediated resistance to grass pathogens challenging. In a recent study, Lehtonen et al. (2006) tested the effects of *N. uncinatum* infection on the frequency of the barley yellow dwarf virus (BYDV) in meadow ryegrass (*L. pratense*). In a common garden experiment within an old pasture with E– and E+ adult meadow ryegrass plants, BYDV-infected aphids (*R. padi*) were released and allowed to reproduce. Plants were harvested and screened for BYDV infection. E+ plants had significantly fewer BYDV infections than E– grasses. The researchers surmised that alkaloids in E+ plants deter aphid feeding, and thus indirectly reduced vectoring of the BYDV virus. *R. padi* is known for being highly sensitive to alkaloids in infected grasses (chapter 6). This experiment demonstrates nicely that other interacting species, in this case the insect vector, can affect outcomes of endophyte-pathogen interactions. Furthermore, the researchers suggested a practical use of their results—intercropping E+ meadow ryegrass with cereal crops that are highly susceptible to BYDV may protect the cereal crop from viral damage.

Pathogen resistance may be particularly crucial at the seed and seedling stage. Increased resistance to herbivores at the adult plant stage may only indirectly affect fitness by altering growth and reproduction over the life span of the host. However, increases in seed or seedling survival directly affect host and endophyte fitness (Faeth & Bultman 2002, Knoch et al. 1993, Wolock-Madej & Clay 1991). Furthermore, alkaloids are often concentrated in seeds (Leuchtmann et al. 2000, Siegel et al. 1990), leading some to suggest that protection of seeds is one of the main routes of mutualistic interaction with the host (e.g., Clay 1988b, Leuchtmann et al. 2000, Siegel et al. 1990). Both E+ and E– grass seeds harbor an array of other endophytic fungi, some of which are known pathogens (Hamilton & Faeth 2005). Tests of the hypothesis that seed-borne endophytes protect seeds against pathogens are scarce for both agronomic and native grasses. Burpee and Bouton (1993) found no difference in resistance of E+ and E– seedlings to *Rhizoctonia* (damping off) disease in tall fescue. Hamilton and Faeth (2005) showed that infected seeds of native Arizona fescue harbor more, not fewer, weedy endophyte pathogens. Moreover, infected seeds had more mortality due to fungal pathogens than E– seeds, and consequently, lower germination and seedling survival rates. They suggested that for systemic endophytes like *Neotyphodium* to persist within the host plant, immunological responses of the host may be suppressed, such that resistance to other fungal endophytes and pathogens is reduced.

Although tests of increased resistance to pathogens of grasses in the adult stage are currently limited (e.g., Schardl et al. 2004), especially for native grasses, relative to studies of resistance to herbivory by invertebrates and vertebrates, the few studies to date suggest wide variability in outcomes of endophyte infection, much like endophyte-mediated resistance to invertebrate and vertebrate "macro" herbivores. Studies of endophyte-mediated resistance to seed pathogens are even rarer. However, the two studies to date suggest either no protective role of endophytes at the seed and seedling stage, or reduced resistance, despite the fact that endophyte alkaloids appear to have antimicrobial activity against potential pathogens in culture, and that alkaloids are often concentrated in seeds. Clearly additional studies are needed in the area of pathogen-resistance via endophyte infections in both agronomic and wild grasses.

3.6 HERBIVORY AND GRASS COMPETITIVE INTERACTIONS

As discussed in chapter 2, endophyte infections potentially alter the competitive abilities in their hosts, sometimes increasing overall competitive ability by increasing resistance to drought and other stress, increasing mineral uptake, and increasing resistance to herbivores and pathogens. Thus alterations in herbivore resistance due to endophyte infection may be considered part and parcel of changes in competitive ability, as improved herbivore resistance may lead to greater growth and thus better competitive ability. However, most studies of herbivore resistance occur in the absence of plant competition, typically in greenhouses on singly grown plants in pots or in the laboratory on excised portions of grass leaf tissues. However, herbivory and competition typically interact to affect the fitness and demography of the grass host (e.g., Strauss & Irwin 2004).

In one of the first comprehensive tests of the interaction between herbivory and plant competition, Clay et al. (1993) grew three species of E+ and E– agronomic grasses (tall fescue, perennial ryegrass, and red fescue) under intra- and interspecific competition (with orchardgrass [*Dactylis glomerata*] and Kentucky bluegrass [*Poa pratensis*]) and with and without the presence of the fall armyworm, *Spodoptera frugiperda*, a generalist pest caterpillar. Generally E+ plants suffered less herbivory and had more biomass than E– plants. However, the outcomes of interspecific competitive interactions were contingent on the identity of the competing grass species, the presence or absence of the endophyte, and the caterpillars. E+ perennial ryegrass grew significantly better in competition with Kentucky bluegrass than E– perennial ryegrass, but only when herbivory was present. Likewise, E+ tall fescue was a poor competitor with orchardgrass when herbivores were absent, but E+ tall fescue outcompeted orchardgrass when fall armyworm was present. Clay et al. (1993) concluded that "competitive hierarchies among grasses are altered by interactions with insect herbivores and fungal endophytes..."

In a more recent greenhouse study, Richmond et al. (2004a) conducted a series of experiments testing the effect of belowground herbivory by the Japanese beetle (*Popilla japonica*) on competitive interactions between E+ and

E– tall fescue and perennial ryegrass with dandelion (*Taraxacum officinale*). Similar to Clay et al. (1993), they found that belowground herbivory reduced biomass of the two turfgrass species. But contrary to Clay et al. (1993), they found no effect of infection on herbivore biomass or survival. Furthermore, unlike the results from Clay et al. (1993), infection had no effect on biomass of grasses under intra- or interspecific competition. In an earlier study, Richmond et al. (2003) showed that E+ perennial ryegrass was a poorer competitor than E– ryegrass when growing in competition with the fast-growing annual grass *Digitaria sanguinalis* (large crabgrass) (see figure 2.10). In two native woodland grasses, *Brachypodium sylvaticum* and *Bromus benekenii* infected with *Epichloë sylvatica* and *E. bromicola*, respectively, infection increased intraspecific competitive abilities of *B. benekenii* but decreased those of *B. sylvaticum* (Brem & Leuchtmann 2002). Thus it seems from these studies that endophyte infection in agronomic and native grasses is certainly not a guarantee of superior competitive abilities either in the presence or absence of herbivores.

Herbivory and competitive ability interact in another indirect fashion related to soil nutrient availability. Assuming that growth and reproduction of many plants are nutrient limited, allocation of nitrogen to nitrogen-rich alkaloids for defensive functions against herbivores may compete with other host growth and reproductive functions. Allocation of nutrients to constitutive or induced allelochemicals is generally assumed to occur in plants that defend against herbivores (e.g., Herms and Mattson 1992). Thus in high-nutrient environments, increased levels of alkaloids may be produced (e.g., Arachevaleta et al. 1992, Lyons et al. 1986) and plant competitive ability increased because herbivores are deterred. Lehtonen et al. (2005a) showed that aphid performance declined on infected and nitrogen-fertilized meadow fescue (*L. pratense*) relative to E– plants, presumably because alkaloids were increased. Similarly Ahlholm et al. (2002) showed that infected *L. pratense* and *F. rubra* plants produced less biomass than E– plants when grown in low-moisture, low-nutrient soils, but the opposite occurred when grown in high-moisture, high-nutrient soils. These results suggest that the costs and benefits of endophyte infection in grasses relative to competition and herbivory are conditional upon resource availability. However, as discussed in chapter 6, it should be remembered that natural populations are not simply a mixture of infected and uninfected hosts of limited genetic background, but rather a mosaic of uninfected plants of various genotypes and infected plants that vary widely in endophyte and plant genotype combinations, and thus in abilities to produce different types and levels of alkaloids. Can one predict which symbiotic combination would be favored by natural selection in different environments?

Faeth and Fagan (2002) attempted to do this with graphic modeling. They used nitrogen flux in the host plant as the common currency in endophyte-host and plant-herbivore interactions because nitrogen is often the key limiting nutrient for plant growth and alkaloids are nitrogen-rich compounds. Their graphical model predicts that low alkaloid-producing endophytes are favored when soil nutrients and herbivory are low. When soil nitrogen increases over some gradient, high alkaloid-producing endophytes are favored over low producers and E– plants. But when nitrogen levels are very high, E– plants may be favored over either low or high alkaloid-producing infected plants. These predictions were supported by patterns of infection and alkaloid production in natural populations as well as by results from a field experiment. Clearly additional observational, experimental, and modeling approaches are needed to explain the maintenance of complex mosaics of E– and E+ plants with widely varying levels of alkaloids in natural populations (e.g., Faeth et al. 2006).

3.7 WHY DON'T ALL ENDOPHYTE-HOST GRASS COMBINATIONS PRODUCE DIVERSE, HIGHLY TOXIC ALKALOIDS?

Given the variation in alkaloids, much of which is heritable either via the host or endophyte genotype, one might expect, as have others (e.g., Clay 1987a, 1990b, Clay & Schardl 2002), strong selection for host-endophyte combinations with high and toxic alkaloid levels. Indeed, endophytes and their alkaloids have been viewed as acquired chemical defenses against herbivores (Cheplick & Clay 1988, Clay 1988b, Vicari & Bazely 1993) and as a way for host plants to keep pace with rapidly evolving herbivores (Carroll 1991, Cheplick & Clay 1988). This should especially hold if alkaloids also act to increase host fitness via other proposed mechanisms, such as increasing resistance to soil nematodes (e.g., Kimmons et al. 1990, Kunkel et al. 2004, Timper et al. 2005) and plant diseases (Siegel & Bush 1996, White & Cole 1985), increasing resistance to seed predation (Knoch et al. 1993, Wolock-Madej & Clay 1991), and as allelopathic agents that enhance competitive abilities (e.g., Matthews & Clay 2001, Peters & Zam 1981). Yet, infected host plants with diverse alkaloids at high concentrations appear as the exception rather than the rule (Faeth 2002, Leuchtmann et al. 2000). There are several reasons why high and diverse alkaloid levels in endophyte-host interactions may be relatively rare.

The first is that alkaloids are costly to produce, similar to other plant allelochemicals that deter herbivores. Metabolic costs of alkaloids can be substantial (e.g., Karban & Baldwin 1997, Ohnmeiss & Baldwin 1994) and requires the use of nitrogen, which is often limiting to plant growth in natural populations. That alkaloids are high in infected agronomic grasses which typically occur in fertilized agricultural pastures or fields is not accidental (Saikkonen et al. 2006). High alkaloid-producing endophyte-host combinations are undoubtedly selected by intense grazing in highly fertile fields. Clay (1998) and Clay et al. (2005) showed that in agronomic tall fescue, infection frequencies increase rapidly with herbivory in old fields. However, infection frequency is much more variable in wild populations (e.g., Faeth 2002, Saikkonen et al. 1998, 2004, Schulthess & Faeth 1998). That alkaloids are costly to produce, even in agronomic grasses, is shown in another experiment by Bultman et al. (2004). Clipping damage induced higher alkaloids in agronomic tall fescue and reduced subsequent susceptibility to *R. padi* aphids. However, regrowth in clipped, infected plants was reduced, and thus increased resistance to aphids via induced alkaloid defenses in infected plants "comes at the cost of tolerance (reduced regrowth)" (Bultman et al. 2004).

The second is that alkaloids, like other plant allelochemicals, are most effective in deterring generalist insect herbivores (Saikkonen et al. 1998, 2004, Strong et al. 1984). Indeed, most cases of insect resistance in agronomic grasses involve generalized insect pests, such as aphids, and bioassays of toxicity or deterrence in infected native grasses likewise use generalist aphids (e.g., Clement et al. 2005, Siegel & Bush 1996). However, even generalist insects may be unaffected by endophyte alkaloids purported as insect deterrents. In a recent study, Krauss et al. (2007) found that *N. lolii* infection in four cultivars of perennial ryegrass (Imp, Nui, Pac, and Sam) did not generally increase resistance to aphid herbivores but did alter attack by their natural enemy parasitoids, despite producing levels of peramine above that known to be toxic to insects (Siegel & Bush 1996). Furthermore, aphid densities were higher on infected, fertilized plants where peramine levels were typically the highest. Notably the only aphid species that was negatively affected by *N. lolii* infection was *R. padi*, while two other species (*Sitobion avenae* and *M. festucae*) were unaffected by infection or associated peramines. *R. padi* is a cosmopolitan generalist aphid species that is widely used in bioassays to test herbivore resistance of infected grasses (e.g., Clement et al. 2005, Davidson & Potter 1995, Hunt & Newman 2005, Siegel & Bush 1996, Siegel et al. 1990). The results from Krauss et al. (2007) suggest that *R. padi* may be hypersensitive to peramine relative to other herbivores and therefore is not a good model herbivore to test endophyte-associated plant resistance.

Third, insect herbivores rapidly evolve resistance to plant allelochemicals (e.g., Fritz & Simms 1992). In fact, specialist insect herbivores, which make up the bulk of herbivorous insect species (Strong et al. 1984), often use these same allelochemicals as attractants to locate hosts, as essential nutrients, and as sequestered defensive compounds against their own natural enemies— predators and parasites. Research by T. Bultman (e.g., Bultman et al. 1997, Faeth & Bultman 2002), P. S. Grewal (e.g., Richmond et al. 2004b), and C. de Sassi (de Sassi et al. 2006) show the same for endophyte-produced alkaloids—alkaloids are sequestered and may reduce attack by predators, parasitoids, and entomopathogenic hosts. Of interest, endophyte-produced alkaloids can also be sequestered by hemiparasitic plants, which may reduce resistance properties of the grass host (Lehtonen et al. 2005b). So, much as it would be erroneous to infer that any given allelochemical in a host plant is a panacea against all herbivores, it is likewise invalid to assume that endophyte-produced alkaloids are effective defensive agents against all herbivores of grasses. This complication of endophytes altering attack by the third trophic level, natural enemies, is explored in more depth in chapter 6.

Fourth, in addition to costs, high levels of allelochemicals such as alkaloids can be autotoxic to the host plant itself (Karban & Baldwin 1997). This may similarly occur in infected host grasses with high alkaloid levels, such as the extremely high levels found in *Achnatherum inebrians* (Miles et al. 1998) and *A. robustum* (Faeth et al. 2006).

Fifth, Faeth (2002) argued that herbivory on native grasses may not be a strong selective pressure on perennial grasses because of inconsistency of herbivore attack and because grasses have evolved adaptations to tolerate herbivory, such as the location of shoot apical meristems close to the soil surface. Survival and Darwinian fitness are not necessarily reduced in grass populations experiencing natural levels of herbivory. Unlike pasture grasses where grazing by livestock is typically consistent and intense, removing large amounts of aboveground biomass, and generalist invertebrate pests dominate, grazing on native grasses by native vertebrates is more sporadic and specialist invertebrate herbivores dominate (Faeth 2002). Many wild grasses have coevolved with native vertebrate grazers and are either adapted to, or are tolerant of, sporadic grazing (e.g., Cullen et al. 2006, McNaughton 1984). Indeed, some grasses benefit from herbivory by overcompensating, especially in resource-rich environments (Hawkes & Sullivan 2001).

3.8 ENDOPHYTES AND TOLERANCE TO HERBIVORY

As noted in section 3.7, grasses in general may be tolerant of herbivory, although the ability to tolerate herbivory may vary among grass species (e.g., Cullen et al. 2006). Tolerance is an evolved plant strategy where plants can compensate for tissue lost to herbivores by regrowing and without a loss of fitness (Del-Val & Crawley 2005, Strauss & Agrawal 1999, Tiffin & Rauscher 1999). In some cases, plants may overcompensate for herbivory (Agrawal 2000, Lennartson et al. 1997, Paige & Whitham 1992) such that fitness increases under herbivory (termed overcompensation) (Maschinski & Whitham 1989). The ability to tolerate or compensate for herbivory is often constrained by plant genotype (Del-Val & Crawley 2005, Tiffin & Rauscher 1999), available resources (Hochwender et al. 2000, Wise & Abrahamson 2007), and ontogeny of the plant (Boege & Marquis 2005). Tolerance to herbivory is generally viewed as an alternative plant strategy to cope with herbivory, as opposed to producing either constitutive or inducible defensive allelochemicals that prevent or deter herbivory. These two plant strategies, tolerance and defense, are usually assumed to be constrained by the energy and nutrient demands of the plant, and plants may use primarily one or the other, or a combination of the two (e.g., Fornoni et al. 2004).

Asexual, seed-borne endophytes and their grass hosts have traditionally been viewed as partnerships whereby grass hosts can acquire indirect chemical defenses as a strategy against herbivory in addition to tolerance (e.g., Cheplick & Clay 1988, Clay 1988b, 1991a). We have seen above that these alkaloid defenses may not always be effective against herbivores and may also incur metabolic costs and nutrient limitations to the hosts. Recent evidence suggests that endophyte infection not only affects chemical defenses, but may also change or constrain tolerance in the host. Endophyte infection alters the ability of grass hosts to regrow following clipping in both agronomic tall fescue (Belesky & Fedders 1996) and perennial ryegrass (Cheplick 1998a). Bultman et al. (2004) showed experimentally that endophyte infection in agronomic tall fescue induces alkaloid defenses that render grass less susceptible to aphid attack, but exacts a cost in terms of reduced tolerance (less regrowth) relative to uninfected plants. In a later study, Sullivan et al. (2007) demonstrate that herbivory on infected tall fescue upregulates the *lolC* gene, which increases levels of lolines and deters further herbivory, but again at the cost of reduced tolerance. Sullivan et al. (2007) concluded that *Neotyphodium* in tall fescue "appears to switch its host's defensive strategy from tolerance to resistance." These studies suggest that there is a trade-off in tolerance versus chemical defensive strategies, as has been assumed for other plants (e.g., Strauss & Agrawal 1999), and that endophytes may mediate changes in defensive strategies.

Of interest, like many effects of endophytes on their host, the capability of endophytes to influence tolerance and chemical defensive strategies appears to depend on host genotype, endophyte haplotype, and environmental factors. Using the same endophyte, but in different genotypes of perennial ryegrass (cv. Yorktown III), Cheplick (1998a) found that both grass genotype and soil nutrients altered the tolerance to simulated herbivory, estimated by daily regrowth rates based on leaf mass or area accumulated over 14 weeks following experimental clipping. Likewise, Bultman et al. (2004) and Sullivan et al. (2007) showed that endophyte strain and nutrient levels altered the effect of the endophyte on tolerance and induced defenses after mechanical and insect damage. Mycorrhizal associations can also alter the ability of the host grass to tolerate herbivory (e.g., Kula et al. 2005), and because endophyte infections can either increase or decrease mycorrhizal abundance (see section 3.9), we should expect either positive or negative interactions between mycorrhizae and endophytes on tolerance of their shared host grass. These results suggest that in wild grass populations, the effects of endophytes on host strategies against herbivory will vary depending on the host genotype and ontogeny, the endophyte strain, the prevailing resource environment, as well as other interacting microbial symbionts.

All the aforementioned studies tested how infection by *Neotyphodium* endophytes in agronomic tall fescue and perennial ryegrass cultivars alter tolerance in terms of vegetative regrowth after artificial clipping or insect damage. These studies provide an important foundation for future studies and a hint of how endophytes affect tolerance and resistance. However, tolerance is defined in terms of host fitness. Therefore conclusive tests of tolerance, notoriously difficult to conduct in plants even without considering microbial symbionts (e.g., Del-Val & Crawley 2005, Tiffin & Rauscher 1999), must include some estimate of host fitness, not just regrowth. In a 3-year field experiment, Faeth (2008) and Faeth and Saikkonen (2007) (described in section 3.10) found that *Neotyphodium* infection increased the ability of wild Arizona fescue to tolerate herbivory from insect herbivores in terms of increased seed output, and infection shifted allocation from vegetative to increased reproduction biomass.

Furthermore, increased tolerance in infected plants increased as host grasses matured and across four different host grass genotypes and three different soil moisture treatments. These results suggest that endophytes can dramatically alter host plant tolerance and may supersede constraints due to resource availability and host grass genetics and age. However, most systemic endophytes inhabit perennial grasses (Clay 1998) that live for decades (Faeth & Hamilton 2006, Olejniczak & Lembicz 2007), so estimates of the effects of endophytes on lifetime fitness of grass hosts relative to tolerance and resistance are currently unavailable.

3.9 ABOVE- AND BELOWGROUND MICROBIAL INTERACTIONS AND HERBIVORY

Pooid host grasses not only often harbor endophytes, but also a variety of belowground or root symbionts, notably the arbuscular mycorrhizae (AM). A few studies have shown that endophytic and AM infections can interact to alter herbivory on the host plant. Barker (1987) showed that the agricultural pest, the Argentine stem weevil (*Listronotus bonariensis*) is deterred from feeding on a perennial ryegrass cultivar infected with *N. lolii*. Perennial ryegrass is often infected with the common AM fungus *Glomus fasciculatum* in New Zealand. In *Neotyphodium*-free plants, the presence of the AM fungus had no effect on feeding behavior by the stem weevil. But in *Neotyphodium*-infected plants, the presence of the AM fungus reduced the deterrency of endophytes to the stem weevil.

In a later study, Vicari et al. (2002) experimentally examined the interaction of the same endophyte in perennial ryegrass under varying levels of soil phosphorous, but with a different AM fungus and herbivore: *Glomus mosseae* and larvae of the noctuid moth, *Phlogophora meticulosa*. Like Barker (1987), they found that the presence of the endophyte alone reduced the damage inflicted by the moth larvae, and when the AM fungus was present, the difference in damage between E+ and E− plants was reduced. In contrast to Barker's (1987) study, however, they found that the AM fungus alone reduced feeding damage by the larvae. Furthermore, there was an interaction between the endophyte and AM fungus in terms of consumption of grass by the larvae. The endophyte increased the relative consumption rate and reduced the efficiency of conversion of ingested food, whereas the AM fungus did not affect various consumption and nutritional measures of the larvae. They concluded that *Neotyphodium* and *G. mosseae* affected larval growth rate and survival, and that some of these effects were additive, whereas others were not, and depended on phosphorous levels in the host (Vicari et al. 2002).

These studies suggest that systemic endophytes and mycorrhizal fungi potentially interact to influence the behavior, physiology, and survival of some herbivores of agronomic grasses. Both herbivores were generalists, and the Argentine stem weevil is an agricultural pest. However, it is unclear how endophytic and AM mycorrhizae interact to influence herbivory at the population or community level in either agronomic or natural systems. Fungal endophytes may inhibit mycorrhizae, at least in agronomic grasses (Bernard et al. 1997, Chu-Chou et al. 1992, Guo et al. 1992, Müller 2003, Omacini et al. 2006, Mack & Rudgers 2008), and thus affect host nutritional quality (chapter 6) and perhaps alkaloid levels, which in turn could alter herbivory and herbivore population dynamics. In a recent experimental study, J. Newman (personal communication) found that aqueous extracts from agronomic tall fescue with two strains of *N. coenophialum* (the common strain and AR452) reduce spore viability of the AM fungus *Glomus intraradices*. He also found that thatch from tall fescue with the common strain reduced AM fungal colonization of another grass, *Bromus inermis*, but thatch from tall fescue with AR542 did not. Mack and Rudgers (2008) showed that *Neotyphodium* infection of the grass *Lolium arundinaceum* reduced arbuscular mycorrhizal colonization but mycorrhizal treatments had no effect on the endophyte. In contrast, Novas et al. (2005) found that a native grass, *Bromus setifolius*, in Argentina that was infected by *Neotyphodium* had increased, rather than decreased, root colonization by AM mycorrhizae. We know of no other studies of even the basic patterns of association between fungal endophytes and AM fungi in wild grasses. Moreover, there have been no experimental studies of which we are aware that have tested the potential four-way interaction between endophytes, AM fungi, host grasses, and herbivores in either agronomic or natural communities. This should provide fertile ground for both basic and applied endophyte researchers because these interactions are probably very common and may change the concepts of how endophytes influence herbivory.

3.10 DO SOME ENDOPHYTES FACILITATE HERBIVORY ON THEIR HOSTS?

There is another, and counterintuitive, reason why high alkaloid production may not often be selected for in infected grasses—herbivory on vegetative parts of host grasses may increase seed production. Seeds are the only mode of transmission for asexual endophytes like *Neotyphodium* and a facultative mode for most *Epichloë* endophytes. If herbivory results in reallocation of host resources to reproduction, as may be the case for some plants (e.g., Anderson & Paige 2003, Trumble et al. 1993), then fitness of seed-borne endophytes may actually be enhanced with some types and levels of herbivory. If so, then we might expect endophytes to "encourage" herbivory through reduced defenses or by increasing the palatability or attractiveness of plant vegetative tissues. The same may hold for endophytes transmitted in ramets (e.g., Pan & Clay 2003) because herbivory typically increases tillering (e.g., Pavlu et al. 2006), and thus propagation of infected clones. If supported, this hypothesis has two important implications.

First, the prevailing concept that endophytes act mutualistically by deterring herbivores via alkaloids is not only rejected but actually reversed. Second, facilitation of herbivory means that endophytes may partially control plant growth and reproduction via herbivores, even if this is at the expense of long-term survival and fitness of the host grass. Increased herbivory usually has negative long-term effects on grass growth and survival, especially when plant-plant competition for resources is severe (e.g., Olff & Ritchie 1998).

What evidence exists for the hypothesis that asexual, seed-borne endophytes facilitate herbivores to increase host allocation to seeds? Nonsystemic endophytes benefit from herbivory because spores and hyphae colonize wounded tissue and dispersal may occur when ingested hyphae and spores pass through insect guts (Devarajan & Suryanarayanan 2006, Faeth & Hammon 1997). Seed-borne endophytes are not thought to be transmitted via spores that could be transmitted by insect herbivores (but see Moy et al. 2000). Nevertheless, the systemic endophytes can certainly alter allocation of host resources to different growth and reproductive functions to increase transmission. For example, *Epichloë* endophytes in the clonal grass *Glyceria striata* are transmitted mainly by growing into stolons of new ramets. *Epichloë* endophytes increase growth and redirect carbon allocation to stolons, thus enhancing their transmission (Pan & Clay 2002, 2003, 2004) (see figure 2.5).

Alternatively, asexual endophytes are usually viewed as "trapped pathogens" (Schardl & Clay 1997) and "evolutionary dead ends" at least without genetic input from occasional hybridization (Wilkinson & Schardl 1997). Asexual endophytes are considered to be entirely dependent on the host grass for resources and transmission (White et al. 2001b) and under control of the host grass (Christensen et al. 2002). Thus it would seem that asexual endophytes would have little control over host reproduction. However, in a recent 3-year field study, Olejniczak and Lembicz (2007) showed that the asexual and vertically transmitted strain of *Epichloe typhina* shifted allocation in a wild grass, *Puccinella distans*, in Poland to early reproduction. This age-specific shift in reproduction likely benefited the endophyte in terms of transmission but reduced lifetime reproductive success and vegetative biomass of the host grass. In this study, herbivory was not controlled.

Might asexual, seed-borne endophytes also indirectly enhance their transmission by attracting or decreasing resistance to vegetative herbivory that redirects resources to increased seed production? Some infected host grasses are more susceptible to herbivory than uninfected hosts (e.g., Clement et al. 2005, Lopez et al. 1995, Saikkonen et al. 1998, 1999, Tibbets & Faeth 1999), but this offers only circumstantial support. A recent experimental study involving the native grass Arizona fescue, where herbivory was controlled, provides support for this intriguing hypothesis.

In a 3-year field experiment, Faeth (2008) used four different plant genotypes of Arizona fescue (*F. arizonica*) infected with *Neotyphodium*. Infection was experimentally removed from clones. Replicates of infected (E+) and uninfected (E−) clones of each plant genotype were grown in a common garden experiment with three water treatments (reduced, ambient, and supplemented water) and two herbivory treatments: ambient herbivory from arthropods and small mammals, and reduced herbivory (via hardware cloth cages and periodic insecticide treatments). Plant growth (aboveground biomass) and reproduction (culms, flowers, and seeds) were measured at the end of the growing season, as well as arthropod abundances and diversity (from the herbivory treatment plants; caged, control plants had no or very few insect herbivores).

The E+ plants had significantly greater aboveground biomass than E– plants, indicating enhanced growth. Of interest, E+ plants had significantly greater numbers and biomass of herbivorous insects, despite the fact that the four infected maternal plants had levels of the alkaloid peramine that averaged 11.05 ± 0.40 ppm (endophyte-removed plants had no peramine). However, alkaloids are clearly not defensive in this case. These levels of peramine are well above those (approximately 3 ppm) known to be detrimental to sucking insects (Seigel & Bush 1996), and most of the insects found on the plants were sucking insects such as thrips, delphacids (planthoppers), and aphids. Levels of alkaloids three orders of magnitude lower than the levels in infected Arizona fescue are known to deter feeding in some insect pests (e.g., 0.1 ppm) (Rowan et al. 1990). The most intriguing result was that infected plants had significantly higher mean average seed mass per plant, but only in the treat-ment where herbivores were present. In other words, herbivory and infection significantly interacted to affect seed mass per plant, and when herbivores were excluded, seed mass did not differ between E+ and E– plants. E+ plants also had greater total reproductive mass (seeds, culms, and inflorescences) and higher reproductive to vegetative mass ratios than E– plants—but again, only when herbivory was present. Furthermore, all experimental grasses (E+ and E–) subjected to herbivory produced five times more seed mass than those free of herbivory, even though the overall size of the plants was reduced by herbivory.

These results suggest that *Neotyphodium* endophytes in Arizona fescue plants benefit from increased herbivory in terms of plants producing more infected seeds. This could explain the puzzling results from previous studies showing that infected Arizona fescue is consistently less resistant to inverte-brate (Lopez et al. 1995, Tibbets & Faeth 1999) and vertebrate (Saikkonen et al. 1999) herbivores than their uninfected counterparts. It is unclear how *Neotyphodium* in Arizona fescue is increasing herbivory, and thus seed pro-duction. However, E+ plants have higher water content (Lopez et al. 1995) and altered physiological properties (Morse et al. 2002, 2007), which may render infected plants more susceptible to sucking insects in semiarid envi-ronments in Arizona. Whatever the mechanism, it appears that in the case of Arizona fescue, infection by *Neotyphodium* increases herbivore abundance, which in turn enhances the production of seeds that house fungal hyphae.

The larger conclusion is that the relationship of endophytes and their host plants relative to herbivory and alkaloid production is still unknown in many host grasses other than well-studied agronomic grasses such as tall fescue and perennial ryegrass. Even in these agronomic grasses, new complexities, such as the effects of alkaloids on the third trophic level (chapter 6) and the non-alkaloidal changes in grass metabolic profiles due to endophyte infection (e.g., Rasmussen et al. 2008), are just now being explored At the very least, we can expect wide variation in these relationships in natural populations and communities, with dependency on host genetics, endophyte strains and hybridization, and environmental context, as evidenced by accumulated studies to date (Saikkonen et al. 2006). Rather than focusing on whether endophytes and their alkaloids are defensive or not, it seems more productive to first recognize and document the wide variability within and among infected grass populations and across host and endophyte species. Then perhaps this variation can be explained based upon the abundances and types of herbivores and their natural enemies, prevailing soil and moisture conditions, and interacting plant species in the community. As with other interspecific interactions (Thompson 1994, 2005), Faeth and Sullivan (2003) and Saikkonen et al. (2004) argued that endophyte-host grass interactions in local populations vary in intensity and direction across geographic and temporal scales based upon relative costs and benefits to the host and endophytes (chapter 1), phylogeny and genetic covariation of the endophyte and grass (chapters 4 and 5), other interacting plant species in the community (section 2.3), and local environmental factors (section 2.2). This added complexity in natural communities renders observations and experiments more challenging in unraveling how endophytes in grasses interact with herbivores. However, the same complexity is inherent in other plant-herbivore interactions and has been successfully incorporated into observations and experiments (e.g., Crutsinger et al. 2006, Linhart et al. 2005) and is thus certainly possible for endophyte-host-herbivore interactions.

4

Genotypic Specificity of Grass-Endophyte Interactions

For continued coevolution of the grass-endophyte symbiosis due to natural selection in variable environments, there must be genetic variation within populations for both the host and endophyte. As will be shown, substantial genotypic variation has been documented for a variety of hosts and their endophytic symbionts. In addition to abiotic selection pressures such as drought or soil fertility, biotic factors such as competitors, herbivores, and pathogens will also impact the coevolutionary relationship between grasses and their endophytes. All else being equal, the breeding systems of both symbionts (asexual versus sexual, outcrossing versus self-fertilizing) will undoubtedly influence levels of genetic variation found in natural populations. Reciprocal interactions between host and endophyte can alter breeding systems in ways that can change the relative distribution of genetic variation within and among populations. Reference has already been made (section 2.1.4) to some of the ways endophytic fungi can change host reproductive systems in grasses such as *Danthonia spicata* (Clay 1994a) and *Glyceria striata* (Pan & Clay 2003). Wherever abiotic, biotic, and reciprocal selection pressures impact the grass-endophyte symbiosis, there is the opportunity for changes in the genetic composition of natural populations which will further influence the coevolutionary dynamics of the interacting symbiotic partners.

The distribution of ecological outcomes for a particular interspecific interaction can be depicted as interaction norms, which Thompson (2005) describes as extensions of traditional reaction norms. He notes how analyses of interactions between grasses and endophytes are an "exemplar" of how to explore interaction norms within natural populations. The symbiotic interaction norm for a grass-endophyte association should depict a dependent variable (e.g., tiller, biomass, or seed production) in relation to a group of host genotypes with and without the endophyte in the same environment. An example is shown in figure 4.1 using data for 10 genotypes of *Lolium perenne* utilized in a drought stress experiment (Cheplick 2004a). The number of tillers are compared for the same genotypes following a 4-week postdrought recovery period. Both host genotype and infection status were highly significant ($P < 0.006$) factors in an analysis of covariance. In addition, there was a significant ($P < 0.01$) genotype × endophyte interaction which is revealed by the crossing of lines in the interaction norm plot (figure 4.1). This implies genotypic variation among hosts in their response to specific endophytic symbionts. As will become apparent later (section 4.2), this type of study is

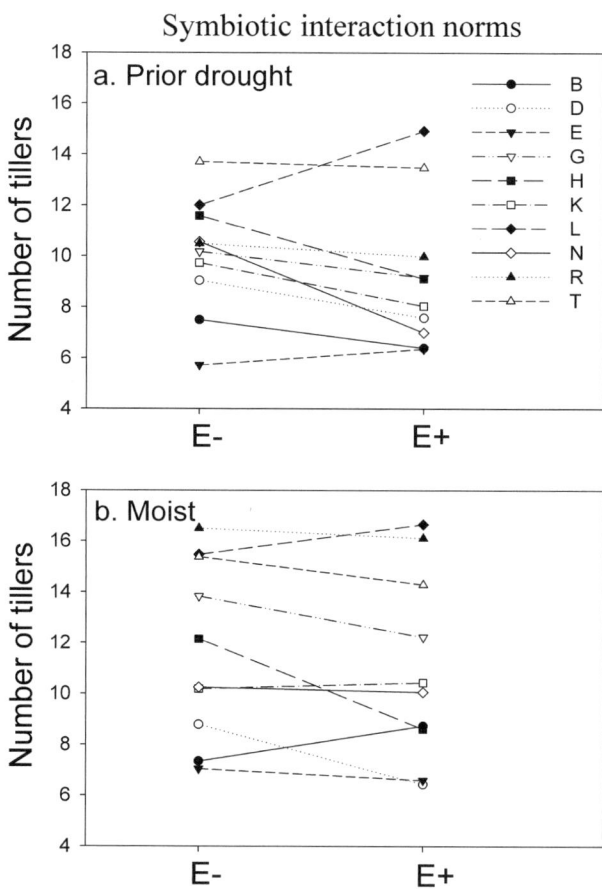

Figure 4.1 Symbiotic interaction norms for the number of tillers produced by 10 genotypes (denoted by capital letters) of perennial ryegrass used in a drought-stress experiment (Cheplick 2004a). Data were recorded for uninfected (E–) and endophyte-infected (E+) plants 4 weeks after (a) a prior drought or (b) moist control conditions.

somewhat problematic, as it typically ignores the potentially confounding effect of variation in endophyte genotypes.

In this same study, a highly significant ($P < 0.0001$) genotype × environment interaction was also detected for tiller production (Cheplick 2004a). This is best visualized in a traditional reaction norm diagram in which the phenotypic responses of a set of genotypes are graphed across a range of environments (Stearns 1992). The same data set used to construct the interaction norm (figure 4.1) was employed to draw a traditional reaction norm (figure 4.2). The crossing of lines in the reaction norm expresses the genotype × environment interaction in both E– and E+ groups. The lack of a significant endophyte × environment (and three-way) interaction term in

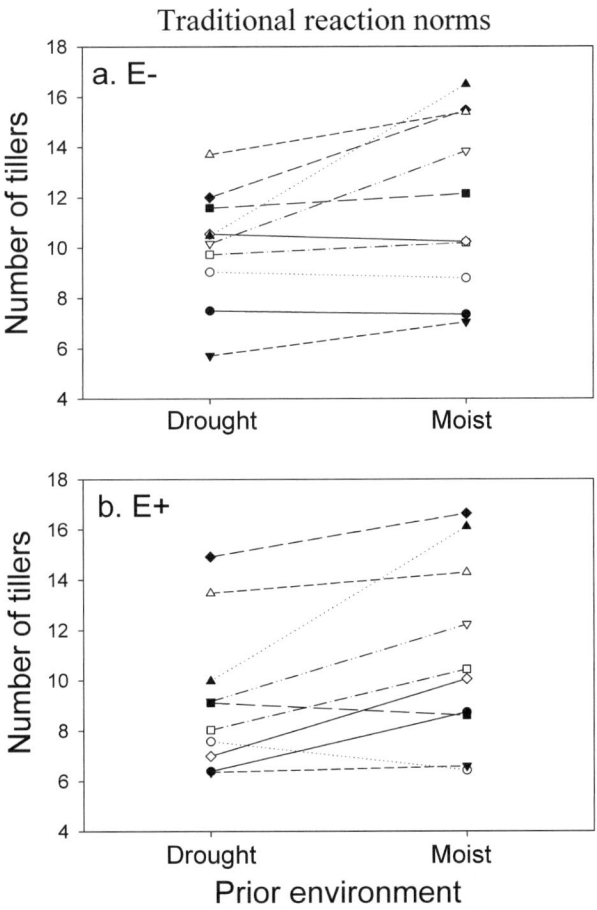

Figure 4.2 Traditional reaction norms for the number of tillers produced by 10 genotypes of perennial ryegrass used in a drought-stress experiment (Cheplick 2004a). Data were recorded for (a) uninfected (E–) and (b) endophyte-infected (E+) plants 4 weeks after a prior drought or moist control conditions. Key to symbols as in figure 4.1.

the analysis indicates that the growth response of these genotypes 4 weeks after exposure to different environmental conditions (moist or drought) was not changed by the presence of the endophyte. Phenotypic plasticity in tiller production across the two prior environments is also indicated for some genotypes by slopes that are substantially above the horizontal (figure 4.2).

The choice of when to use an interaction or reaction norm to depict the phenotypic responses of a set of host genotypes will, of course, depend on the specific objectives of a particular experiment. It will also likely depend on which specific interaction terms were significant in the analysis of variance. The symbiotic interaction norm is probably best to use when it is desired to show the range of endophyte-mediated effects on a set of host genotypes reared in the same environment. The traditional reaction norm is more useful when it is desired to show genotype × environment interactions for the same set of host genotypes when infected or uninfected.

The following sections will explore the genotypic specificity of grass-endophyte associations from the standpoint of both host and endophyte, with attention given to the role of genetic variation in the coevolutionary process.

4.1 HOST GENOTYPES AND INTERACTIONS WITH ENDOPHYTES

The title of this section is somewhat misleading, as it is really a three-way interaction (Thompson 2005) that is probably most appropriate for full analysis of the grass-endophyte symbiosis:

$$\text{Genotype}_{host} \times \text{Genotype}_{endophyte} \times \text{Environment.}$$

Unfortunately, many studies of endophyte interactions with different host genotypes have only compared the performance of the host genotypes with and without the endophyte, without explicitly considering endophyte genotypes (West 2007). Thus in such studies the caveat must be kept in mind that there may be an undetermined effect of different endophyte genotypes residing within different host genotypes. That is, reported genotype$_{host}$ × endophyte status interactions (based on the presence or absence of the endophyte) indicate that specific genotype$_{host}$ and genotype$_{endophyte}$ combinations are what really differ among host genotypes when infected versus when not infected. Furthermore, when there is a significant three-way interaction (genotype$_{host}$ × endophyte status × environment), this denotes that there is variation in the phenotypic response of specific host genotype-endophyte genotype combinations to changes in environmental conditions.

With these caveats in mind, host genotypic variation in response to endophyte infection will be considered for the two most widely studied, agronomically important grass-endophyte symbioses: tall fescue (*Lolium arundinaceum*) infected by *Neotyphodium coenophialum* and perennial ryegrass (*L. perenne*) infected by *Neotyphodium lolii*.

4.1.1 Agronomic grasses

Many of the studies of tall fescue and perennial ryegrass have shown that when the responses of individual host genotypes are examined both with and without the endophyte, there is usually statistically significant variation among genotypes and often a significant genotype × infection interaction. In light of the fact that these two grasses are predominantly outcrossing, it is not surprising to find significant genotypic variation in quantitative and molecular traits.

In certain cases, while there may not be any notable differences between E– and E+ plants in the phenotypic values for a trait averaged across a set of genotypes, individual genotypes can still show significant differences in phenotypic responses to endophyte infection. For example, in a study of five tall fescue (cv. Kentucky 31) genotypes grown in a benign controlled environment with ample soil water and nutrients (Belesky et al. 1987), mean (± SE) tiller production was 35.4 ± 4.4 for the genotypes when infected and 31.2 ± 3.4 when uninfected ($t = 0.76$, not significant). However, there was a major genotype × infection interaction ($F = 3.36$, $P = 0.02$). Inspection of the reaction norm (figure 4.3a) reveals that this interaction was due to a single genotype which showed a negative growth response to infection.

A more pronounced interaction of genotype × infection is shown in two reaction norms drawn from data collected over 2 years for 13 genotypes of perennial ryegrass in a field trial in Germany (Hesse et al. 2004). Although seven phenotypic variables were recorded in this detailed study, only the number of reproductive tillers is depicted here for the genotypes in the first (figure 4.3b) and second year (figure 4.3c). The mean (± SE) number of reproductive tillers per plant in the first year was 112.8 ± 10.3 for E– and 116.1 ± 12.3 for E+ plants ($t = 0.20$, not significant). These averages mask the marked genotypic variation found for this trait (figure 4.3b). In the second year, there was a trend toward greater reproductive tiller production in E– plants (192.4 ± 19.1) compared to E+ plants (155.7 ± 16.0), although these grand means (averaged across the means of the 13 genotypes) were not significantly different ($t = 1.47$). Again there was considerable crossing of lines in the reaction norm (figure 4.3c), emphasizing how simple averages across E– and E+ groups do not reveal the complexities of genotype-specific reactions to endophyte infection.

Clearly these two examples underscore the need to recognize that the lack of a significant difference in a phenotypic trait between heterogeneous sets of E– and E+ plants does not necessarily mean that there is no impact of the endophyte on the evolutionary ecology of the host population. Anytime there are host genotype × endophyte interactions, there is the possibility for selection to implement a unique sorting of host genotypes that depends on infection status and probably endophyte genotype as well (Cheplick & Cho 2003). This genotypic sorting process might be expected to apply to the dynamics of a newly established stand of a grass population as it develops over time (see section 5.1.3 for further discussion).

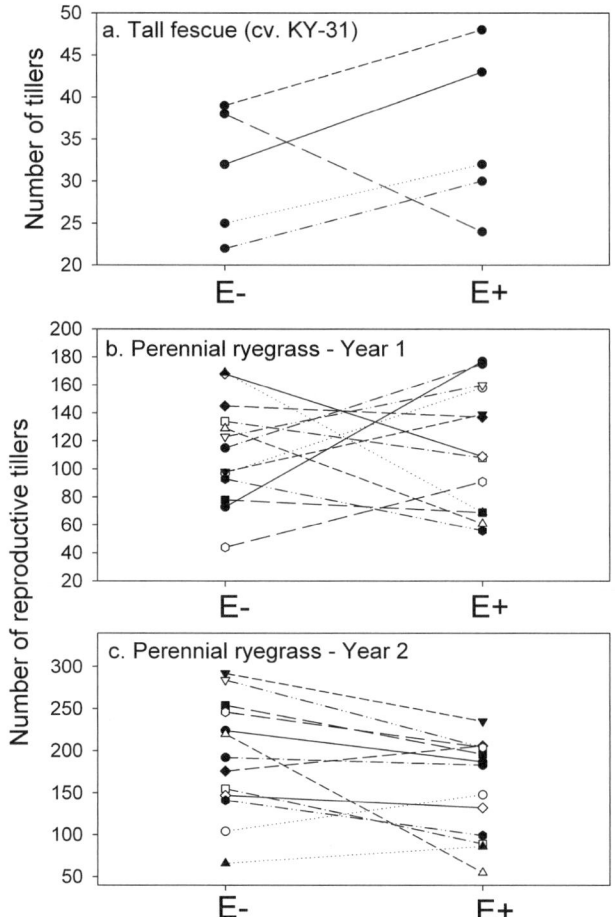

Figure 4.3 Symbiotic interaction norms for the number of tillers produced by five genotypes of tall fescue (cv. Kentucky 31) in (a) a benign, controlled environment and (b, c) for the number of reproductive tillers produced by 13 genotypes of perennial ryegrass over 2 years in a field trial in Germany. Data are presented for uninfected (E–) and endophyte-infected (E+) plants. Data for (a) from Belesky et al. (1987) and for (b, c) from Hesse et al. (2004).

The pattern of phenotypic responses to a set of environments can be illustrated by reaction norms drawn separately for each host genotype. Such a representation reveals the genotypic specificity of endophyte effects and the potential for the plasticity of phenotypic changes to be modified by endophyte infection. In tall fescue, genotype × endophyte interactions have been detected for a diversity of traits, such as carbon exchange rates (Marks & Clay 1996), competitive ability (Malinowski et al. 1999a), mineral nutrient uptake (Malinowski et al. 2000, Rahman & Saiga 2005), relative growth rate (Belesky & Fedders 1996), and seed production (Rice et al. 1990).

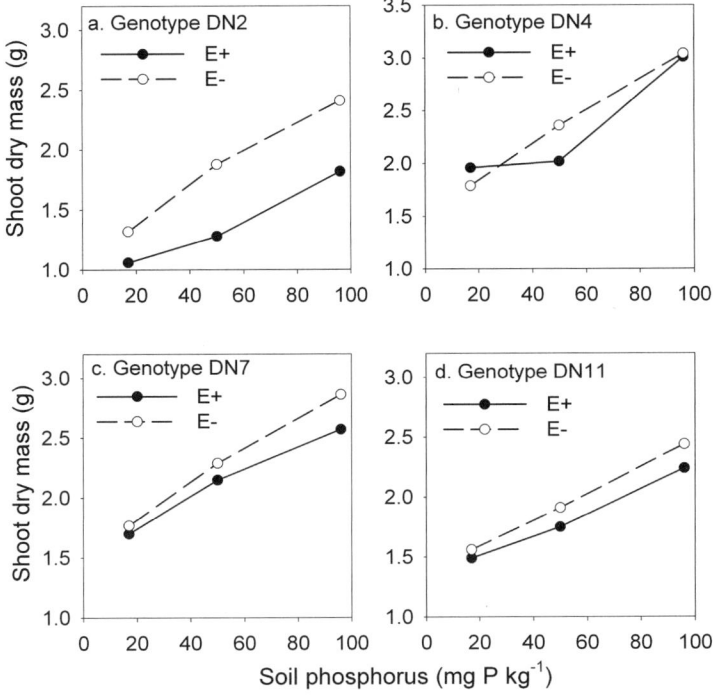

Figure 4.4 Traditional reaction norms for shoot dry mass of four tall fescue genotypes in relation to soil phosphorus. Data are presented for uninfected (E–) and endophyte-infected (E+) plants grown under greenhouse conditions (from Malinowski et al. 1998).

As an illustration of how both the endophyte and the experimental environment can interact with host genotype to influence phenotype, reaction norms for shoot mass are shown for four tall fescue genotypes in relation to soil phosphorus (figure 4.4) (Malinowski et al. 1998). In this greenhouse study, both E– and E+ replicates of the four genotypes were grown for 5 months at three phosphorus levels: 17, 50, and 96 mg/kg soil. Although variance ratios (F values) from analyses of variance were not supplied in the original paper, the effects of phosphorus treatment, genotype, and endophyte status were all significant ($P < 0.05$). In addition, the endophyte × genotype and the phosphorus × genotype interaction terms were significant (Malinowski et al. 1998). The reaction norms (figure 4.4) show how the endophyte reduced shoot dry mass in three genotypes, but did not in the fourth (genotype DN4). The slope of the reaction norm represents the change in shoot mass relative to the change in P level from 17 mg/kg to 96 mg/kg soil and thereby indicates phenotypic plasticity across the soil phosphorus gradient (Stearns 1992). For all genotypes, this slope is greater for E– plants. The difference between E– and E+ plants is most apparent for genotype DN2 (figure 4.4a), where the slope is 0.0138 for E– and 0.0096 for E+ plants. Thus in this study the endophyte may be reducing phenotypic plasticity in its host over the range in soil conditions examined. However, this reduction in plasticity is not necessarily a negative effect if stability in phenotypic traits improves fitness over a range of conditions within a heterogeneous environment.

Like tall fescue, perennial ryegrass and its endophyte have been extensively examined from the perspective of individual host genotypes. Genotype × endophyte interactions have been reported for many traits measured under a variety of conditions. These include tiller production (Cheplick 1997a, 2004a, Cheplick & Cho 2003, Hesse et al. 2004), net photosynthesis (Spiering et al. 2006a), carbohydrate storage (Cheplick & Cho 2003), aboveground dry mass (Cheplick 2004a, 2007, Hesse et al. 2004, Lewis 2004), leaf area (Cheplick 1997a, 2004a, Cheplick & Cho 2003), regrowth rate following defoliation (Cheplick 1998a), and seed yield (Hesse et al. 2004).

4.1.2 Native grasses

Clearly more research on the genotypic specificity of grass-endophyte interactions is desirable, especially for host grasses that are native to the areas from which they are collected and studied, and that are not necessarily of agronomic importance (see section 7.1.1). Within the past decade there have been a few genotypically based investigations of other grass-endophyte symbioses. Meijer and Leuchtmann (2000) investigated "disease expression" in nine genotypes of the caespitose woodland perennial *Brachypodium sylvaticum*, a grass native to temperate Eurasia. Disease expression was defined as the proportion of tillers that were choked by *Epichloë sylvatica*. In the first year, 74.1% of the total variance in choke rate was explained by host genotype, although this variance component declined to 27.5% in the second year. Disease expression in this system is likely to be controlled in part by particular combinations of host and endophyte genotypes. Host genotype × endophyte interactions were also detected for dry mass in both years. Although environmental variables (carbon dioxide [CO_2], light, and soil fertility) were experimentally manipulated and plants showed phenotypic responses to some of them, the authors concluded that particular host-endophyte genotype combinations were "more important for growth variation of *B. sylvaticum* than the environment" (Meijer & Leuchtmann 2000). The prominent host genotypic component to the observed variation in disease expression and growth in conjunction with similar effects of fungal genotype (Meijer & Leuchtmann 2001) is likely to support continued coevolution of this particular grass-endophyte symbiosis.

Faeth and Fagan (2002) replicated E+ and E– ramets of four host genotypes of Arizona fescue (*Festuca arizonica*), a caespitose perennial native to the American southwest. Plants were grown in an outdoor plot in Arizona and had ambient water and soil nutrients, or supplemented water and soil nutrients (four treatment combinations). The change in plant volume over one growing season, used as a measure of host growth rate, was significantly different among genotypes ($F = 12.67$, $P < 0.001$) and there was also a significant genotype × endophyte interaction ($F = 3.95$, $P < 0.01$). Two genotypes showed a greater growth rate when infected, while the others were unaffected by endophyte presence. Another study of the same grass-endophyte symbiosis (four host genotypes) also revealed significant genotype × endophyte interactions for growth and tiller production over a 2-year period, but the endophyte mostly had a negative effect (Faeth & Sullivan 2003). The authors remarked that host genotype was "paramount in elucidating the interactions between host and endophyte" in this system.

In a study of the effect of the endophyte *Epichloë glyceriae* on clonal growth of the native perennial *G. striata*, two host genotypes were utilized from each of three populations in southern Indiana (Pan & Clay 2003). Host genotype effects nested within population were apparent for tiller and stolon production, and clonal growth biomass, but there were no significant genotype × endophyte interactions. The effect of infection was consistent across genotypes, resulting in fewer tillers, but significantly more stolons than E– plants. Thus selection should favor infected genotypes whenever enhanced clonal growth aids survival or genet persistence (Pan & Clay 2003).

From the natural grass-endophyte systems examined to date, it appears that host genotype and its interaction with endophytes as determinants of the expression of quantitative traits are as pervasive or more so than in the agronomic grasses tall fescue and perennial ryegrass. However, there still remain far too few detailed studies of native grasses in nature, especially of multiple host genotypes from geographically widespread populations. The lack of information makes it difficult to draw any firm conclusions, but host genotype × endophyte interactions are likely to be relatively common in grass-endophyte symbioses. Sexual reproduction is rarely completely occluded in grass populations infected by endophytes, even in *Epichloë*-induced choke disease at high infection frequencies (Bucheli & Leuchtmann 1996, Chung & Schardl 1997a, Groppe et al. 1999, Meijer & Leuchtmann 2000, Schardl & Leuchtmann 2005). This is because some tillers of an infected host may escape infection and produce viable seeds on unchoked inflorescences. In addition, 100% infection frequency is probably not common within most grass populations (section 5.2). Hence, as long as some level of gene exchange can occur among some grass genotypes, new genetic combinations will arise. Host genotypic variation in conjunction with possible variation in endophyte genotypes can provide part of the raw material (Thompson 1999, 2005) necessary for continued coevolution of the grass-endophyte symbiosis.

4.2 HOST PHENOTYPIC PLASTICITY

The possibility that endophytes can modulate phenotypic plasticity in their grass hosts is intriguing and worthy of study in other grass-endophyte systems (Cheplick 1997a). One evolutionary model of host-pathogen interactions showed how plasticity promoted greater levels of cooperation between the symbionts and led to reductions in pathogen virulence (Taylor et al. 2006).

Because plasticity is a genetically based characteristic, it may evolve by natural selection (e.g., Schlichting & Pigliucci 1998). If plasticity in a particular trait is positively related to Darwinian fitness, as evidenced by improved competitive vigor or seed production, for example, then plasticity may be envisioned as adaptive in a heterogeneous environment. When particular endophyte-host genetic combinations enhance positive host responses to improved environmental conditions, selection may favor them. However, plasticity may not always be advantageous under all conditions. For example, reduced sensitivity of the host to environmental variation (e.g., when endophyte-infected) might

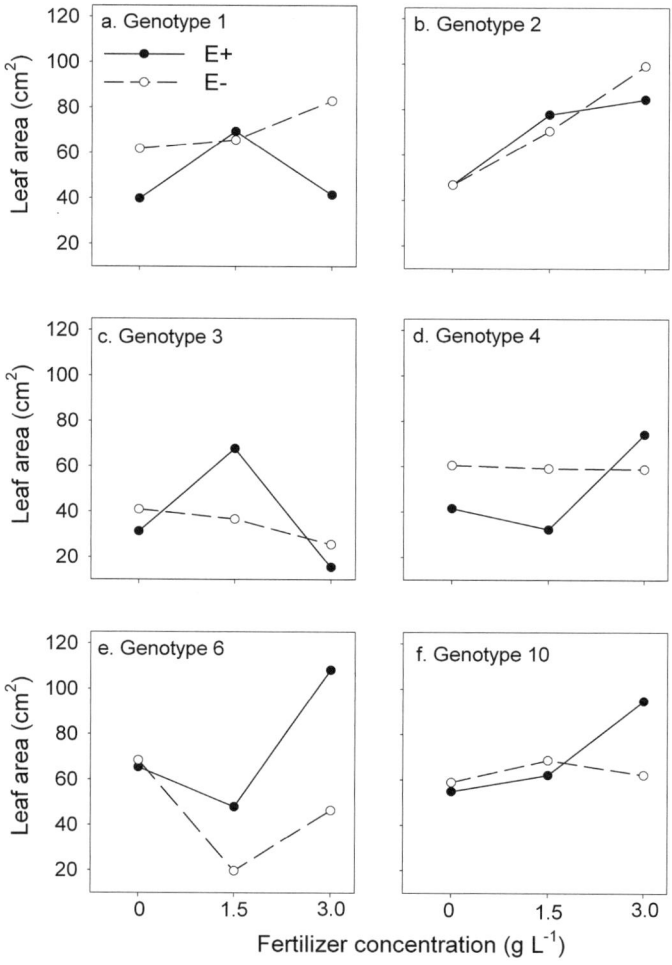

Figure 4.5 Traditional reaction norms for leaf area of 11 perennial ryegrass genotypes in relation to mineral fertilizer concentration. Data are presented for uninfected (E−) and endophyte-infected (E+) plants grown under greenhouse conditions (from Cheplick 1997a).

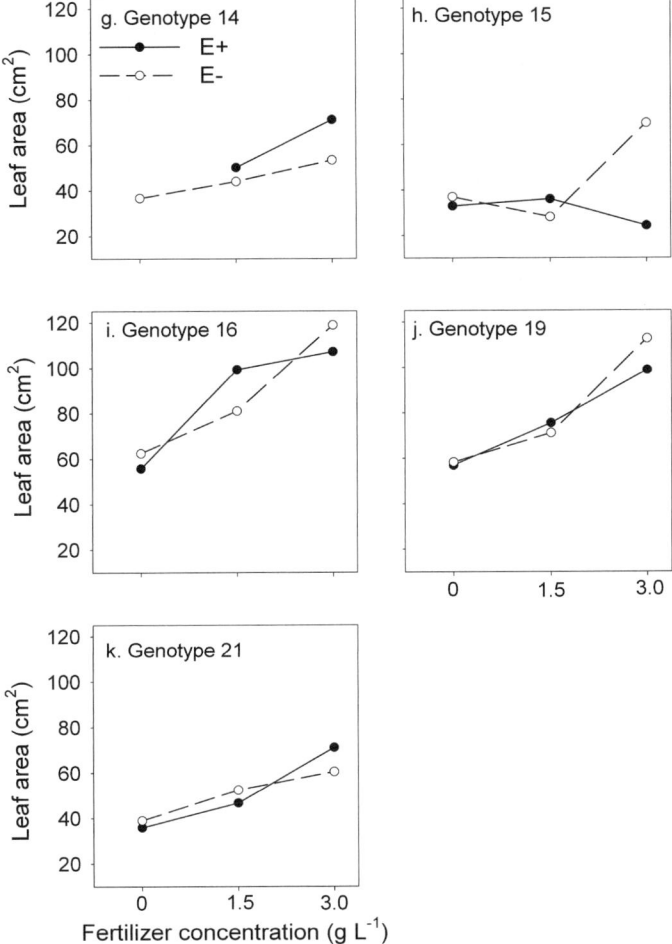

Figure 4.5 *Continued*

lead to selection against developmental instability and plasticity if high sensitivity to environmental changes has a negative impact on fitness (Schlichting & Pigliucci 1998). Clearly the potential influence of endophytic fungi on their hosts, revealed by genotype × endophyte interactions, adds a new level of complexity to the evolution of populations that contain a mixture of infected and uninfected host genotypes.

The study by Cheplick (1997a), designed to investigate the effects of endophytes on phenotypic plasticity of perennial ryegrass, will be used to demonstrate the complex nature of genotypic specificity in these effects and their dependence on environmental conditions. Twelve infected genotypes of *L. perenne* (cv. Yorktown III) were cloned manually into ramets. The endophytes were eliminated from half the ramets by fungicide treatment (see section 2.1.1); the remaining half remained infected. All plants were grown in the greenhouse and subjected to three treatments that provided variation in general soil fertility: N-P-K (20–20-20) fertilizer at 3 g/L (high nutrients) or 1.5 g/L (medium nutrients), and a control that received water (low nutrients). There were five replicates per each genotype-endophyte status treatment combination. Plants were clipped to soil level at 11 weeks and allowed to regrow until 25 weeks. At both 11 and 25 weeks, tiller number, leaf mass, and leaf area were obtained.

In a three-way analysis of covariance (using initial ramet fresh mass as the covariate), the effects of nutrient treatment and genotype were significant for all three variables at both 11 and 25 weeks (Cheplick 1997a). More important to this discussion, genotype × endophyte, and genotype × endophyte × treatment interactions were found for tiller number, leaf mass, and leaf area at 25 weeks. The three-way interaction demonstrates that phenotypic plasticity in response to improved environmental conditions (i.e., enhanced mineral nutrient availability) is dependent on specific host genotype-endophyte combinations. Figure 4.5 shows reaction norms of leaf area at 25 weeks separately for each host genotype (there are 11 genotypes because most replicates of one genotype died after clipping at 11 weeks) (Cheplick 1997a). Although the overall effect of endophyte infection in the analysis of covariance was not significant, it can be seen that the endophyte reduced plasticity in some genotypes (e.g., genotype 15) and increased plasticity in others (e.g., genotype 10). For other genotypes (e.g., genotypes 19 and 21), plants responded similarly to changed nutrient environments whether they were infected or not (figure 4.5).

The coefficient of variation (CV) calculated among nutrient treatments is an effective way to compare the phenotypic plasticity of genotypes when infected versus uninfected. The CV is the ratio of the standard deviation to the mean for a particular trait, often expressed as a percentage. For E− perennial ryegrass, the CVs of leaf area ranged from 1.5% (genotype 4) to 54.5% (genotype 6), with a mean (± SE) of 27.7 ± 5.4%. Note that the extremely low CV of genotype 4 when uninfected reflects the horizontal appearance of the reaction norm plot and indicates a lack of plasticity (figure 4.5d). For E+ plants of the same ryegrass genotypes, the CVs of leaf area ranged from 19.5% (genotype 15) to 70.2% (genotype 3), with a mean (± SE) of 36.2 ± 4.4%. At least for some genotypes, leaf area plasticity in response to changes in soil nutrient levels appears to be greater in E+ relative to E− plants.

Selection may favor increases in the leaf area of perennial ryegrass whenever soil conditions improve, as this would enhance the light-harnessing ability of host plants. Greater leaf area might improve vegetative vigor, competitive ability, or tolerance to herbivory. If this is true, then in a hypothetical "mixed" population of these 11 perennial ryegrass genotypes in a natural community where soil fertility varies, selection should favor the particular host-endophyte combinations represented by genotypes 6 and 10 (figure 4.5e, f), but not that of genotypes 1 and 15 (figure 4.5a, h). It is interesting to further speculate that some host genotypes (e.g., genotypes 2, 16, and 19) may be selectively favored simply because they show a strong response of leaf area to increased nutrient availability, regardless of whether or not they are infected (figure 4.5b, i, j). In these genotypes where the host-endophyte relationship is essentially commensalistic, endophyte genotypes may effectively be hitchhiking along with their selectively favored host genotypes.

To date, for most grass-endophyte symbioses it is unclear how endophytic fungi influence the observable phenotypic variation in a genotypically variable host population and the impact this might have on ecological and coevolutionary processes. In a heterogeneous, disturbed environment, increased host plasticity should permit rapid adjustments to changes in habitat conditions (Bazzaz 1996). Such plasticity can be adaptive when a growth, survival, or reproductive advantage is conferred on plastic genotypes compared to developmentally canalized genotypes (West-Eberhard 2003). It is interesting to speculate that seed-borne endophytes may be able to exploit phenotypic plasticity of the host in response to a changing environment if it results in increased tillering and seed production, which would help to maintain and propagate the endophyte. This could occur even if increased plasticity was not in the long-term interest of the host. In some stable or severe habitats, a more tightly integrated, developmentally canalized host phenotype may be advantageous. In grasses common in such an environment, the endophyte should then act as a stabilizing factor, reducing phenotypic responses to rare, unusual conditions. Further studies on phenotypic plasticity, developmental trajectories, and phenotypic integration (*sensu* Pigliucci 2004) in endophyte-infected grasses are clearly in order (see also section 7.1.3).

4.3 ENDOPHYTE GENOTYPE

Genotypic variation among endophytes isolated from different hosts both within and among species should be expected, especially for those fungi that have retained sexual reproduction and horizontal transmission. In addition to quantifying genetic diversity within and between endophyte populations (e.g., Arroyo García et al. 2002, Brem & Leuchtmann 2003, Leuchtmann & Clay 1989b, 1990, 1996), endophyte researchers have begun to examine how the variation among endophyte genotypes can impact alkaloid production (Faeth et al. 2002a, 2006) (see chapter 3) and sexual reproduction by the fungus (Bucheli & Leuchtmann 1996, Meijer & Leuchtmann 2001,Tintjer & Rudgers 2006). In addition, variable effects of different endophyte genotypes have been explored from the perspective of the host in a few grass-endophyte systems (Hill et al. 1996, Johnson-Cicalese et al. 2000, Morse et al. 2007).

Although examples of genotype-specific effects of both plant pathogens and mutualists (e.g., mycorrhizal fungi) on the growth and evolutionary fitness of their hosts have been reported (Koch et al. 2006, Parker 1995, Salvaudon et al. 2005), to date there is not much known about whether or not genetic variation in endophytic fungi significantly impacts host grass populations. However, endophyte genotypic variation may be influential to associated communities, impacting other trophic levels. For example, variation among isolates of *Epichloë elymi* from the native grass *Elymus hystrix* significantly impacted the patterns of arthropod herbivory on the host (Tintjer & Rudgers 2006). In this section, after defining several commonly used terms, we will explore genetic diversity in fungal endophytes and its role in determining the ecological outcomes of grass-endophyte interactions.

4.3.1 Haplotypes, strains, and races

As with any living entity, genetic variation in endophytic fungi has a hierarchical structure, with increasing genetic differentiation along the ecological hierarchy from individuals to populations to species. However, there are a variety of terms used in endophyte research to depict the various intermediate levels of genetic distinctiveness within the ecological hierarchy. The terms are not often defined and ambiguity exists over which terms are appropriate in particular circumstances.

An *isolate* is a fungus brought into pure culture from nature (Carlile et al. 2001). It is considered to be an individual genotype that may or may not differ from other isolates. Because the fungal genome is haploid, a genetically distinct isolate from an endophyte-infected tiller is commonly referred to as a haplotype and represents a distinct, multilocus genotype. In most endophyte research, an isolate is the mycelium that grows directly out of a piece of surface-sterilized stem or leaf tissue of an infected host. Genetic differences among isolates determined by molecular techniques, sometimes coupled with morphological features of the cultured fungal colony (e.g., Christensen & Latch 1991), are used to delimit haplotypes and to quantify genetic variation within and among endophyte populations (e.g., Arroyo García et al. 2002, Leuchtmann & Clay 1989b, Sullivan & Faeth 2004).

A *strain* has been defined as a genetic variety of a fungus isolated from nature (Carlile et al. 2001) and the term has been mostly used by endophyte researchers to denote genetically distinct isolates. As such, a strain is mostly synonymous with a particular genotype or haplotype that can be identical for several individual isolates. In contrast, a *race* of a pathogen or endophyte comprises a broader range of genetically distinct individuals that are typically specialized by occurring on or in specific populations, cultivars, or host species (Burdon 1987). If such groups of isolates are still interfertile in experimental crosses, then the individual groups are characterized as races, not biological species, even though they may occur on distinct host species (Brem & Leuchtmann 2003, van Horn & Clay 1995). Surveys of widespread pathogens infecting agricultural crops have typically reported substantial racial variations (Burdon 1987).

Clearly there is no simple rule for partitioning genetic variation among endophyte isolates that are genetically distinct. However, the level of genetic analysis achieved will depend on the choice of molecular technique, the number of isolates available, and the number of gene loci or DNA markers examined. Some of the DNA-based techniques tend to be more sensitive than others (e.g., isozymes) in detecting fine-scale genetic variation. In addition, the extent of genotypic diversity observed should be expected to depend on both host and endophyte species and their primary breeding system (sexual versus asexual or a mixed strategy where both are possible).

4.3.2 Assessing endophyte genotypes

To experimentally tease apart all the components of the three-way interaction— $genotype_{host} \times genotype_{endophyte} \times environment$—is a daunting task. Endophytic fungi must be isolated and cultured from different infected host genotypes. Genetic differences among isolates, when they exist, must be elucidated by molecular markers (isozymes, DNA) that can be used to define multilocus genotypes (Brem & Leuchtmann 2003, Bucheli & Leuchtmann 1996, Groppe et al. 1995, Leuchtmann & Clay 1989b, 1990, 1996, Sullivan & Faeth 2004, van Horn & Clay 1995). For example, molecular markers based on simple sequence repeat (SSR) loci are now available for *Neotyphodium* endophytes and provide a valuable tool for assessing genetic variation among endophyte isolates (van Zijll de Jong 2003, 2005). If there are distinct endophyte genotypes (also known as haplotypes due to the haploid nature of the fungal genome), these can be used to inoculate different host genotypes using the carefully controlled, aseptic techniques described elsewhere (Latch & Christensen 1985, Leuchtmann & Clay 1988b). Although most individuals arising from bulk-collected seeds of outcrossing host species such as perennial ryegrass or tall fescue are likely to be distinct genotypes, if host genotypes have not been previously defined, it may be necessary in some grass species to denote genotypes based on molecular characterization of host DNA.

Once distinct endophyte and host genotypes are available, inoculations can then be carried out (Johnson-Cicalese et al. 2000, Tintjer & Rudgers 2006), preferably replicates of all host genotypes being inoculated with each of the endophyte genotypes. Note that successful manipulation of endophyte status in inoculation experiments depends greatly on the compatibility of host and endophyte genotypes; thus there is some bias as to which particular host-endophyte combinations can actually be investigated. Furthermore, this protocol requires that endophyte-free (E−) ramets of the host genotypes be generated prior to inoculation. Thus treatment of some ramets with a systemic fungicide, or heat treatment to remove endophytes from seeds and the emerging seedlings (see section 2.1.1), will be required to provide E− plants for inoculation. It is important to also inoculate each host genotype with the original endophyte genotype isolated from it for comparison with uninoculated, infected control plants to control for possible effects (on the host) of the inoculation procedure itself (Leuchtmann & Clay 1988b).

If one has performed the cross-inoculations, then genotype-specific effects of both endophyte and host on plant growth (or other) variables can be teased apart, as well as potential $genotype_{host} \times genotype_{endophyte}$ interactions. If all plants were reared in a common environment, this is as far as the analysis can proceed. However, if the researcher had the time, resources, and inclination to replicate the cross-inoculations and grow the same host-endophyte genotypic combinations in multiple environments, then the magnitude of the environmental component of the three-way interaction could be determined. When cross-inoculations are not undertaken, one can still partition some of the components of the interaction using quantitative genetics techniques. For example, one can use the seeds from naturally infected and uninfected plants to generate half-sib families that can be grown across multiple environments (Ahlholm et al. 2002). This affords the opportunity to quantify the proportion of the variation in responses caused by host genotype, infection status, and environment. However, any effect of endophyte genotype is unfortunately subsumed within the host genotype effect.

As can be seen, a complete elucidation of the three-way interaction of genotype$_{host}$ $\times$ genotype$_{endophyte}$ $\times$ environment is a laborious undertaking, and to date we know of no study that has thoroughly characterized any grass-endophyte symbiosis in this way. These types of studies are sorely needed to pinpoint the genetic (host and endophyte) and environmental effects on endophyte-host interactions and coevolution.

4.3.3 Effects on host growth and physiology

There is some evidence that endophyte genotype can influence host growth. Three endophyte genotypes (N2, N11, N12) previously isolated from three tall fescue genotypes (PDN2, PDN11, PDN12) were employed in a drought stress experiment as reported in Hill et al. (1996). Endophytes were inserted into hosts developed from tissue culture techniques. Uninfected plants of the same host genotypes were used as controls. Although not all possible cross-inoculations among endophyte and host genotypes were carried out, replicates of host genotype PDN2 were inoculated with each of the endophyte genotypes. This allowed an assessment of whether or not endophyte genotype could differentially influence growth of a single host genotype under the conditions of the experiment. Aboveground dry mass of the host when infected by its own endophyte isolate (N2) was significantly greater than in the same host when endophyte-free (figure 4.6). When infected by the other two endophytes (N11, N12), genotype N11 significantly improved growth of the same host genotype, while the effect of N12 was no different than that of N2 (figure 4.6). Leaf water potential and turgor pressure were greater for host genotype PDN2 when infected with N11 and N12 than when infected with its usual endophytic symbiont, N2 (Hill et al. 1996). The authors speculated that different quantities of hormones produced by each endophyte isolate could lead to the differential responses of the same host genotype to short-term drought stress. Regardless of the mechanism, these data provide some of the best evidence that it is specific genotypic combinations of both host and endophyte that determine the morphology and physiology of endophyte-infected grasses, as well as dictating how selective pressures act on specific endophyte-host combinations.

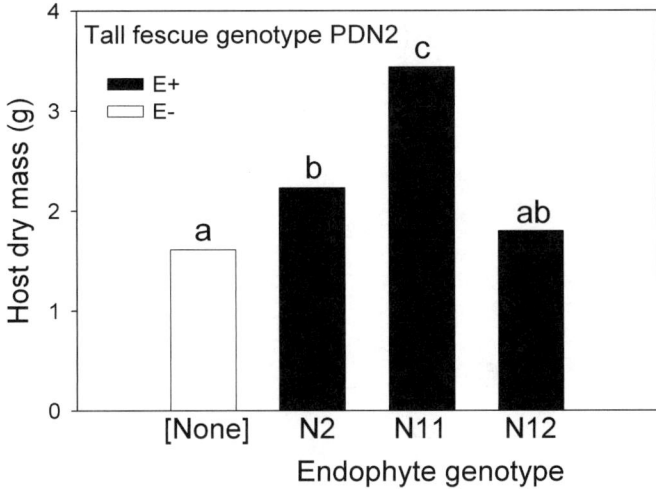

Figure 4.6 Dry mass of tall fescue genotype PDN2 when uninfected [none], or when infected with one of three different endophyte genotypes (N2, N11, and N12). N2 is the endophyte originally isolated from PDN2. Drawn with data from table 5 of Hill et al. (1996). Bars with different letters are significantly different at $P < 0.05$.

Cross-inoculations of seedlings of two cultivars of tall fescue (Maris Kasba [MK] and El Palenque [EP]) were performed for another drought stress experiment using two distinct endophyte strains designated AR501 and KY31 (Assuero et al. 2000). Although separate host genotypes were not retained, for most measured variables (tiller number, growth rate, photosynthesis) there was a significant cultivar × endophyte strain interaction. Symbiotic interaction norms for the rate of photosynthesis are plotted for both cultivars when uninfected or infected with each of the two endophyte strains (figure 4.7). In the MK cultivar, photosynthesis was increased by endophyte strain KY31, while in cultivar EP, photosynthesis was greatest when infected by endophyte strain AR501 (figure 4.7). If these tall fescue cultivars are relatively genetically homogeneous, the cultivar × endophyte strain interaction becomes analogous to a genotype$_{host}$ × genotype$_{endophyte}$ interaction for the measured variables.

Another inoculation experiment employed three endophyte genotypes of *Epichloë festucae* isolated from two subspecies of *Festuca rubra*, and a fourth endophyte genotype (*Neotyphodium* sp.) isolated from *Poa ampla* (Johnson-Cicalese et al. 2000). These different isolates were cultured on sterile media and then used to individually inoculate tillers of several genotypes of *F. rubra* subsp. *fallax* (Chewings fescue) and *F. rubra* subsp. *rubra* (creeping red fescue). Endophyte-free plants of each genotype were grown as well. Although only 11% of 340 inoculated tillers were successfully infected, endophyte effects on host size and seed production depended on both endophyte and host genotype (Johnson-Cicalese et al. 2000). For example, the endophyte isolated from *P. ampla* enhanced panicle and seed production in one Chewings fescue

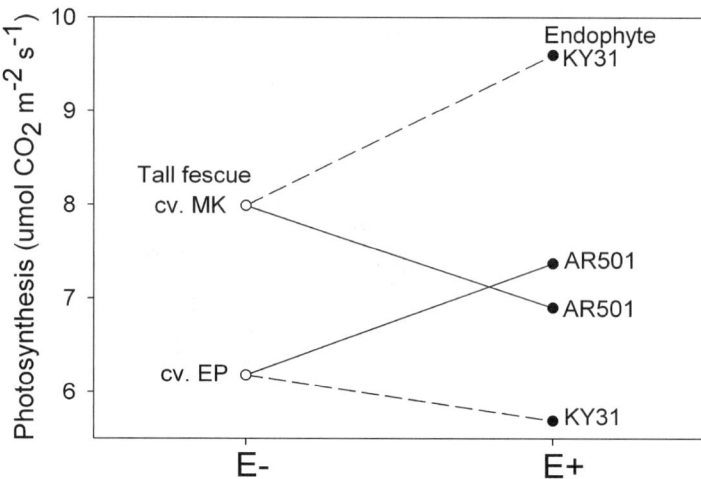

Figure 4.7 Symbiotic interaction norms for the rate of photosynthesis of two tall fescue cultivars (MK and EP) when uninfected (E–) or infected (E+) with endophyte strain KY31 or AR501. Drawn with data from table 4 of Assuero et al. (2000).

genotype (relative to the uninfected control), but had the opposite effect in a different host genotype of the same subspecies. This research provides solid evidence that endophyte isolates can have genotype-specific effects on host growth and reproduction.

To further understand the ecological genetics of grass-endophyte symbioses in natural communities it would be prudent to examine native grasses infected by variable endophyte genotypes. Based on genetic variation in microsatellite markers, Sullivan and Faeth (2004) demonstrated that the *Neotyphodium* endophyte of the native grass *F. arizonica* (Arizona fescue) exists as several distinct haploid genotypes (i.e., haplotypes) of both hybrid and nonhybrid origin. In a recent study of the growth and physiology of this grass, four host genotypes were examined, each of which contained one of two possible nonhybrid haplotypes denoted H1 and H2 (Morse et al. 2007). Specific haplotype-host genotype combinations differed in physiological parameters, such as photosynthetic rate, respiration, leaf water potential, and the density of stomata. For example, genotypes harboring endophyte H1 showed greater photosynthesis over a 49-day growth period than those harboring H2. Plants with endophyte H1 also had greater dry mass and higher relative growth rate than those with H2 (Morse et al. 2007). Although each endophyte haplotype was associated with only two particular host genotypes, and genetically identical replicates of the same host genotype infected by different haplotypes were not available, the results strongly imply that endophyte genotypes, in combination with different host genotypes, collectively determine growth and physiology in this native grass. Future work on this and other grass-endophyte systems will undoubtedly uncover more instances of the ecological relevance of genetic variation among endophytes inhabiting diverse host genotypes, populations, or species.

4.3.4 Genetic diversity

Across a broad range of endophytes and grasses, genotypic diversity of endophyte isolates has been reported, but has often been relatively low compared to the genotypic diversity of their hosts (table 4.1). Whether using isozymes, mitochondrial DNA (mtDNA), SSR markers, or microsatellites, often fewer than 20% of the isolates from a particular host are genetically distinct. Sometimes there is no detectable genetic variation. For example, 20 isolates of *Atkinsonella hypoxylon* from four populations of *S. leucotricha* were all genetically identical (Leuchtmann & Clay 1989b).

Based on isozymes, other isolates of *A. hypoxylon* from four *Danthonia* spp. varied as to the proportion that were genetically distinct from 8.6% in *D. epilis* to 22.4% in *D. spicata* (table 4.1) (Leuchtmann & Clay 1996). For 192 isolates from the same four host species, 20.3% were distinguishable as separate genotypes based on mtDNA (van Horn & Clay 1995). The proportions of the isolates that represented distinct genotypes were greater for each of the *Danthonia* spp. (table 4.1) compared to the isozyme study. Using the more sensitive technique of random amplified polymorphic DNA (RAPD) markers at 11 putative loci, Kover et al. (1997) found more than 50% of the isolates from *D. spicata* were genotypically distinct. The authors maintained that *A. hypoxylon* exhibited high genotypic diversity because most clones were derived from sexually produced, horizontally transmitted spores.

Genotypic diversity for *Epichloë* spp. endophytes appears to be within the range of that reported for *A. hypoxylon* (table 4.1). From 7% to 17% of the isolates from four wild grass hosts were identified as distinct genotypes, based on isozymes of 10 (Leuchtmann & Clay 1990) or 11 enzyme systems (Bucheli & Leuchtmann 1996).

Genetic variation in *Epichloë* spp. endophytes based on DNA has also been examined. Amplified fragment length polymorphism (AFLP) markers were used to explore the genetic structure of two populations of *E. festucae* inhabiting red fescue (*F. rubra*) in natural grasslands in Spain (Arroyo García et al. 2002). Although *E. festucae* can show a mixed breeding system (Schardl & Leuchtmann 2005), in these red fescue populations the primary method of endophytic reproduction appeared to be asexual, as stromata were rarely observed. Using a combination of primers, 84 AFLP loci could be scored in 18 isolates from each population. Average gene diversity over all loci was 0.25 in one populations and 0.30 in the other (Arroyo García et al. 2002). Fully 80% of the molecular variance was found within populations.

In another study that employed AFLP fingerprinting, 26 isolates of *Epichloë bromicola* were obtained from three *Bromus* spp. (Brem & Leuchtmann 2003). Based on 500 polymorphic restriction fragments, average gene diversities were 0.03, 0.12, and 0.20 for *B. ramosus*, *B. erectus*, and *B. benekenii*, respectively. About 30% of the total genetic variation was within populations. The rather

Table 4.1 Genotypic diversity of fungal endophytes isolated from various grass hosts. Distinct endophyte genotypes are given as a percentage of the number of isolates, except for *Neotyphodium coenophialum* and *Neotyphodium lolii*, where it is the percentage of single sequence repeat (SSR) loci that were unique (van Zijll de Jong et al. 2003). "None" denotes that all isolates were of the same genotype.

Endophyte (technique)	Host(s)	No. of populations	No. of isolates	Distinct endophyte genotypes (%)	Reference
Atkinsonella hypoxylon (isozymes)	*Danthonia compressa*	8	122	7.4	Leuchtmann & Clay 1989b
	Danthonia sericea	1	6	None	
	Danthonia spicata	11	143	7.7	
	Stipa leucotricha	4	20	None	
Atkinsonella hypoxylon (isozymes)	*Danthonia compressa*	24	71	9.9	Leuchtmann & Clay 1996
	Danthonia sericea	9	22	13.6	
	Danthonia spicata	23	49	22.4	
	Danthonia epilis	13	58	8.6	
Atkinsonella hypoxylon (mtDNA)	*Danthonia compressa*	25	60	11.7	van Horn & Clay 1995
	Danthonia sericea	9	30	16.7	
	Danthonia spicata	26	57	45.6	
	Danthonia epilis	14	44	18.2	
Atkinsonella hypoxylon (RAPDs)	*Danthonia spicata*	3	141	51.1	Kover et al. 1997
Epichloë festucae (microsatellites)	*Festuca rubra*	12	189	13.2	Wäli et al. 2007
Epichloë sylvatica (isozymes)	*Brachypodium sylvaticum*	10	266	7.1	Bucheli & Leuchtmann 1996
Epichloë brachyelytri (isozymes)	*Brachyelytrum erectum*	3	18	16.7	Leuchtmann & Clay 1990
Epichloë elymi (isozymes)	*Elymus hystrix*	2	40	15.0	Leuchtmann & Clay 1990
	Elymus virginicus	3	40	10.0	
Neotyphodium coenophialum (isozymes)	*Lolium arundinaceum*	14	52	3.8	Leuchtmann & Clay 1990
Neotyphodium coenophialum (SSR markers)	*Lolium arundinaceum*	5	5	9.7	van Zijll de Jong et al. 2003
Neotyphodium lolii (SSR markers)	*Lolium perenne*	5	5	6.3	van Zijll de Jong et al. 2003
Neotyphodium starrii (isozymes)	*Festuca obtusa*	5	14	42.9	Leuchtmann & Clay 1990
Neotyphodium starrii (microsatellites)	*Festuca arizonica*	4	123	6.5	Sullivan & Faeth 2004

high diversity of isolates from *B. benekenii* is surprising, as the endophyte is completely asexual in this host (and in *B. ramosus*) (Leuchtmann 2003). In contrast, on *B. erectus* the endophyte causes choke disease and reproduces sexually (Schardl & Leuchtmann 2005), giving rise to an expectation of relatively high genetic diversity for the isolates from this host.

Whether using isozymes (Leuchtmann & Clay 1990) or DNA markers (van Zijll de Jong et al. 2003), the genotypic diversity of the *Neotyphodium* endophytes of tall fescue and perennial ryegrass appears to be relatively low. From 14 population sources for tall fescue, some of which were commercial cultivars, 47 of 52 isolates showed the same isozyme profile, and the 5 that did not had a different profile, but were from a single cultivar (cv. Triumph) (Leuchtmann & Clay 1990). In the same study, all six isolates from perennial ryegrass (cv. Repell) had identical isozyme profiles. Furthermore, of 123 isolates of *Neotyphodium starrii* from Arizona fescue, only 6.5% represented distinct haplotypes (Sullivan & Faeth 2004). This low genetic diversity of *Neotyphodium* spp. may be a reflection of the exclusively asexual reproduction of these endophytes, which have no known sexual phase (Schardl & Leuchtmann 2005).

To identify and quantify genetic diversity within and among *Neotyphodium* and *Epichloë* endophytes, van Zijll de Jong et al. (2003) developed molecular markers based on SSR loci. The percentage of SSR loci that were unique was relatively low for the endophytes from tall fescue (9.7%) and perennial ryegrass (6.3%). They also used AFLP profiles to identify 879 polymorphic bands. The majority (>65%) of the polymorphism was distributed between *Neotyphodium* and *Epichloë* species; probably due to their asexual reproductive mode, low levels of polymorphism (5%) were found in each of the two hosts (van Zijll de Jong et al. 2003).

Genotypic diversity of endophytes within host populations has been documented in other grass-endophyte symbioses. In the isozyme studies of Leuchtmann and Clay (1989b, 1990), there were sufficient isolates obtained from individual host plants in some populations to permit quantification of the percentage of isolates that represented distinct genotypes (table 4.2). Within three populations of *Danthonia compressa* infected by *A. hypoxylon*, from 9% to 25% of the isolates were distinguishable based on isozyme profiles. Comparable values for four populations of *D. spicata* infected by the same endophyte species ranged from 8% to 21%. It is interesting to note that many more endophyte genotypes from the same host species could be discerned with RAPD markers (Kover et al. 1997): from 43% to 63% of the isolates were different genotypes (table 4.2).

In *Epichloë elymi* isolates from one population of *Elymus hystrix*, no isozyme variation was found, while in one population of *Elymus virginicus* infected by the same endophyte, only two variants out of 36 isolates were detected (Leuchtmann & Clay 1990). *E. elymi* has a mixed strategy of reproduction, including both asexual and sexual phases (Schardl & Leuchtmann 2005). Within four populations of *B. sylvaticum*, isolates of *Epichloë sylvaticum*, a species that appears to be predominantly asexual, were of the same genotype

Table 4.2 Within-population genotypic diversity of fungal endophytes isolated from various grass hosts. "None" denotes that all isolates were of the same genotype in that population. Only populations with at least 15 isolates are shown.

Endophyte (technique)	Host	Population	No. of isolates	No. endophyte genotypes (%)	Reference
Atkinsonella hypoxylon (isozymes)	Danthonia compressa	8840	3	3 (9.4)	Leuchtmann & Clay 1989b
		8841	16	4 (25.0)	
		8844	53	6 (11.3)	
Atkinsonella hypoxylon (isozymes)	Danthonia spicata	8847	22	4 (18.2)	Leuchtmann & Clay 1989b
		8848	38	3 (7.9)	
		8849	51	7 (13.7)	
		8850	19	4 (21.1)	
Atkinsonella hypoxylon (RAPDs)	Danthonia spicata	FR	46	25 (54.3)	Kover et al. 1997
		TS	32	20 (62.5)	
		GR	63	27 (42.9)	
Epichloë sylvatica (isozymes)	Brachypodium sylvaticum	A	129	14 (10.8)	Bucheli & Leuchtmann 1996
		B	19	6 (31.6)	
		C	36	2 (5.6)	
		D	32	1 (none)	
		E	26	6 (23.1)	
Epichloë elymi (isozymes)	Elymus hystrix	8864	34	1 (none)	Leuchtmann & Clay 1990
	Elymus virginicus	8862	36	2 (5.6)	
Epichloë festucae (microsatellites)	Festuca rubra	P3	67	12 (17.9)	Wäli et al. 2007
		P9	15	3 (20.0)	
		P11	18	2 (11.1)	
Neotyphodium starrii (microsatellites)	Festuca arizonica	Flagstaff	30	1 (none)	Sullivan & Faeth 2004
		Clint's Well	30	7 (23.3)	
		Merritt Draw	35	2 (5.7)	
		Buck Springs	28	2 (7.1)	

in one population, while multiple genotypes (6–32% of the isolates) were found in the other populations (Bucheli & Leuchtmann 1996) (table 4.2). While it is clear that genotypic variation among *Epichloë* endophytes can and does occur within host populations, it is by no means clear whether or not levels of genetic diversity are tied to endophyte breeding systems. In this regard, it is notable that Leuchtmann and Clay (1990) stated that there was no evidence across the 17 host grass species they studied that sexual endophytes were electrophoretically more variable than asexual ones.

4.3.5 Host-adapted endophyte races

As currently defined, one endophyte species can clearly be found in more than one host species (Schardl & Leuchtmann 2005), but is there evidence, as predicted by coevolutionary theory (e.g., Kaltz & Shykoff 1998, Thompson 2005), for local adaptation or maladaptation of endophytic fungi to specific host species or populations? Some of the best evidence for local adaptation of endophyte populations comes from experimental cross-inoculations of endophytes isolated from different host populations or species. For instance, researchers interested in the applied aspects of grass-endophyte symbioses have performed deliberate inoculations of endophytic isolates from one host species into another potential host species that is normally inhabited by a different endophyte. Such experiments are performed in an effort to create new grass-endophyte associations that might have agronomic value (Easton 2007, Johnson-Cicalese et al. 2000).

After inoculating perennial ryegrass with *N. coenophialum*, the endophyte normally found in tall fescue, Koga et al. (1993) reported that the leaf sheaths of the new host contained collapsed, degenerative hyphae. Thus the endophyte of *L. arundinaceum* did not appear to be compatible with the closely related congener *L. perenne*. Variable levels of compatibility among congeneric grass hosts were also found when representative isolates comprising six "taxonomic groups" of *Neotyphodium* spp. endophytes were inoculated into seedlings of perennial ryegrass (cv. Nui), tall fescue (cv. Roa), and meadow fescue (cv. Ensign) (Christensen 1995). In that study, most isolates could form associations with all three hosts, suggesting that a fair degree of cross-compatibility existed among these closely related grasses. For nine isolates from tall fescue inoculated into perennial ryegrass, 0–33% of the inoculations resulted in endophyte establishment (mean ± SE = 13.5 ± 3.3%). In contrast, when the same isolates were inoculated into their usual host (tall fescue), 11–54% resulted in endophyte establishment (mean ± SE = 32.5 ± 5.9%). When the same isolates from tall fescue were inoculated into meadow fescue (*Lolium pratense*), 3–67% resulted in endophyte establishment (mean ± SE = 32.3 ± 7.0%). Note that this level of inoculation success for meadow fescue receiving the tall fescue endophyte is the same as that obtained when tall fescue was inoculated with isolates of its own endophyte species. However, some host-endophyte combinations were unstable and endophyte-free tillers began to be produced as the host grew (Christensen 1995). This was especially true for perennial ryegrass inoculated with isolates from the other two hosts, and for tall fescue inoculated with isolates from meadow fescue. In a related study (Christensen et al. 1997), some isolates of *E. festucae* from *F. rubra* and *F. longifolia* resulted in increased mortality and stunted growth when inoculated into tall fescue, meadow fescue, and perennial ryegrass.

Pronounced variation in host-endophyte compatibility has been described between *A. hypoxylon* and three host grasses: *D. spicata* (three populations), *D. compressa* (two populations), and *S. leucotricha* (three populations) (Leuchtmann & Clay 1989a). Isolates were obtained from each population and used for experimental cross-inoculations. Within the *Danthonia* genus, cross-inoculations were highly successful, while there was complete incompatibility between isolates across *Danthonia* and *Stipa* (figure 4.8a). Although all of these isolates are of the same endophyte species, there has evidently been some level of evolutionary divergence to the point that different races inhabit the two taxonomically divergent host genera.

Compatibility relationships have also been explored among *Epichloë* endophytes and their host grasses (Brem & Leuchtmann 2003, Christensen et al.

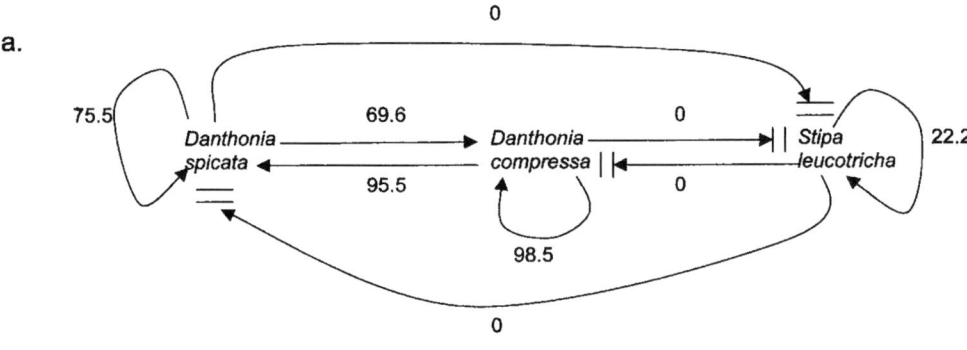

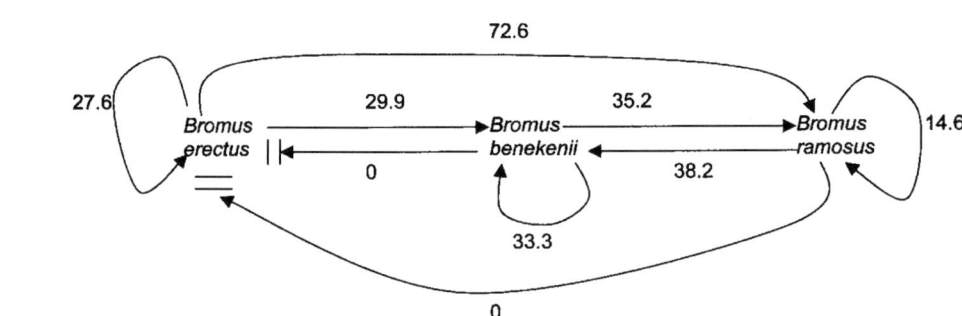

Figure 4.8 Reciprocal cross-inoculations of endophytic isolates across various grass hosts. Numbers represent the percentage of seedlings successfully infected following inoculation. Data for (a) are from tables 3 and 4 of Leuchtmann and Clay (1989a) and for (b) are from table 4 of Brem and Leuchtmann (2003).

1997, Chung et al. 1997, Johnson-Cicalese et al. 2000, Leuchtmann & Clay 1993). The host range of *E. elymi* was examined by reciprocal inoculations using isolates from three *Elymus* host spp. (Leuchtmann & Clay 1993). Isolates of what is now classified as *Epichloë brachyelytri* (Schardl & Leuchtmann 2005) were obtained from *Brachyelytrum erectum* [note that at the time of the original study, all isolates from the four grasses were assigned to *Epichloë typhina*]. Isolates of all three *Elymus* species could reciprocally infect all four grass host species. However, isolates from *B. erectum* could only infect seedlings of *B. erectum*. The authors noted that their study was the first example of successful cross-inoculation between hosts within different taxonomic tribes of the grass family (Leuchtmann & Clay 1993). That is, endophytes from *E. virginicus* of the tribe Triticeae were able to establish and persist for almost 4 years after being placed into a member (*B. erectum*) of the tribe Brachyelytreae.

Solid molecular evidence for host-adapted races of *E. bromicola* isolated from three *Bromus* species is available (Brem & Leuchtmann 2003). The isolates were genetically differentiated according to their host species and host specificity was indicated by reciprocal inoculations (figure 4.8b). Isolates from *B. erectus* could readily infect seedlings of *B. ramosus*, but the reverse was not true. Isolates from *B. benekenii* were unable to infect seedlings of *B. erectus* (figure 4.8b). The endophytic fungi of *B. ramosus* and *B. benekenii* are asymptomatic and vertically transmitted through seeds, while that of *B. erectus* causes choke disease and does not transmit through seeds (Schardl & Leuchtmann 2005). Brem and Leuchtmann (2003) suggested that the host specificity of the asexual isolates from *B. ramosus* and *B. benekenii* could be a consequence of reproductive isolation following the loss of the ability to produce choke-causing stromata. These asexual forms presumably evolved following a host shift from *B. erectus*. Thus these seed-transmitted isolates of *E. bromicola* exist as distinct races, or perhaps incipient species. Over a long enough time, the lack of gene flow among the asexual races should promote genetic drift, and as mutations accumulate, might lead to the eventual evolution of new *Epichloë* species on *B. ramosus* or *B. benekenii*.

4.3.6 Multistrain infections and fungal hybridization

Given the genotypic variation found among isolates of fungal endophytes, the question arises as to whether two or more strains of systemic endophytes can inhabit the same host individual. This is of more than academic interest: as discussed later, the occurrence of two endophyte strains or genotypes or "species" within the same host is a necessary precondition to fungal hybridization (Meijer & Leuchtmann 1999, Selosse & Schardl 2007) which has apparently occurred multiple times during the evolutionary history of clavicipitacean fungi (Gentile et al. 2005, Moon et al. 2004, Schardl & Moon 2003, Tsai et al. 1994).

Wille et al. (1999) artificially inoculated all possible double-strain mixtures of four strains of *E. bromicola* into seedlings of *B. erectus*. Although individual tillers only harbored one endophyte genotype, 8% of the host plants showed evidence of double infection, albeit in different tillers of the same individual. In a study of *E. sylvatica* infecting the grass *B. sylvaticum*, isozyme polymorphism at seven loci was used to characterize endophyte genotype in plants sampled at natural sites (Meijer & Leuchtmann 1999). Of 63 host plants examined, 18 (= 28.6%) of them contained two endophyte genotypes in separate tillers and 7 (= 11.1%) contained three endophyte genotypes. Finally, Christensen et al. (2000) were able to successfully inoculate tall fescue plants with both *N. coenophialum*, its usual endophyte, and *N. lolii* isolated from perennial ryegrass. Over time the incidence of double infection decreased, but tillers infected with the two endophytes were identified in three tall fescue individuals. The authors speculated that hyphae of a dominant strain were most likely to colonize newly emerging tillers, and over time all tillers of most plants would be likely to contain but a single endophyte genotype (Christensen et al. 2000).

All three of the above studies provide evidence that multiple endophyte strains can co-occur, at least for a limited time, within the same host plant. In addition, all show that the individual tiller may be the basic modular unit of the grass-endophyte symbiosis, as multiple endophyte strains were generally not found within the same tiller. Nonetheless, this does not rule out the possibility of hyphal fusion leading to hybridization among endophytes, at least during the early stages of tiller formation. Indeed, Moon et al. (2004) mentioned that the most likely means of fungal hybridization was by somatic fusion of endophytic hyphae via parasexuality. In endophytes with horizontal transmission, the opportunity for hybridization could occur during the sexual cycle (e.g., transfer of spermatia by insects from one stroma to another, enabling cross-fertilization) (Bultman et al. 1995). However, despite in vitro mating studies carried out with *Epichloë* endophytes (Chung & Schardl 1997b), direct experimental evidence of hybridization in vivo among endophytes has not been forthcoming. Nevertheless, there is much molecular and phylogenetic evidence that hybridization among endophytes is a potent force in the evolution of new endophyte lineages.

Phylogenetic analyses have demonstrated that about two-thirds of asexual, vertically transmitted endophytes are interspecific hybrids with heteroploid genomes (Moon et al. 2004). A few examples are briefly noted here, but much more detail is available for the interested reader in the various reviews by Christopher Schardl and colleagues (Moon et al. 2004, Schardl & Craven 2003, Schardl & Leuchtmann 2005, Schardl & Moon 2003, Schardl & Wilkinson 2000, Selosse & Schardl 2007). In tall fescue, molecular data on multiple copies of genes from different clades of *Epichloë* have revealed that three hybridization events were likely to have been involved in the evolution of *Neotyphodium* endophytes (Tsai et al. 1994). Another form of the *Neotyphodium* endophyte found in perennial ryegrass appears to be a hybrid of *N. lolii* and *E. typhina* (Schardl et al. 1994). More recently, most of the isolates of *Neotyphodium* spp. from native grasses in Argentina were likewise shown to be interspecific hybrids of *E. festucae* and *E. typhina* (Gentile et al. 2005).

For hybridization to occur, co-occurring endophyte strains or species must merge and fuse nuclei; the resulting hybrid is presumed to be favored by selection (Schardl & Moon 2003, Schardl & Wilkinson 2000). However, other than the observation of the commonness of hybrids (Hamilton & Faeth 2007), there have been virtually no experimental tests of the notion that hybrid endophytes are more fit than nonhybrid ones (see section 5.3.1). After heteroploid hybrids are formed, genome size may be reduced and novel hybrid genotypes can be readily maintained through successive generations by vertical transmission (Schardl & Craven 2003, Selosse & Schardl 2007). For asexual endophytes that cannot generate new genetic recombinants by ordinary sexual means, there may be reduced potential for evolutionary diversification and long-term survival (Holland et al. 2007). The parasexual cycle may mitigate this problem as hyphal fusion between two fungal strains (or species) can rapidly yield unique recombinant strains in asexual species (Carlile et al. 2001). Hybridization has important consequences for reducing gene flow differentials between sexually reproducing hosts and asexual endophytes. The consequences of this change in gene flow differential at the population level for coevolution are explored in chapter 5.

Schardl and Moon (2003) suggested that asexual, nonhybrid endophytes (*Neotyphodium* spp.) arise most often from *Epichloë* spp. by host jumps and are then subjected to hybridization events when host plants become coinfected by other *Epichloë* genotypes. New recombinants formed in this manner may combine the characteristics of the hybridizing parental genotypes and may be selectively favored on some hosts or in particular environments (Hamilton & Faeth 2007). Genetic variation in both endophyte and host will expedite the continued coevolution of the symbiosis. During the evolutionary history of endophytic fungi, interspecific hybridization has likely been important in both reducing the accumulation of deleterious mutations (i.e., Muller's ratchet) and promoting genotypic diversity (Schardl & Craven 2003). In addition, hybridization among endophytes and the resulting meiotic sterility of the hybrids may have been important selection pressures for the evolution of vertical transmission and mutualistic symbiosis (Selosse & Schardl 2007).

5

Evolutionary Ecology of
Grass-Endophyte Interactions

The consequences of endophyte infection to the evolutionary ecology of grass populations are difficult to assess. In addition to investigations of the dynamics of ramets and genets for infected and uninfected segments of the population, it is important to analyze the host population as a collection of individual host genotype-endophyte combinations interacting with the biotic and abiotic environment. Studies at the interface of demography and genetics, which are probably the most illuminating in terms of the underlying coevolutionary process, should integrate the population dynamic consequences of genotypic variation in both partners, as detailed for plant-pathogen interactions by Burdon and Thrall (2001). In this chapter, we first describe experimental research on the population dynamics of endophyte-infected grasses, including results of a 3-year experiment with replicated *Lolium perenne* genotypes (infected and uninfected) growing in an outdoor plot (Cheplick 2008). This is followed by a detailed section that considers various explanations for the persistence and high frequency of endophytes in natural populations. Later we examine how the coevolution of endophytes and their hosts fits within the framework of Thompson's (2005) geographic mosaic theory.

5.1 DYNAMICS OF HOST POPULATIONS

Models developed specifically to describe population dynamics in mutualistic symbioses (e.g., Holland et al. 2002, Wolin 1985) may not be especially useful for grass-endophyte systems. This is because (1) the grass-endophyte interaction ranges from parasitic to mutualistic within and between host species, as discussed throughout this book; (2) the nature of the symbiosis is highly contingent on environmental conditions; (3) the genetic background of both host and endophyte impacts the grass-endophyte relationship (chapter 4); (4) one partner (the endophyte) is an obligate symbiont, while the other (the plant) is not (although this feature can be incorporated into a simple mutualism model) (Dean 1983); and (5) the majority of endophyte-infected grasses are caespitose perennials (Clay 1998) with populations composed of potentially long-lived genets in closed communities. In such communities, establishment of new genets from seeds may be a relatively rare event. This last consideration is important because if recruitment into the plant community is relatively rare, endophyte-mediated effects on sexual reproduction, whether positive or negative (section 2.1.4), are liable to only change the dynamics of host genet populations very slowly, if at all. However, one simple computer simulation showed that even with a modest fitness advantage, the frequency of infected individuals should gradually increase over time (Clay 1993) provided there are at least some opportunities for seedling establishment.

Despite the complexities inherent in the development of theoretical models to describe the dynamics of grass-endophyte interactions, several structured metapopulation models have depicted how endophyte-infected plants can persist within host populations (Gyllenberg et al. 2002, Saikkonen et al. 2002). Although it may be easy to envision the persistence of endophytes that are exclusively vertically transmitted within host seeds, Saikkonen et al. (2002) showed how endophytes could persist even if vertical transmission was imperfect and the endophyte reduced host fitness in some environments. Their model demonstrated that mutualism was not a necessary prerequisite for the survival and maintenance of vertically transmitted endophytes, as long as the environment was a heterogeneous, patchy one.

Only a small number of studies on grass-endophyte ecology have provided moderately long-term demographic data on hosts that span several years under field conditions (Clay 1990a, Clay & Holah 1999, Faeth & Hamilton 2006, Fowler & Clay 1995, Hill et al. 1998, Olejniczak & Lembicz 2007). In his review of the "population dynamics" of endophyte-infected grasses, Clay (1998) summarized several categories of endophyte-mediated effects on host seed germination, plant vigor and reproduction, and host stress tolerance, all of which were covered earlier in chapter 2. Although it is recognized that endophytes can influence host population dynamics through their general effects on individual host growth and physiology, here a more restricted definition of population dynamics will be used that emphasizes the "change in numbers with time" (Silvertown & Charlesworth 2001). In the summary to the earlier review, Clay (1998) stated that "the population dynamics of many grasses are intimately coupled with infection by systemic fungal endophytes." In this section, the veracity of this contention will be evaluated by presenting information on the dynamics and survival of individual host plants (i.e., genets) with and without endophytes, and on the dynamics of the tillers (i.e., ramets) that comprise each genet. Factors that can impact host population dynamics, such as resource storage, seed production, seedling establishment, and endophyte persistence within plant parts, will also be examined.

5.1.1 Genet and ramet dynamics

Changes in the number of genetically distinct individuals (genets) within a grass population may occur by mortality of established plants over time, or by the recruitment of new seedlings following an episode of sexual reproduction. An excellent example of a long-term study of genet dynamics for a perennial grass (*Bouteloua gracilis*) in a natural ecosystem is provided by Fair et al. (1999).

One of the most detailed, long-term field studies of endophyte-infected hosts involved a demographic comparison of E− and E+ plants of two grasses (*Lolium arundinaceum* and *Sporobolus poiretii*) and one sedge (*Cyperus virens*) over a 3-year period under field conditions in Louisiana (Clay 1990a). Ramets of uniform size of the two grasses were planted into a grassland, while those of the sedge were planted into a disturbed, seasonally flooded site. Data were recorded each year on the percentage of the planted population that survived and flowered, and on the per capita number of tillers and inflorescences. At the end of the third growing season, aboveground dry mass was obtained.

A strong mutualistic association was evident for *L. arundinaceum* (tall fescue, cv. Kentucky 31) infected by *Neotyphodium coenophialum*, as shown by improved genet survival (table 5.1) and greater tiller production over the

Table 5.1 Long-term survival of uninfected (E−) and endophyte-infected (E+) plants of seven perennial grasses and one sedge (*C. virens*). + denotes a statistically significant positive effect of endophyte infection, while an O denotes no significant effect.

			Survival (%)			
Host	Endophyte	Duration (years)	E−	E+	Effect of infection	Reference
Cyperus virens	*Balansia cyperi*	3	3	22	+	Clay (1990a)
Danthonia spicata	*Atkinsonella hypoxylon*	1.5	54.6	65.3	O	Clay (1984)
Festuca arizonica	*Neotyphodium starrii*	5	96.2	93.2	O	Faeth & Hamilton (2006)
		7	100	98.3	O	
Lolium arundinaceum	*Neotyphodium coenophialum*	3	56	84	+	Clay (1990a)
Lolium perenne	*Neotyphodium lolii*	2.7	90.4	87.4	O	Cheplick (2008)
Puccinellia distans	*Epichloë typhina*	3	100	100	O	Olejniczak & Lembicz (2007)
Sporobolus poiretii	*Balansia epichloë*	3	74	81	O	Clay (1990a)
Stipa leucotricha	*Atkinsonella texensis*	2.5	["no significant effect"]		O	Fowler & Clay (1995)

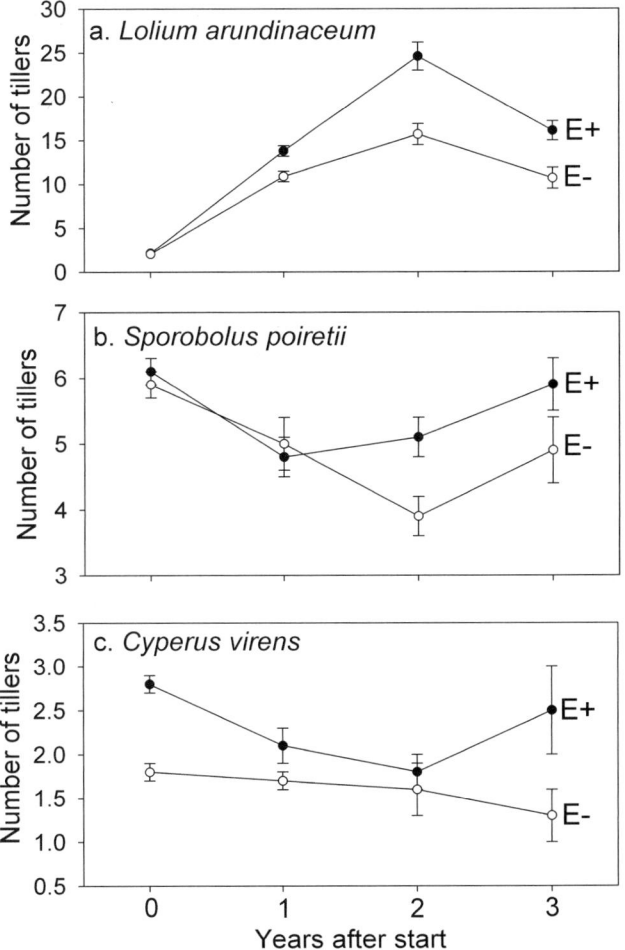

Figure 5.1 Tiller dynamics of uninfected (E–) or endophyte-infected (E+) plants of (a) *Lolium arundinaceaum* (tall fescue), (b) *Sporobolus poiretii* (smut grass), and (c) *Cyperus virens* (green sedge) over 3 years in the field. Data from Clay (1990a).

3 years (figure 5.1a). By the end of the study, the mean (± SE) dry mass of E+ plants was 5.2 ± 0.5 g ($n = 141$), significantly greater than that of E– plants (3.7 ± 0.5 g, $n = 94$). Bearing in mind that this grass was an agronomic cultivar (Kentucky 31) of tall fescue well-known for its positive growth responses to endophyte infection under a variety of conditions (see chapter 2) and its antiherbivore properties (Ball et al. 1993, Saikkonen 2000), it is perhaps no surprise that greater growth and survival were found under field conditions. Later research has revealed that infected individuals of this same cultivar of tall fescue become increasingly dominant in field plots over several years (Clay & Holah 1999, Clay et al. 2005, Hill et al. 1998).

In contrast to tall fescue, survival of the other caespitose grass, *S. poiretii*, did not depend on whether or not plants were infected by *Balansia epichloë* (table 5.1) (Clay 1990a). Tiller production was slightly greater in E+ plants (figure 5.1b), but final dry masses of E+ and E− plants were not significantly different (1.5 ± 0.1 g, $n = 125$ and 1.3 ± 0.2 g, $n = 59$, respectively). Sexual reproduction of *S. poireti* was drastically impaired by this endophyte which produces stromata and is capable of horizontal transmission. Only 5% of E+ ramets made an inflorescence, compared to 70% of E− ramets. Thus recruitment of new genets into the population would mostly be possible only for endophyte-free individuals. Clay (1990a) reasoned that this grass-endophyte relationship was therefore a pathogenic one.

For the perennial caespitose sedge *C. virens*, infection by the choke-causing endophyte *Balansia cyperi* improved genet survival over the 3-year study, although overall survivorship was quite low for this species (table 5.1). Although infected individuals tended to have more tillers (figure 5.1c), by the end of the experiment dry mass did not differ between the few surviving E+ (1.9 ± 0.5 g, $n = 22$) and E− plants (2.2 ± 0.8 g, $n = 3$). In this grass-endophyte symbiosis, infected plants always show inflorescence abortion and are incapable of producing seeds (Clay 1990a). Nevertheless, E+ individuals can sometimes produce asexual, viviparous plantlets that are also infected (Clay 1986). Thus, despite the inability of E+ individuals to generate new genotypes via sexual reproduction, it is possible for existing genotypes (and their endophytes) to be asexually propagated and thereby increase numerically over time.

A glance at the column "effect of infection" in table 5.1 reveals that for the majority of host grass populations that have been monitored for at least a few years, survival is generally very high and mostly unaffected by endophytes. In the longest study, survival of Arizona fescue genets exceeded 98% over a 7-year period (Faeth & Hamilton 2006). None of the native grasses exhibited a beneficial effect of endophytes on survival. Furthermore, if there is some fitness cost to the host of endophyte infection, it was not manifested as a reduction in long-term genet survival for any grass species. Of course, this does not discount the possibility that endophyte effects on host survival occur over even longer time frames (e.g., >5 years), or that effects occur only during rare, environmentally severe events such as prolonged drought. Such episodes of hard selection can impact the population dynamics of a wide range of species (Saccheri & Hanski 2006).

Changes in the numbers of tillers (ramets) produced over time can indirectly impact the long-term dynamics of genets. A genet that readily adds tillers under the prevailing environmental conditions will inevitably increase in size. Large size can be selectively advantageous for a variety of reasons. The compact spatial arrangement of tillers within the genet of a caespitose grass allows the consolidation of resources from the immediate environment (Briske & Derner 1998) and this may be more effective for large, multitillered genets. A compacted phalanx of ramets may be selectively favored in grasses that inhabit the competitive environment typical of successional fields (Cheplick 1997b, 2003, Cheplick & Chui 2001, Humphrey & Pyke 1998). Increases in size by tiller addition should improve competitive ability, genet persistence, and perhaps survival under adverse conditions. In addition, a greater accumulation of tillers may correlate with greater inflorescence, spikelet, and seed production (e.g., Cheplick & White 2002) due to allometry of sexual reproduction with vegetative size (Cheplick 2005 and section 5.1.2.2). In turn, greater sexual reproduction would open more opportunities for the recruitment of new genets into the population from seed. In short, it is not difficult to envision how particular host genotype-endophyte combinations that enhance tillering ability could be selectively favored in many communities where grasses grow.

5.1.2 A common garden experiment with perennial ryegrass

We have already reviewed some of the evidence that tiller production can be greater in E+ relative to E− plants in a number of grasses (section 2.1.2). In addition, in the Clay (1990a) study reviewed earlier, infected plants mostly produced greater numbers of tillers over the 3-year period for the three study species (figure 5.1). A few studies have also shown that tiller production and population persistence of perennial ryegrass can be greater for E+ plants under some conditions (Eerens et al. 1998b, Francis & Baird 1989, Popay 1997, Ravel et al. 1995, 1999). But do the ramet dynamics of genets vary when grown for a few years in a common garden under ambient field conditions? And, more importantly, are ramet dynamics of individual genets affected by endophyte infection?

An attempt was made to examine these questions for 10 replicated genotypes of perennial ryegrass (*L. perenne* cv. Yorktown III) grown in a field plot in central New Jersey, over three growing seasons (Cheplick 2008). Relative interaction intensities based on final aboveground dry mass were mostly positive, as reported earlier (section 1.3, figure 1.6), suggesting that endophytes were of some benefit in terms of host growth. In this experiment, 10 to 12 replicate ramets were planted for each host genotype-endophyte status combination. Survival of ramets was high (>87%) and did not differ significantly between E+ and E− plants (table 5.1). The number of tillers was recorded on 13 occasions throughout the 3 years of the experiment. At the time of planting (mid-November 2001), mean (± SE) size between E+ and E− plants did not differ based on either the number of tillers per ramet (1.14 ± 0.05 for E+ and 1.26 ± 0.05 for E−) or on total tiller length (25.87 ± 0.96 cm for E+ and 26.41 ± 0.98 cm for E−; $F_{1, 186} = 0.15$, $P = 0.70$). Thus there was no initial size advantage to E+ versus E− replicate plants at the start of the field trial.

5.1.2.1 Tiller dynamics

Repeated-measures analysis of variance (ANOVA) was performed on the log_{10}-transformed number of tillers, using host genotype, infection status, and their interaction as between-subject effects. Genotype had a highly significant effect on tiller production ($F_{9,\ 162} = 7.51$, $P < 0.0001$), but endophytes did not ($F_{1,\ 162} = 1.37$, $P = 0.24$). However, time showed a significant interaction with both genotype ($F_{108,\ 1944} = 4.02$, $P < 0.0001$) and infection status ($F_{12,\ 1944} = 2.12$, $P = 0.01$), indicating that tiller (ramet) dynamics varied with host genotype (genet) and depended on whether plants were infected.

Contrasting patterns of tiller dynamics are presented for three genotypes in figure 5.2. The peaks in these curves correspond to the rapid growth that

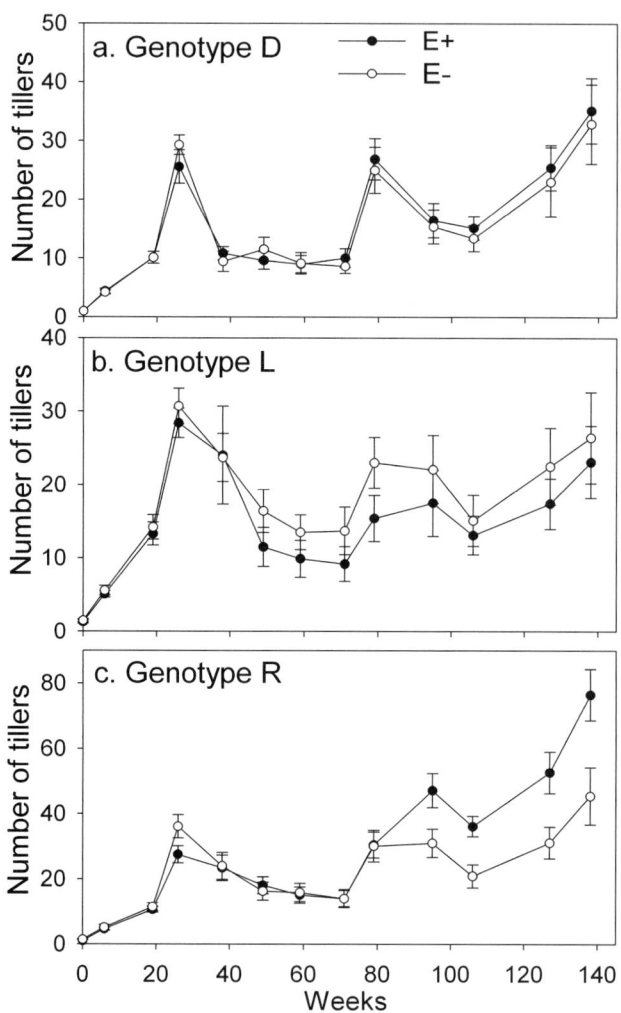

Figure 5.2 Tiller dynamics of uninfected (E–) or endophyte-infected (E+) plants of three genotypes of perennial ryegrass over three growing seasons in a common garden (Adapted from Cheplick, 2008).

occurred during the early spring (May) of each growing season. The subsequent drop-offs that follow the first two peaks in tiller numbers correspond to the senescence of flowering tillers and tiller mortality that occurred over the ensuing winter. The pattern shown by genotype D (figure 5.2a), in which tiller production was congruent for E+ and E− plants, was found in four of nine genotypes (all replicated plants of a tenth genotype were unfortunately determined to be infected and therefore could not be analyzed in this way). For genotype R, the tiller dynamics of E+ and E− plants were similar until the final growing season, when E+ plants showed a greater increase in tiller numbers relative to E− plants (figure 5.2c). Three of the nine genotypes showed this pattern, with the result that the final mean number of tillers by the end of the experiment was significantly greater for E+ plants (38.58 ± 2.35, $n = 97$) relative to E− plants (33.60 ± 2.63, $n = 85$) according to a univariate ANOVA ($F_{1, 163} = 6.22$, $P = 0.01$). Only two genotypes showed consistently greater tiller production in E− plants (e.g., genotype L in figure 5.2b).

The number of flowering tillers was also recorded during each flowering season, which was June at the study site. Each flowering tiller in perennial ryegrass produces a single elongate inflorescence (a spike) with a group of sessile, alternating spikelets in which multiple florets can mature seeds following outcrossing. The percent of the E+ and E− segments of the common garden population that flowered did not differ significantly in any year, but was typically high (E+ = 98.2%, 78.2%, and 87.7% and E− = 97.9%, 84.1%, and 83.5% in years 1, 2, and 3, respectively). Repeated-measures ANOVA on the $\log_{10}$-transformed number of spikes revealed a significant between-subjects effect of genotype ($F_{9, 163} = 3.51$, $P = 0.0005$), but not endophyte ($F_{1, 163} = 0.91$, $P = 0.34$). Furthermore, although spike production differed significantly among years ($F_{2, 326} = 95.66$, $P < 0.0001$) (table 5.2) and genotypes varied among years ($F_{18, 326} = 6.80$, $P < 0.0001$), there was no interaction of time with infection status ($F_{2, 326} = 6.80$, $P = 0.10$) or with the genotype × infection interaction ($F_{16, 326} = 1.06$, $P = 0.39$). For this mixed field population of perennial ryegrass, there was clearly no effect of endophyte infection on flowering tiller production (table 5.2).

Table 5.2 Mean (± SE) number of flowering tillers for uninfected (E−) and endophyte-infected (E+) plants of perennial ryegrass (averaged across all genotypes). Plants grew in a field plot in central New Jersey, over 3 years (Cheplick 2008).

Year	n	E−	n	E+
1	94	10.5 ± 0.7	107	9.4 ± 0.6
2	88	4.1 ± 0.4	101	4.2 ± 0.5
3	85	3.3 ± 0.3	97	4.1 ± 0.4

5.1.2.2 The allometry of reproduction

Despite the lack of differences in flowering behavior of the E− and E+ plants in the common garden population, recall that E+ plants had a significantly greater number of tillers overall. In addition, the final dry aboveground mass of E+ plants was 1.15 ± 0.08 g ($n = 97$), which was significantly greater ($F_{1, 163} = 5.41$, $P = 0.02$) than that of E− plants (0.93 ± 0.07 g, $n = 85$). This leads naturally to the question of whether or not there is a potential fitness payoff (in terms of sexual reproduction) to larger vegetative size (Cheplick 2005). To address this issue, the allometry of spike production was analyzed using the number of vegetative tillers at the time of flowering as a surrogate for plant size in the first 2 years (all variables were log_{10} transformed). For the third year when plants were harvested, in addition to the number of vegetative tillers, aboveground dry mass of all vegetative tillers was also used as a more precise metric for size.

In the spring of the first year after transplantation, spike production was not correlated with vegetative size for either E− ($F_{1, 92} = 0.01$, $P = 0.94$) or E+ plants ($F_{1, 105} = 0.60$, $P = 0.44$). In contrast, spike production the second year was positively correlated with the number of vegetative tillers for both E− ($F_{1, 86} = 5.58$, $P = 0.02$) and E+ plants ($F_{1, 99} = 5.90$, $P = 0.02$). However, the best and most statistically significant relationship between spike production and vegetative size was found in the final year of the study. For E− plants, log[no. spikes] $= -0.52 + 0.36$ log[mass] ($F_{1, 83} = 17.00$, $P < 0.0001$). The scaling relation was very similar in E+ plants: log[no. spikes] $= -0.49 + 0.37$ log[mass] ($F_{1, 95} = 15.47$, $P = 0.0002$). As might be expected given the positive relationship between tiller production and vegetative mass, spike production in the third year was also positively correlated with the number of vegetative tillers for E− ($F_{1, 83} = 13.99$, $P = 0.0003$) and E+ plants ($F_{1, 95} = 14.98$, $P = 0.0002$).

It is evident from the foregoing analyses that larger vegetative size does translate into increased spike production over time, but presently it is unclear if endophytes might indirectly increase sexual reproduction as plants age due to their positive effects on the size of some genets (e.g., figure 5.2c). It is also uncertain whether increased spike production leads to greater seed set. However, in another field study of perennial ryegrass conducted at an experimental station in Halle, Germany (Hesse et al. 2004), data were collected on the number of reproductive tillers and seed yield (i.e., dry mass of all seeds produced) for each of 13 replicated genotypes. Across the genotypes, mean seed yield was positively correlated with the mean number of flowering tillers for both E− ($r^2 = 0.65$, $P < 0.01$) and E+ plants ($r^2 = 0.43$, $P < 0.05$). The slope of the line depicting this relation was somewhat greater for E− plants (figure 5.3). Thus genotypes of perennial ryegrass that produce more flowering tillers will mature more seed.

If it is assumed that greater seed output results in a higher likelihood of successful seedling establishment, then the promotion of flowering spike production by endophyte infection could, at least for some host genets, result in

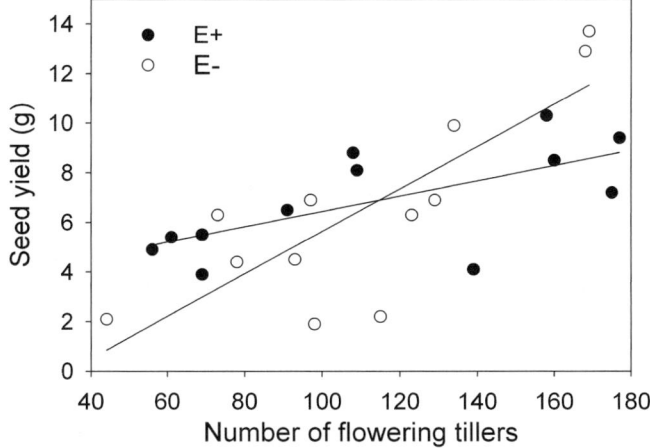

Figure 5.3 The relationship between seed yield (dry mass of seeds per genet) and the number of flowering tillers per genet for 12 genotypes of perennial ryegrass grown in a plot in Halle, Germany. Means are presented for uninfected (E−) or endophyte-infected (E+) plants for each genotype. Data from table 2 of Hesse et al. (2004).

improved representation of both endophyte and host genotypes in future generations. Obviously, for an exclusively vertically transmitted endophyte such as *N. lolii*, selection should favor increased seed production by the host. This could be accomplished by (1) greater spike production, (2) more seeds set per spike, (3) production of larger, heavier seeds by E+ plants, or (4) greater germination of E+ seeds. All of these variables were recorded for the 13 genotypes in the field experiment of Hesse et al. (2004), but both positive, negative, and no effects of endophytes were found, depending on host genotype.

The fact that it took three growing seasons to detect endophyte-mediated effects on host tiller dynamics and vegetative size in the common garden experiment of Cheplick (2008) underscores the importance of long-term field studies when examining the dynamics of endophyte-infected grasses, most of which are long-lived perennials. Another 3-year common garden study of the perennial grass *Puccinellia distans* and its asexual endophyte *Epichloë typhina* reinforces this point (Olejniczak & Lembicz 2007). In the first 2 years of the study, E+ plants produced more flowering tillers with longer inflorescences than E− plants. However, by the third year E+ plants had fewer and shorter inflorescences than E− plants and had reduced aboveground and belowground dry mass. Seed size was significantly lower on infected inflorescences for all 3 years. The authors suggested that stimulation of host reproduction early in life incurs a future cost to the host (Olejniczak & Lembicz 2007). Also, they speculated that the reduction in the dry mass of roots, which store resources for spring regrowth, might result in the persistence of endophyte effects beyond the third year of the host's life.

5.1.2.3 Carbohydrate costs of endophyte infection

Is there any evidence for a carbohydrate cost of endophyte infection for grass hosts? As detailed in chapter 2, endophyte-related reductions in growth and photosynthesis have been detected in some grasses, especially under low-resource conditions (e.g., Cheplick 2007, Cheplick et al. 1989). Because the endophyte must obtain the carbohydrate substrates necessary for its catabolism from its photosynthetic host, might the cost of endophyte infection be manifested by a reduced carbohydrate pool in infected hosts? This idea matches the explanation of "host-fungus competition" for carbohydrates that has been suggested as a reason for the depression of aboveground yield in perennial ryegrass infected by mycorrhizal fungi (Buwalda & Goh 1982). However, studies with tall fescue and perennial ryegrass have not reported a consistent influence of endophyte infection on carbohydrate storage (in tiller bases) or on regrowth following defoliation (Belesky & Fedders 1996, Belesky et al. 1989, Cheplick 1998a, Cheplick & Cho 2003, Rasmussen et al. 2008).

In general, very few studies have addressed the potential costs of endophyte infection in native grass species. However, there is limited evidence that costs can be detected, especially when resources such as soil water or minerals are low. For example, in the native grasses *Festuca rubra* and *Lolium pratense*, Ahlholm et al. (2002) found possible costs of infection, manifested as reduced tillering, biomass, or inflorescence production. This was especially true for *L. pratense* infected by *Neotyphodium uncinatum* and growing under resource-limited conditions (Ahlholm et al. 2002). However, carbohydrate costs were not specifically measured in that study.

In their defoliation experiment with perennial ryegrass (cv. Yorktown III), Cheplick and Cho (2003) reported a significant interaction effect of host genotype with infection status on the percent total nonstructural carbohydrates (TNC) stored in tiller bases at the time of defoliation. Although not reported in that publication, TNC was also assessed in the same 10 genotypes following 4 weeks of regrowth (five E– and five E+ replicate plants per genotype). These TNC data were arcsine, square-root transformed and subjected to ANOVA, with host genotype, infection status, and their interaction as the effects. Endophyte infection had no effect on final percent TNC ($F_{1,\ 79} = 0.02$, $P = 0.89$), but there were significant effects of genotype ($F_{9,\ 79} = 19.34$, $P < 0.0001$) and genotype × infection ($F_{9,\ 79} = 2.43$, $P = 0.02$). This interaction is readily apparent in table 5.3, where relative interaction intensities (Armas et al. 2004) have been provided to express the degree to which endophytes positively or negatively impacted TNC for each host genotype. Again, this example illustrates how it can appear that endophytes are having no overall effect on a particular host population (note the very similar grand means in table 5.3 for E– and E+ plants) even though individual host genets are "reacting" quite differently to their particular endophytic symbiont. It remains to be determined whether these genet-based differences in endophyte effects on TNC actually affect demographic parameters in the field such as tiller dynamics, flowering behavior, or overwinter survival.

Table 5.3 Total nonstructural carbohydrates (%) in 5-cm samples of the tiller bases of 10 genotypes of *Lolium perenne* (cv. Yorktown III) after 4 weeks of regrowth following defoliation. Values are the means (± SE) for four to five uninfected (E–) or four to five endophyte-infected (E+) plants per genotype. Previously unreported data from Cheplick and Cho (2003). Relative interaction intensities (RIIs) based on these data are indicated (see section 1.3).

Genotype	E–	E+	RII
B	11.01 ± 2.81	9.24 ± 1.48	−0.0877
D	14.92 ± 1.91	12.56 ± 1.92	−0.0859
E	19.66 ± 2.17	15.29 ± 3.01	−0.1250
G	24.89 ± 3.78	32.16 ± 5.48	+0.1274
H	11.11 ± 3.35	4.70 ± 0.98	−0.4055
K	13.74 ± 2.71	11.71 ± 1.54	−0.0800
L	4.04 ± 0.62	10.91 ± 1.51	+0.4600
N	10.27 ± 2.21	9.51 ± 1.12	−0.0386
R	4.74 ± 1.02	8.88 ± 2.49	+0.3040
T	3.63 ± 0.70	3.42 ± 0.52	−0.0302
Grand mean	11.80 ± 2.18	11.84 ± 2.51	+0.0038 ± 0.0763

5.1.2.4 The cost of reproduction

Another cost to endophyte infection that has not been previously explored is the potential cost of reproduction. Although not applicable to choke-causing endophytes that occlude inflorescence development, asexual, vertically trans-mitted endophytes like *N. lolii* that permit sexual reproduction could entail a future cost. Such a cost is due to a physiological trade-off: resources allocated to sexual reproduction divert from the resource pool available for growth or storage (Stearns 1992). The short-term cost of a current bout of sexual repro-duction could include reduced growth or reproduction in the subsequent sea-son (Obeso 2002).

One commonly used technique to detect costs utilizes the naturally occur-ring variation in reproduction of plants in the field and develops phenotypic correlations between reproduction and the putative cost (Obeso 2002). In the common garden experiment with perennial ryegrass (Cheplick 2008), annual spike (table 5.2) and tiller production (figure 5.2) were recorded over three growing seasons. Growth following a reproductive event was calculated as the weekly relative growth rate based on tiller addition during the ensuing spring. This growth measure was used as the dependent variable in a regres-sion employing the prior year's spike production as the independent variable. Despite much variation in spike production within and between genets, these regressions were insignificant in both the first ($F_{1, 99} = 2.34$, $P = 0.13$) and sec-ond spring ($F_{1, 95} = 0.22$, $P = 0.64$) following reproduction for E+ plants and for the second spring for E– plants ($F_{1, 83} = 1.49$ $P = 0.23$). However, the regres-sion of growth during the first spring on spike production of the prior year was significant and negative for E– plants ($F_{1, 86} = 4.36$, $P = 0.04$), indicating a possible growth cost to sexual reproduction. Nevertheless, the relationship was very weak, with prior spike production explaining only 4.8% of the variance in spring growth the following year. In can be concluded that growth costs associated with earlier sexual reproduction are not too important to tiller dynamics in this perennial ryegrass population, with the possible exception of endophyte-free plants following a relatively high bout of spike production (table 5.2).

Future reproduction, expressed as the log(number of spikes + 0.5) in the year following a reproductive episode, was regressed onto the prior year's reproduction, also expressed as the log(number of spikes + 0.5). When individual values for each plant were used to describe overall phenotypic correlations (Obeso 2002), the number of spikes in year 2 made by E+ plants was significantly negatively related to the number of spikes they made in year 1 ($F_{1, 99} = 5.43$, $P = 0.02$). This relationship, which suggests a cost to reproduction, was relatively weak, explaining only 5.2% of the variance in spike production the second year. However, the same relationship for E− plants was not significant ($F_{1, 86} < 0.01$, $P = 0.97$). When genet means were used to describe genetic correlations, results were insignificant for both E+ and E− plants due to the low number of genotypes available, but did suggest a negative relationship (figure 5.4a, b).

No cost of reproduction was detected at all between the second and third years of spike production. A significant positive correlation was evident between log(number of spikes + 0.5) in year 3 and log(number of spikes + 0.5) in year 2 for both E− ($F_{1, 83} = 7.77$, $P = 0.007$) and E+ plants ($F_{1, 95} = 13.89$, $P = 0.0003$). Again, genetic correlations were insignificant, but suggestive of a positive relationship (figure 5.4c, d).

The dramatic difference in the results of the two cross-year comparisons of sexual reproduction (figure 5.4) may be due to the relative amount of spike production that occurred in the different years. Notice in table 5.2 that the greatest reproductive episode occurred in the first year of the study for both E− and E+ plants. This high production of spikes in the first year would be more likely to result in a future cost of reproduction compared to the much lower spike production in the second year. Low levels of reproduction in long-lived perennial plants may have only minor or no effects on future growth and reproduction. Indeed, this is listed as one of the "difficulties in detecting the costs of reproduction" in the review by Obeso (2002). Apparently in this perennial ryegrass plot, plants able to produce a small number of spikes in the second year (table 5.2) were able to produce about the same number of spikes in the third year.

Despite the importance of cost-benefit considerations to various models of mutualism (Bacon & Hill 1996, Foster & Wenseleers 2006, Keeler 1985, Thompson 2005), it has been a challenge for endophyte researchers to detect costs of infection with population dynamic consequences in any grass-endophyte symbiosis. Measuring the costs of mutualism under field conditions is notoriously difficult for any symbiotic system (Bronstein 2001). While the costs of endophyte infection theoretically would include host

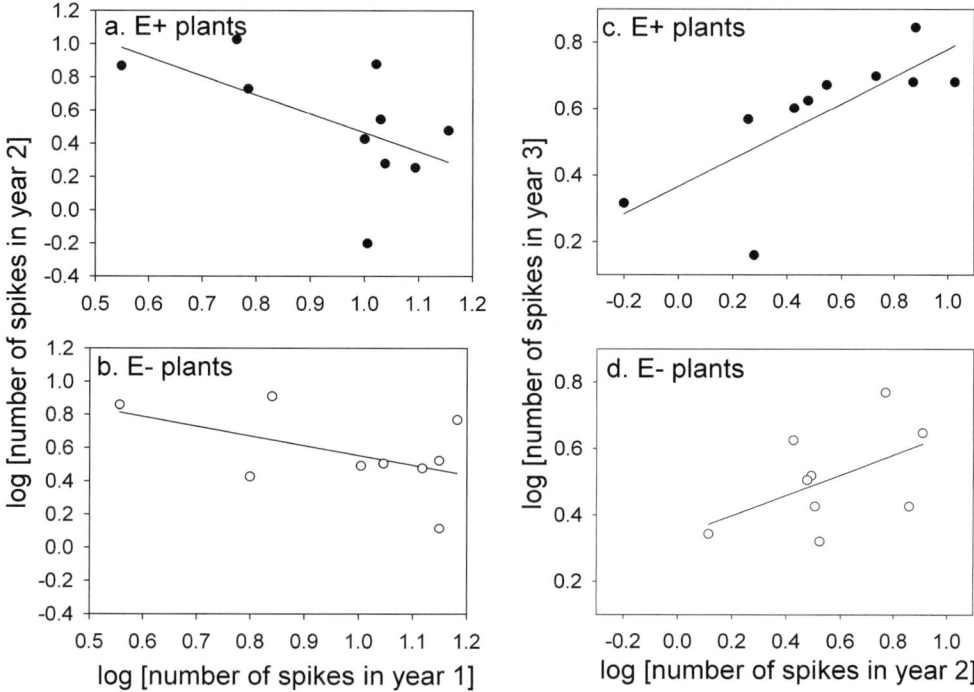

Figure 5.4 The relationship between sexual reproduction (assessed as the number of flowering spikes) in subsequent years for uninfected (E−) or endophyte-infected (E+) plants of genotypes of perennial ryegrass over three growing seasons in a common garden (data from Cheplick, 2008). (a, b) Relationship between log spike production in the second year to log spike production in the first year in E+ and E− plants. (c, d) Relationship between log spike production in the third year and log spike production in the second year in E+ and E− plants.

carbohydrate losses and reduced growth and reproduction of hosts, at least under some conditions, the long-term role of endophytes in the dynamics of ramets and genets in natural populations of their host grasses remains to be determined.

5.1.3 Genotypic sorting by natural selection

Following an episode of sexual reproduction, seed maturation, and dispersal in a mixed grass population composed of uninfected and infected individuals, seeds may become dormant for a time and enter the soil seed bank (Baskin & Baskin 1998). At some future time when conditions are suitable, seeds will germinate and new seedlings will establish. Many grasses have transient seed banks in which seeds survive only until the first germination season following their maturation; other grasses form persistent seed banks in which seeds survive for two or more seasons (Baskin & Baskin 1998). For asexual fungal endophytes, the longevity of seeds in the soil bank could impact their ability to exhibit vertical transmission and selection should favor the maintenance of both fungal and seed viability.

In their table showing "examples of grasses whose seeds have been found in persistent seed banks," Baskin and Baskin (1998) include several species known to be potential hosts for endophytes (*Brachypodium sylvaticum*, *F. rubra*, *L. perenne*), as well as genera known to contain endophyte-infected species (*Agrostis, Andropogon, Bromus, Danthonia, Panicum, Poa, Puccinellia, Stipa*). The fact that the longevity of grass seeds buried in soil typically is relatively short (Lewis 1973, Priestley 1986), often less than 5 years, means that endophytes may be lost from a host seed population as seeds die or are consumed by predators.

Endophyte infection can also be lost from aging seeds if the fungus does not remain viable over time (Rolston et al. 1986, Welty et al. 1987). Environmental conditions during seed storage can have major effects on seed and endophyte survival. As shown by Welty et al. (1987), the loss of both endophyte and seed viability was greatest over 18 months when seeds were stored at relatively warm temperatures (20°C–30°C) and high relative humidity (50–80%) for tall fescue and perennial ryegrass. Lower temperatures and humidity greatly slowed the loss of endophytes from seeds. In a study of two cultivars (Nui and Ellett) of *L. perenne* (Latch & Christensen 1982), endophyte-infected seeds were stored at 0°C to 5°C for up to 7 years. After 6 or 7 years, 40% of cv. Nui remained infected, while after 6 years, 80% of cv. Ellett remained infected. Clement et al. (2004) reported that 100% of the seedlings of *L. perenne* and 96% of the seedlings of *Hordeum brevisubulatum* (wild barley) arising from E+ seeds that had been stored for 8 years at 4°C contained viable endophytes. Comparable values for seeds stored for 8 years at 20°C were only 7.1% and 0% for perennial ryegrass and barley, respectively.

One of us (G.P.C.) has stored the seeds of several endophyte-infected perennial ryegrass cultivars within plastic bags in a 4°C refrigerator for 12 years. Seeds of one cultivar (Yorktown III) were examined periodically for the frequency and level of endophyte infection by growing seedlings from subsamples of seeds after 0, 3, and 12 years of cold storage. After 12 years, 80% to 88% of these subsamples germinated. The frequency of infection had declined from 95% to 80%, but the level of infection (assessed by hyphal counts at 400×) remained consistently high (table 5.4). A second cultivar (Palmer II) examined after 12 years of storage was 87% infected. Infection levels were slightly lower than for the other cultivar (table 5.4), but unfortunately initial infection levels are not available for comparison. Given that the longevity of *L. perenne* seeds in the soil may be only 3 to 4 years (Baskin & Baskin 1998, Priestley 1986), it would appear that the endophyte could persist indefinitely within dormant seeds.

Following habitat disturbance, grass seedlings may recruit from transient or persistent seed banks. Endophytes are, of course, only one of many

Table 5.4 Longevity of endophyte infection in seeds of two cultivars of *Lolium perenne*. Data are infection frequency (percentage of infected plants) and infection level (mean ± SE number of hyphae at 400× in infected plants). All seeds were placed into storage in closed plastic bags at 4°C beginning in 1994.

Years of storage	cv. Yorktown III			cv. Palmer II		
	n	Frequency (%)	Level	n	Frequency (%)	Level
0	55	94.5	5.54 ± 0.34	—	NA	NA
3	20	95.0	6.30 ± 0.50	—	NA	NA
12	30	80.0	6.08 ± 0.41	30	86.7	4.69 ± 0.44

n = number of plants sampled, NA = data not available
From G. P. Cheplick (unpublished data).

biotic factors that could impact seedling establishment (Cheplick 1998b). Unfortunately there are no studies that have examined seedling establishment and subsequent survival for mixed populations of E− and E+ seeds in any grass species. This represents an important area for future research on the evolutionary ecology of grass-endophyte interactions because greater depth of understanding of the underlying coevolutionary dynamics requires details on the unique genotypic sorting that must go on in infected host populations.

Natural selection within the mixed host population will depend not only on plant genotype and the extent of phenotypic differences that are due to variation among these genotypes. It will also depend on whether or not a particular host is endophyte-infected, and how endophyte and host genotype interact to influence variation in phenotypic traits that are subject to selection. Examples of such traits are growth and photosynthetic rates, tillering ability, and inflorescence production. Although endophyte genetic variation may be modest or even nonexistent, in most grass-endophyte symbioses it is probably safest to assume that each host genet with its endophyte is a unique entity. Thus any agent of natural selection, whether abiotic (e.g., drought) or biotic (e.g., herbivores), may select for some particular host-endophyte combinations, but not others. In addition, under some environmental circumstances and for some endophyte species, selection may favor those host genets that happen to be endophyte-free. The genotypic sorting that occurs is probably quite slow in long-lived, adult populations of perennial grasses. Also, selection differentials between E− and E+ segments of the host population may be relatively small, resulting in slow microevolutionary responses to whatever selection pressures may be present, at least in relation to endophyte infection.

The genotypic sorting process could be predicted to occur at any stage of the host species life cycle. Consider a mixed host population that consists of both E− and E+ genets as it undergoes sexual reproduction, assuming the endophyte is asexual and transmitted vertically within seeds. If seeds are mostly the products of outcrossing events, recombinant host genotypes can be expected in the new generation. This means that an outbred E+ maternal parent can produce genetically diverse offspring, all containing the same endophyte genotype as the mother host (assuming no new somatic mutations). Some host genotype-endophyte combinations in the offspring pool could be favorable for some siblings in respect to germination and seedling establishment, but may not be favorable for others. Siblings would, at a minimum, differ in 50% of their genes if they are full siblings (i.e., they share the same father and mother), but would be more likely to differ in 75% of their genes because most will probably have different paternal parents (presumed here to be unrelated to the mother). Of course, if selfing occurs, "complete siblings" (Cheplick 1993b) would be produced that are genetically more similar to one another than are full siblings. In short, there exists a continuum of genetic relatedness among the offspring produced by a maternal parent within a plant population (Cheplick 1993b). Thus the general type of breeding system exhibited by a host grass species will greatly influence the level of genetic variation among the myriad offspring produced following an episode of flowering in a large, mixed population. It is this genetic variation among offspring that provides the underlying support for the genotypic sorting process.

Although the host and endophyte genotype in vertically transmitted symbioses are sometimes considered to coexist as a "lineage" over long time periods, this is not necessarily the case. Not every seed produced by an E+ mother will necessarily be infected or always give rise to an infected seedling. For example, Kover & Clay (1998) found that seedling "escape" from seeds infected by *Atkinsonella hypoxylon* can reduce vertical transmission by 25% in *Danthonia compressa*. There is also the possibility that the endophyte could fail to infect some tillers of the maternal parent, which will therefore produce some E− seeds if the endophyte-free tillers flower. Furthermore, an endophyte within a seed or seedling might die at some point, without the death of the seed or seedling. Finally, genetic diversity among the offspring of an outcrossing host could potentially disrupt a tightly coevolved host-endophyte genetic combination at any time. The relative extent to which any of these phenomena occur is presently not known for any infected host population.

In short, variation exists among host genotypes and in their infection status. If conditions are conducive to seedling establishment and many new individuals have recently recruited into a population, genotypic sorting can occur as some plants die or fail to reproduce or simply produce fewer seeds than others can when they do eventually reproduce. The influence of endophytic fungi on the complex dynamics expected during this sorting process has not yet been investigated. However, the potential selective agents that mediate the sorting process are likely to be diverse, ranging from abiotic pressures such as drought to biotic pressures commonly found in grass-dominated communities such as competition and herbivory.

5.2 EXPLAINING THE PERSISTENCE AND INFECTION
FREQUENCY OF ENDOPHYTES IN GRASS POPULATIONS

The notion that systemic endophytes in grasses, especially asexual ones, are strongly mutualistic with their host grass is a deeply ingrained idea (e.g., Clay 1988b, 1990b) that still persists in the grass endophyte literature (e.g., Clay & Schardl 2002). This original mutualistic concept was largely supported by empirical studies with two agronomic grasses, tall fescue and perennial ryegrass, even though some early studies indicated variability in the direction of the interaction in these systems (e.g., Cheplick 2004a, West et al. 1995) Assuming that asexual endophytes are mutualistic, some predicted that endophyte-infected populations should become 100% infected over time (Clay 1998, Leuchtmann & Clay 1997, Wilkinson & Schardl 1997) because natural selection via herbivore resistance, competitive abilities, or resistance to stresses such as drought should favor infected plants. If asexual, vertically transmitted endophytes are not at 100% in populations, then this is explained as frequencies that have not yet reached equilibrium through natural selection (Clay 1993, 1998). Indeed, a recent study (Clay et al. 2005) and previous studies (table 10.5 in Clay 1998) indicated a shift to higher infection frequencies in agronomic tall fescue over time via natural selection by herbivores, although apparently this increase in infection frequency is not always the case even for tall fescue (e.g., Spyreas et al. 2001). This same type of argument has also been applied to *Epichloë* frequencies in a slightly different way. Brem and Leuchtmann (2002) argued that 100% infection frequency of sexual *Epichloë sylvatica* in natural populations of *B. sylvaticum* must be due to some other benefit, such as increased herbivore resistance, because experimental tests show infection reduced intraspecific competitive abilities. In other words, endophytes that are largely parasitic (e.g., disease-causing, stromata-forming *Epichloë*) should not persist in populations at high frequencies because E– plants would be selectively favored. However, as we discuss below, using observations of infection frequencies in nature to infer the direction of endophyte-host interactions may be misleading and confound correlation with causation (see chapter 2).

The idea that infection frequencies should increase over time and remain at high levels for vertically transmitted endophytes is grounded in evolutionary theory. Microbial symbionts that lose sexuality and are strictly vertically transmitted in their hosts should be strong mutualists because microbial fitness is closely linked to that of their hosts (e.g., Ewald 1994, Frank 1994, Law 1985, Lewis 1985, Massad 1987). Conversely, if asexual symbionts are parasitic, then infected hosts should be selected against relative to uninfected hosts, and frequencies should decline over time. The third alternative, that asexual endophytes are neutral in their interactions with the grass host, would also result in eventual declines in infection frequencies if the endophyte fails to grow into some seeds (imperfect transmission) or if the endophyte is randomly lost through hyphae inviability in seeds or adult plants (e.g., Siegel et al. 1984a). If imperfect transmission and hyphal inviability occur, then this suggests that the mutualistic benefits of endophyte infection must also compensate for this mechanism of infection decline in frequency. Although imperfect transmission has been hypothesized (do Valle Ribeiro 1993, Kover & Clay 1998, Ravel et al. 1997b and loss of viability can be experimentally induced (see chapter 2), there are not yet, to our knowledge, comprehensive studies of the rate of imperfect transmission across infected grass species other than the observation that seed lots from some infected agronomic grasses are often less than 100% infected (e.g., Meister et al. 2006).

Sullivan et al. (2007) proposed another explanation for the less than 100% infectivity of agronomic tall fescue (*L. arundinaceum*) based on an experimental study of the regulation of genes for loline production in infected plants after herbivore and mechanical damage. They found that previous herbivore and mechanical damage negatively affected reproduction of the bird-oat cherry aphid (*Rhopalosiphum padi*), an aphid known to be highly sensitive to loline alkaloids. Furthermore, they showed that damaged infected tall fescue increased expression of the *lolC* gene, one of several genes essential in endophyte loline biosynthesis. However, damaged E+ hosts had lower foliar nitrogen levels than E− plants and higher carbon:nitrogen (C:N) ratios, suggesting that E+ tall fescue allocated more nitrogen to defense than growth. Sullivan et al. (2007) surmised that the presence of *N. coenophialum* shifted the strategy of the host in response to herbivory from tolerance to defense. The differences in costs and benefits of these strategies of E+ and E− grasses, depending on the availability of nitrogen, may explain the persistence of E− grasses in populations. Note that this explanation is akin to that in section 5.2.2, where the effect of endophytes on their host's fitness is variable, depending on biotic and abiotic factors.

More recent observations, however, suggest that, first, the frequency of asexual endophytes, such as *Neotyphodium*, in natural populations and even in agronomic old fields is highly variable and not always near or at 100% infectivity (e.g., Bony et al. 2001, Piano et al. 2005, Ravel et al. 1997b, Saikkonen et al. 2000, Vinton et al. 2001, White 1987) (table 5.5). In some populations of both cultivars and wild grasses, infection levels may be quite low (e.g., Bazely et al. 2007, Faeth et al. 2006, Latch et al. 1987, Lewis et al. 1997, Saikkonen et al. 2000, Spyreas et al. 2001) (table 5.5). Second, interactions of asexual endophytes with host grasses span the spectrum from parasitic to neutral to mutualistic, depending on host and endophyte genotype and prevailing environments (e.g., Faeth & Sullivan 2003, Müller & Krauss 2005, Saikkonen et al. 1998). These recent observations therefore beg two important questions. First, if asexual endophytes are strongly mutualistic, then how do we explain less than 100% frequencies in grass populations? And second, if asexual endophytes are often parasitic or neutral in their interactions with grass hosts, how do we explain the persistence of infections in populations when they should be selected against, or randomly lost over time? This section addresses these two dilemmas regarding the persistence of endophytes in natural populations. Then we address frequencies of other horizontally transmitted endophytes, whose dynamics are very different.

Table 5.5 Variation in systemic endophyte infection frequencies among populations of some wild grass species on different continents. Studies involving composite herbarium specimens (e.g., White & Cole 1985, White 1987) or not collected in a systematic fashion by population (e.g., Clay & Leuchtmann 1989) were not included because among-population differences could not be determined.

Grass species	Endophyte species	Continent*	No. of populations sampled	Range of infection frequency within population	No. of populations infected, among-population infection frequency	References
Festuca arizonica	Neotyphodium starrii	NA	5	77–98%	5/5, 100%	Schulthess & Faeth 1998
Festuca arizonica	Neotyphodium starrii	NA	33	64–100%	33/33, 100%	Faeth et al. 2002b
Elymus canadensis	Epichloë spp.	NA	36	0–100%	32/36, 89%	Vinton et al. 2001
Achnatherum (= Stipa) robustum	Neotyphodium	NA	17	48–100%	17/17, 100%	Faeth et al. 2006
Lolium perenne (cultivars)	Neotyphodium lolii	EUR	52	?	14/52, 27%	Lewis & Clement 1986
Lolium spp. (wild)	Neotyphodium lolii	N. & S. EUR	523	0–100%	324/523, 61%	Lewis et al. 1997, Ravel et al. 1997b
Lolium perenne (wild)	Neotyphodium lolii	S. EUR.	11	33–58%	11/11, 100%	Zabalgogeazcoa et al. 2003
Lolium perenne (wild)	Epichloë	S. EUR.	11	0–5%	3/11, 27%	Zabalgogeazcoa et al. 2003
Lolium perenne (wild)	Neotyphodium lolii	S. EUR.	18	?	2/18, 11%	Guillaumin et al. 2001
Lolium (= Festuca) arundinaceum	Neotyphodium coenophialum	S. EUR	60	0–100%	58/60, 97%	Piano et al. 2005
Agrostis capillaris	Neotyphodium or Epichloë	N. EUR	9	67%	1/9, 67%	Saikkonen et al. 2000

(continued)

Table 5.5 *Continued*

Grass species	Endophyte species	Continent*	No. of populations sampled	Range of infection frequency within population	No. of populations infected, among-population infection frequency	References
Agrostis capillaris	Neotyphodium or Epichloë	N. EUR	44	17%	1/44, 2%	Bazely et al. 2007
Agrostis castellana	Epichloë	S. EUR	12	0–5%	3/12, 25%	Zabalgogeazcoa et al. 2003
Alopecuris pratensis	Neotyphodium or Epichloë	N. EUR	3	0	0	Saikkonen et al. 2000
Alopecuris arundinaceus	Epichloë	S. EUR	18	0–1%	1/18, 6%	Zabalgogeazcoa et al. 2003
Brachypodium phoenicoides	Epichloë	S. EUR	4	0–1%	1/4; 25%	Zabalgogeazcoa et al. 2003
Brachypodium sylvaticum	Epichloë	S. EUR	1	?	1/1; 100%	Zabalgogeazcoa et al. 2003
Bromus erectus	Epichloë	S. EUR	27	?	3/27, 11%	Guillaumin et al. 2001
Bromus tectorum	Epichloë	S. EUR	5	?	1/5, 20%	Guillaumin et al. 2001
Calamagrostis epigejos	Neotyphodium or Epichloë	N. EUR	5	0	0	Saikkonen et al. 2000
Calamagrostis lapponica	Neotyphodium or Epichloë	N. EUR	10	0	0	Saikkonen et al. 2000
Dactylis glomerata	Neotyphodium or Epichloë	N. EUR	11	17%	1/11, 17%	Saikkonen et al. 2000
Dactylis glomerata	Epichloë	S. EUR	9	?	9/9, 100%	Zabalgogeazcoa et al. 2003
Deschampsia flexuosa	Neotyphodium or Epichloë	N. EUR	28	10–20%	5/28, 13%	Saikkonen et al. 2000

Host	Endophyte	Region	n	Infection range	Infection	Reference
Deschampsia flexuosa	*Neotyphodium* or *Epichloë*	N. EUR	51	14–33%	3/28, 21%	Bazely et al. 2007
Deschampsia caespitosa	*Neotyphodium* or *Epichloë*	N. EUR	21	9–30%	2/21, 19%	Saikkonen et al. 2000
Elymus repens	*Neotyphodium* or *Epichloë*	N. EUR	8	9%	1/8, 9%	Saikkonen et al. 2000
Lolium (= Festuca) arundinaceum	*Neotyphodium coenophialum*	N. EUR	15	96–100%	13/15, 98%	Saikkonen et al. 2000
Lolium (= Festuca) arundinaceum	*Neotyphodium coenophialum*	N. EUR	29	5–100%	29/29, 100%	M. Helander, K. Saikkonen, S. Faeth (unpublished data)
Lolium (= Festuca) arundinaceum	*Neotyphodium coenophialum*	S. EUR	56	?	20/20, 100%	Clement et al. 2001
Lolium (= Festuca) arundinaceum	*Neotyphodium coenophialum*	S. EUR	9	62–100%	9/9, 100%	Zabalgogeazcoa et al. 2003
Lolium (= Festuca) arundinaceum	*Neotyphodium coenophialum*	S. EUR	19	?	12/19, 63%	Guillaumin et al. 2001
Festuca ampla	*Neotyphodium* or *Epichloë*	S. EUR	1	?	1/1, 100%	Zabalgogeazcoa et al. 2003
Festuca ovina	*Neotyphodium*	S. EUR	1	?	1/1, 100%	Zabalgogeazcoa et al. 2003
Festuca ovina	*Neotyphodium*	S. EUR	53	?	10/53, 19%	Guillaumin et al. 2001
Festuca ovina	*Neotyphodium*	N. EUR	27	10–50%	5/27, 29%	Saikkonen et al. 2000
Lolium (= Festuca) pratense	*Neotyphodium*	N. EUR	25	10–100%	8/25, 42%	Saikkonen et al. 2000

(continued)

Table 5.5 *Continued*

Grass species	Endophyte species	Continent*	No. of populations sampled	Range of infection frequency within population	No. of populations infected, among-population infection frequency	References
Lolium (= *Festuca*) *pratense*	*Neotyphodium*	S. EUR	4	?	4/4, 100%	Guillaumin et al. 2001
Festuca rubra	*Epichloë festucae*	N. EUR	50	14–80%	11/50, 32%	Saikkonen et al. 2000
Festuca rubra	*Epichloë festucae*	N. EUR	49	10–100%	27/49, 53%	Bazely et al. 2007
Festuca rubra	*Epichloë festucae*	S. EUR	60	?	24/60, 40%	Guillaumin et al. 2001
Festuca rubra	*Epichloë festucae*	S. EUR	27	44–92%	27/27, 100%	Zabalgogeazcoa et al. 1999
Festuca rubra	*Epichloë festucae*	S. EUR	28	0–100%	27/28, 96%	Zabalgogeazcoa et al. 2003
Holcus lanata	*Epichloë*	S. EUR	12	0–1%	3/12, 25%	Zabalgogeazcoa et al. 2003
Melica ciliata	*Neotyphodium* or *Epichloë*	S. EUR	15	?	12/15, 80%	Guillaumin et al. 2001
Phalaris arundinacea	*Neotyphodium* or *Epichloë*	N. EUR	11	0	0	Saikkonen et al. 2000
Phleum pretense	*Neotyphodium* or *Epichloë*	N. EUR	5	33%	1/5, 33%	Saikkonen et al. 2000
Poa trivialis	*Neotyphodium* or *Epichloë*	N. EUR	45	11–67%	11/45; 39%	Bazely et al. 2007
Festuca rubra subsp. *pruinosa*	*Epichloë festucae*	S. EUR	4	58–81%	4/4,100%	Zabalgogeazcoa et al. 2006b
Lolium (= *Festuca*) *arundinaceum*	*Neotyphodium coenophialum*	S. EUR/N. AF	104	0–100%	99/104, 95%	Clement et al. 2001

Host	Endophyte	Region	Number	Infection	Range	Reference
Lolium (= Festuca) arundinaceum	Neotyphodium coenophialum	N. AF	56	55/56, 98%	0–100%	Clement et al. 2001
Lolium (= Festuca) arundinaceum	Neotyphodium coenophialum	N. AF	56	24/, 86%	0–100%	Clement et al. 2001
Bromus setifolius	Neotyphodium tembladerae	SA	13	8/13, 42%	0–100%	White et al. 2001
Bromus setifolius	Neotyphodium tembladerae	SA	36	15/36, 43%	0–100%	M. V. Novas, M. Collantes, D. Cabral (unpublished data)
Festuca argentina	Neotyphodium	SA	60	95–100%	?	Bertoni et al. 1993
Festuca hieronymi	Neotyphodium	SA	60	95–100%	?	Bertoni et al. 1993
Triticum spp.	Neotyphodium	AS	411	19/411, 5%	0–15%	Marshall et al. 1999
Achnatherum inebrians	Neotyphodium gansuense	AS	20	83/83, 100%	80–100%	Nan & Li 2001
Agropyron cristatum	Neotyphodium	AS	6	1/6, 17%	?	Nan & Li 2001
Bromus inermis	Neotyphodium	AS	6	1/6, 17%	?	Nan & Li 2001
Bromus magnus	Neotyphodium	AS	1	1/1, 100%	?	Nan & Li 2001
Deschampsia caepitosa	Neotyphodium	AS	4	1/4, 25%	?	Nan & Li 2001
Elymus cylindricus	Neotyphodium	AS	2	1/2, 50%	?	Nan & Li 2001
Elymus dahuricus	Neotyphodium	AS	3	1/3, 33%	?	Nan & Li 2001
Elymus nutans	Neotyphodium	AS	23	2/23, 9%	?	Nan & Li 2001
Elymus tangutorum	Neotyphodium	AS	21	12/23, 52%	?	Nan & Li 2001
Elytrigia dahuricus	Neotyphodium	AS	1	1/1, 100%	?	Nan & Li 2001
Festuca alatavica	Neotyphodium	AS	1	1/1, 100%	?	Nan & Li 2001
Festuca modesta	Neotyphodium	AS	1	1/1, 100%	?	Nan & Li 2001

(continued)

Table 5.5 Continued

Grass species	Endophyte species	Continent*	No. of populations sampled	Range of infection frequency within population	No. of populations infected, among-population infection frequency	References
Festuca rubra	Neotyphodium	AS	5	?	3/5, 60%	Nan & Li 2001
Festuca sinensis	Neotyphodium	AS	8	?	8/8, 100%	Nan & Li 2001
Hordeum bogdanii	Neotyphodium	AS	3	?	2/3, 67%	Nan & Li 2001
Hordeum violaceum	Neotyphodium	AS	2	?	2/2, 100%	Nan & Li 2001
Leymus chinensis	Neotyphodium	AS	1	?	1/1, 100%	Nan & Li 2001
Poa alpina	Neotyphodium	AS	2	?	1/2, 50%	Nan & Li 2001
Poa pratensis	Neotyphodium	AS	4	?	1/4, 25%	Nan & Li 2001
Poa sphondylodes	Neotyphodium	AS	11	?	1/11, 9%	Nan & Li 2001
Poa tibetan	Neotyphodium	AS	2	?	1/2, 50%	Nan & Li 2001
Polypopon monspeliensis	Neotyphodium	AS	1	?	1/1, 100%	Nan & Li 2001
Roegneria stricta	Neotyphodium	AS	1	?	1/1, 100%	Nan & Li 2001
Stipa purpurea	Neotyphodium	AS	7	?	1/7, 14%	Nan & Li 2001
Bromus tomentellus	Neotyphodium	AS	50	0–100%	45/50, 90%	Mirlohi et al. 2006

Echinopogon ovatus	Neotyphodium	AU/NZ	17	17–90%	17/17, 100%	Miles et al. 1998
Echinopogon phleoides	Neotyphodium	AU/NZ	1	0	0/1, 0%	Miles et al. 1998
Echinopogon nutans var. major	Neotyphodium	AU/NZ	1	100%	1/1, 100%	Miles et al. 1998
Echinopogon nutans var. nutans	Neotyphodium	AU/NZ	4	0	0/4, 0%	Miles et al. 1998
Echinopogon mckiei	Neotyphodium	AU/NZ	5	0	0/5, 0%	Miles et al. 1998
Echinopogon intermedius	Neotyphodium	AU/NZ	6	0–100%	1/6, 17%	Miles et al. 1998
Echinopogon caepitosus	Neotyphodium	AU/NZ	6	0–100%	1/6, 17%	Miles et al. 1998
Echinopogon cheelii	Neotyphodium	AU/NZ	6	0–100%	3/6, 50%	Miles et al. 1998

* NA = North America, EUR = Europe, AF = Africa, SA = South America, AU/NZ = Australia/New Zealand, AS = Asia.

5.2.1 Explaining low infection frequencies when endophytes are mutualistic

If one assumes that endophytes in grasses are generally mutualistic, then natural selection should favor infected grasses relative to uninfected ones, and infection frequencies should increase (Clay 1988b, 1990b, Clay et al. 2005, Francis & Baird 1989, Ravel et al. 1997b, Wilkinson & Schardl 1997), eventually reaching 100% infectivity (Clay 1990b, 1993, 1998). Why then are most grass populations not 100% infected?

One explanation for less than 100% infectivity often found in both agronomic and wild grass populations is that infection frequencies have not yet reached equilibrium (Clay 1988b, 1993). Clay (1993) estimated from a computer simulation that if E+ plants have twice the fitness advantage of E− grasses in a population, starting infection frequencies of 50% of a type III (asexual, obligate vertical transmission) would become fixed at 100% in less than 10 generations. Data from some agronomic grass studies support this increase of infection frequencies over time (Clay 1998: table 10.5, Clay et al. 2005, Siegel et al. 1985) and older populations can have higher infection rates (Cunningham et al. 1993, Lewis & Clement 1986). However, apparently infection rates in agronomic grasses, like tall fescue, do not necessarily increase through time in old fields (e.g., Spyreas et al. 2001), and agronomic fields of tall fescue appear to equilibrate at about 70% infection, although starting infection rates may be higher (T. Phillips, personal communication). Disruption or disturbance of selective pressures or migration of E− plants from other populations is the mechanism commonly provided for failure to reach 100% equilibrium (Clay 1993, Ravel et al. 1997b).

Ravel et al. (1997b) provided another and nonmutually exclusive explanation for failure to reach high infection frequencies, assuming that the interaction is mutualistic. They proposed that imperfect transmission may account for less than 100% infection frequencies. Hyphae may fail to reach all tillers of an infected plant or may fail to grow into culms and seeds of inflorescences of the host grass. Furthermore, infected seeds and possibly seedlings and adult plants may lose infection if hyphae become inviable (do Valle Ribeiro 1993). A common practice to "cure" seeds of infection is to apply heat in humid conditions which kills the endophyte or to store seeds in warm conditions to achieve E− seeds from infected lines (e.g., Siegel et al. 1984a, Welty & Azevedo 1985) (chapter 2). Ravel et al. (1997b) constructed a simple model that incorporated the life cycle of the host grass, including seed germination, seedling competition, flowering, and mortality, and assuming no seed dormancy, with some proportion of E− seeds produced by E+ plants (imperfect transmission). With a relatively low rate of imperfect transmission of about 10%, uninfected plants can be maintained at equilibrium in the population. The work of Ravel et al. (1997b), while only a mathematical model, suggests that biologically reasonable assumptions of imperfect transmission alone could account for the persistence of E− plants in populations, even when the infected plants are selectively favored because they confer benefits to the host.

Saikkonen et al. (2004) proposed another explanation. Because asexual endophytes often occurred in outcrossing grasses, genetic mismatches should often occur between the recombining host plant and the relatively static endophyte genome. The asexual endophyte may be "mismatched" with the sexually recombinant seed and the seedlings may be less fit than well-matched endophyte-host combinations of the previous generation. This may explain why some systemic endophyte infections reduce sexual reproduction of their hosts in favor of clonal growth to maintain favorable endophyte-host combinations (e.g., Pan & Clay 2002).

We know little of how disturbance or migration may maintain E− plants in a mutualistic endophyte-host system. According to population genetics models, however, relatively small amounts of gene flow can disrupt Hardy-Weinberg equilibria, and changes to selective regimes because of fluctuating environments may do likewise (Thompson 1994, 1999) such that a mixture of E− and E+ grasses are maintained. The latter would suggest that mutualistic interactions between host and endophyte vary with time and space, and may switch back and forth between mutualism and parasitism. This, however, is the very essence of the geographic mosaic theory of species interactions (Thompson 1994, 2005). If this is the case, then explanations for why infected populations vary in infection frequency over time and among populations when endophytes are sometimes mutualists become moot.

Finally, in a recent model for an annual grass, Gundel et al. (2008) showed that asexual *Neotyphodium* infections can persist in populations even if the mutualistic effect is very small and vertical transmission efficiency is sufficiently high. They conclude that, in natural systems, seed immigration and environmental fluctuations generally prevent infection levels from reaching local equilibrium. In their model, environmental disturbances tend to increase infection frequencies whereas immigration reduces representation of infected annual grasses.

5.2.2 Explaining high infection frequencies when endophytes are nonmutualistic or variable in their effects

In their paper, Ravel et al. (1997b) stated, or perhaps understated, another alternative explanation for the persistence of E− grasses in populations: "An extremely pessimistic version...would be that the beneficial effects of endophytic infection appear only in agricultural systems, while they could be absent in wild populations. Measures of endophyte effects in wild populations are badly needed." This alternative hypothesis was perhaps more prophetic than pessimistic. The observation of Ravel et al. (1997b) that studies of endophyte effects in wild populations are very scarce still rings true today (see chapter 7) (Faeth & Saikkonen 2007). However, accumulating evidence from wild populations suggests that endophyte effects in wild populations are much more variable (e.g., Saikkonen et al. 2006) and often run counter to results found in agroecosystems. In some cases, asexual systemic endophytes act parasitically or neutrally with their hosts, depending on the host and endophyte genotypes and environments, and interactions with the third trophic level (e.g., Faeth & Sullivan 2003, Lehtonen et al. 2005b, Müller & Krauss 2005, Saikkonen et al. 2004, 2006, Tibbets & Faeth 1999) (chapters 4 and 6).

In one of the few well-studied infected wild grasses, Arizona fescue, infection frequencies remain high despite the fact that multiple experiments show that E+ Arizona fescue is more susceptible to herbivory (Lopez et al. 1995, Saikkonen et al. 1999, Tibbets & Faeth 1999), generally grows and reproduces worse (Faeth & Sullivan 2003), and is a poorer competitor (Faeth et al 2004, Morse et al. 2007) than E− conspecifics. Furthermore, long-term survival of E+ and E− Arizona fescue plants does not differ (Faeth & Hamilton 2006) (table 5.1). Yet infection frequencies remain high (60–100%) in natural populations of Arizona fescue (Schulthess & Faeth 1998). Evolutionary theory predicts that symbiotic parasites that decrease host fitness cannot persist if they are exclusively vertically transmitted in populations (e.g., Ewald 1987, 1994, Fine 1975, Lipsitch et al. 1995, 1996). We then must ask the reciprocal question: if *Neotyphodium* in Arizona fescue is, on average, detrimental to the fitness of the host, how do endophytes persist and how are high frequencies in natural populations maintained?

Faeth et al. (2007), Faeth and Sullivan (2003), and Saikkonen et al. (2002) proposed four explanations for the persistence of endophytes in natural populations.

Mutualistic interactions may occur infrequently but have large effects on population dynamics

Rare benefits even with persistent costs of infection may maintain infection frequencies via fluctuating selection pressures. The rare mutualistic benefits may occur at critical points in the life span of the grass host. This may be particularly evident in perennial grass hosts which may live for decades or even centuries (e.g., Faeth & Hamilton 2006). Severe droughts, for example, occurring once or twice during the life span of a perennial grass that favors E+ plants may result in a selective bottleneck through which few E− plants survive. Thus infection frequency may increase after population contractions during severe drought periods but then decline during the more average or favorable intervening years. In a theoretical analysis of the persistence of less fit genotypes of the desert annual *Linanthus parryae*, Turelli et al. (2001) proposed that genotypes with a mean or geometric relative fitness of less than one can persist if their relative fitness in high-yield years is greater than one. In addition, if the seed bank is long lived (i.e., each generation contributes only a small fraction of the overall seed bank), then less temporal variation in relative fitness is required to compensate for an overall fitness disadvantage (Turelli et al. 2001). Because *Neotyphodium* is a vertically transmitted symbiont, *Neotyphodium* frequencies may be analogous to allele frequencies.

In addition, natural selection may act more strongly on vulnerable life stages such as seeds and seedlings. However, experimental evidence, at least with Arizona fescue (Faeth et al. 2002b, Faeth & Sullivan 2003, Neil et al. 2003) does not show any differences in E+ and E− plants in germination or seedling survival success. Likewise, Faeth and Hamilton (2006) also showed no difference in long-term survival of E+ and E– Arizona fescue, although E+ seedlings had higher survival in the early seedling stage, but eventually E+ and E– grasses had equivalent survival over a 3- to 5-year period. Moreover, genotypic interactions of host and endophyte also influence the outcomes of fluctuating selection. At each transmission event (seed production), endophyte haplotypes are reshuffled into different plant genotypes during sexual repro-duction of the host (e.g., Sullivan & Faeth 2004). For any given year, different endophyte-host genotypic combinations may be at a fitness disadvantage rela-tive to E– grasses, but an E+ lineage having a high relative fitness year may be enough not only to ensure that lineage's continued presence in the popu-lation, but also to maintain the observed high frequencies of *Neotyphodium* infections in natural populations. As Ravel et al. (1997b) noted, there is a dearth of any demographic studies of E+ and E– wild grasses, not even con-sidering various endophyte-host combinations, so conclusions about life stage survival and lifetime fecundity are at this time impossible. Studies of the life-time fitness of E+ and E– grasses that incorporate genetic differences are sorely needed to determine how fluctuating selection affects the persistence of E+ and E– plants in natural populations.

Spatial structuring increases persistence and coexistence

Thompson's (1994, 2005) geographic mosaic model of coevolution synthesized how the direction and strength of interspecific interactions varies across broad spatial scales depending upon phylogeny, morphology, and genetic covariation of the interacting species. The interaction between local selection pressures and gene flow between communities forms the core of this model. One of the predictions of the model is that due to gene flow, in some communities, interac-tions will be maladapted or selective pressures will be thwarted. Experimental and theoretical analysis of plant-pathogen interactions (e.g., Burdon & Thrall 1999) between populations appears to support this prediction. In addition, a recent model (Nuismer et al. 1999) examining the evolutionary dynamics of an obligate symbiont in two populations under varying selection regimes for and against mutualism has demonstrated that local adaptations can be sub-verted by migration, and that this subversion of local selection pressures can occur over a wide range of selection and migration values. The net result is a stable polymorphism of interaction types (Nuismer et al. 1999).

The evolutionary dynamics of grasses and endophytes are more complex because, instead of two interaction types (mutualistic symbiont and antago-nistic symbiont), there are also symbiont-free grasses. Furthermore, perennial grass populations may be highly structured because adult plants are obviously immobile and seed dispersal is relatively low, and usually by wind (Cheplick 1998b). So their evolutionary dynamics may be highly influenced by migra-tion and exhibit high degrees of patchiness of E– and E+ individuals that vary by plant genotype, and for E+ grasses, by endophyte haplotype (e.g., Faeth 2002, Faeth & Fagan 2002). Using a metapopulation model, Saikkonen et al. (2002) and Gyllenberg et al. (2002) demonstrated that mixed populations of E+ and E− individuals can be formed under a range of ecological conditions, including scenarios incorporating patches where the *Neotyphodium* endophyte functions as a parasite. In homogeneous populations of identical patches, mutualistic interactions are necessary for coexistence (Gyllenberg et al. 2002). However, in heterogeneous environments with nonidentical patches, coexistence of E+ and E− can occur, even if the endophyte is non-mutualistic or parasitic (Saikkonen et al. 2002). Thus if wild grass populations are highly structured in space, which most likely are, then we might expect persistence of infected and uninfected individuals in local patches, even if the endophyte is not strictly mutualistic (Faeth 2002, Saikkonen et al. 2002). In a more recent mathematical analysis, Faeth et al. (2007) demonstrated that a vertically transmitted but parasitic endophyte (e.g., *Neotyphodium* species or strain that had net negative effects on its host grass) could coexist with a more pathogenic horizontally transmitted endophyte (like *Epichloë*) by protecting or inhibiting infection by the more pathogenic endophyte. While these mathematical models provide insight into possible dynamics of spatially structured grass populations with E+ and E− plants, to date we know of no empirical studies that incorporate spatial structure into studies of the dynamics of endophyte-infected grass populations.

Asexual endophytes are occasionally transmitted horizontally

White et al. (1996) and Moy et al. (2000) suggested the possibility that strictly asexual *Neotyphodium* endophytes may be occasionally transmitted horizontally because of the presence of epiphyllous nets and conidia in the leaves of some infected grasses, much like the ancestral *Epichloë*. If transmitted horizontally to adult plants, then infection frequencies could be maintained even if the interaction is neutral or parasitic. Horizontal transmission of *Neotyphodium* has yet to be unequivocally demonstrated. Furthermore, mathematical models show that horizontal transmission of asexual endophytes could be easily overlooked and could maintain infection frequencies even with imperfect transmission in a population with both vertically and horizontally transmitted endophytes, such as *Neotyphodium* and *Epichloë* (Faeth et al. 2007). From a biological perspective, we might expect that if *Neotyphodium* can be readily inoculated into E− plants in the laboratory with simple techniques (e.g., Christensen 1995) then infection may occasionally be transmitted horizontally in nature, possibly by root-to-root contact or herbivorous insect vectors. Other horizontally transmitted endophytes can be vectored by insect herbivores (e.g., Faeth & Hammon 1997), and systemic ones like *Neotyphodium* may be as well. T. Phillips (personal communication) reports that accessions of E− tall fescue (Kentucky 31) appear to become infected over time, although the mechanism is unclear. However, horizontal transmission rates would have to equal or exceed loss of the endophyte via hyphae inviability and imperfect transmission for maintenance or an increase in infection frequencies.

Asexual endophytes control aspects of host plant reproduction

Epichloë endophytes are well known for altering reproductive function and allocation in host grasses. *Epichloë*, as the causative agent of choke disease, destroys inflorescences in some grass hosts and produces stromata in their stead (Clay 1986, 1991b). Spores from stromata are then transported by specialized flies in the genus *Botanophila* that also use the stromata as larval food sources (Bultman et al. 1995, Bultman & Leuchtmann 2003). *Epichloë* may alter host reproduction in other, more subtle ways. For example, Meijer and Leuchtmann (2001) reported that asexual, vertically transmitted *Epichloë* in a native woodland grass, *B. sylvaticum*, promotes self-fertilization that results in continuation of favorable plant genotypes for the fungus. Pan and Clay (2002, 2003) showed that *Epichloë glyeriae* increases clonal growth of its grass host *Glyceria striata* under some conditions and thus increases transmission to tillers of the host grass.

Neotyphodium, unlike its ancestral *Epichloë*, however, is thought to be strongly mutualistic and under control of the host plant as a "trapped pathogen" (White et al. 2001a, Wilkinson & Schardl 1997). However, other vertically transmitted symbionts are not mutualistic and yet are evolutionarily stable. For example, Werren and O'Neill (1997) showed that maternally inherited, asexual symbionts need not act as host mutualists to persist in host populations, contrary to conventional wisdom (e.g., Ewald 1994, Law 1985, Wilkinson & Schardl 1997). Maternally inherited cytoplasmic microbes, such as *Wolbachia*, persist in many invertebrate species by acting as reproductive parasites. *Wolbachia* reduces male function, increases feminization, reduces male:female sex ratios by producing parthenogenetic daughters, and reduces the fitness of uninfected hosts (by cytoplasmic incompatibility). *Wolbachia* does not increase, and often decreases, host fitness, yet is usually vertically transmitted in the cytoplasm of eggs (Werren 1997; but see Huigens et al. 2000).

Neotyphodium may similarly act as a reproductive parasite and promote its transmission to host offspring by increasing or decreasing allocation to female and male function, respectively, or by increasing vivipary or pseudovivipary, which are well-known in grasses (e.g., Elmqvist & Cox 1996). The hypothesis that asexual endophytes may retain the capacity to alter reproduction of their hosts has been largely untested. However, as discussed in chapter 3, Olejniczak and Lembicz (2007) recently demonstrated experimentally that an asexual and vertically transmitted strain of *E. typhina* shifted reproduction in the wild grass *P. distans* to an earlier age of reproduction, apparently to the benefit of the endophyte, but at the expense of growth of the host grass. Moreover, this notion is not without parallel for other vertically transmitted and specialized fungal symbionts. Poulsen and Boomsa (2005) showed that vertically transmitted clonal fungi grown by leaf-cutting ants actively rejected mycelia from other colonies even though mixed fungal colonies produced more biomass, and therefore would be more beneficial to the ant colonies. Thus these vertically transmitted and domesticated fungi retain control over the growth and species composition of the fungal garden even after millions of years of domestication by the ants and even though control may be detrimental to the host. Similarly, as we shall see in chapter 6, the presence of *Neotyphodium* increases herbivory, which results in increased allocation to seed production at the expense of vegetative growth, at least for one native grass, Arizona fescue. In addition, Rice et al. (1990) showed that N. *coenophialum* infection increased seed production in tall fescue. These results suggest that *Neotyphodium* may indirectly control host reproduction and increase its transmission, even if this may result in lower long-term fitness of the host grass. Certainly additional studies, especially in native grasses, are needed before we can rule out that *Neotyphodium* is strictly vertically transmitted and is only a passive partner in the symbiosis.

Another way that vertically transmitted, but parasitic symbionts can be maintained in host populations has been explored theoretically by Faeth et al. (2007) and Lipsitch et al. (1996). In their models, vertically transmitted strains can persist in populations if they provide protection against more virulent, horizontally transmitted strains (Faeth et al. 2007, Lipsitch et al. 1996). In the case of endophytes of grasses, vertically transmitted *Neotyphodium* could theoretically persist, even though they have negative effects on the host, if they provide protection or immunity against virulent and sexual forms of *Epichloë* (Faeth et al. 2007).

Of course, as emphasized throughout this book, the direction of the endophyte-grass interaction is often conditional, like other apparent mutualisms (e.g., Bronstein 1994a), on plant and endophyte genotype, and other biotic interactions and abiotic factors. Thus all of the aforementioned and more specific hypotheses to explain variation in endophyte frequencies fall under this larger umbrella of conditionality of the grass-endophyte interaction.

5.3 COEVOLUTION OF ENDOPHYTE AND HOST POPULATIONS

The concept of coevolution itself has evolved. Thompson (1994) nicely traces the roots of the notion of coevolution from Darwin through the evolutionary synthesis in the first half of the 20th century to modern concepts today. Ehrlich and Raven's (1964) seminal paper describing specialization of butterflies and their host plants from a phylogenetic perspective was one approach to how species interactions become specialized and how this specialization can lead to cospeciation and adaptive radiation of insects and their host plants. Other approaches have focused on gene-for-gene changes in plant pathogens and their hosts (e.g., Flor 1955), while later researchers (e.g., Pimentel 1988) incorporated population dynamics and genetics for antagonistic interactions. These approaches predicted genetic changes in the host and pathogen or parasite and a reduced oscillation amplitude of host and pathogen population numbers over time (Thompson 1994). Thompson (1994) notes a fourth approach based upon interspecific competitive interactions in communities, where reciprocal changes in competing species results in rapid evolutionary changes with consequences for patterns of species coexistence, character displacement, limiting similarity, size ratios, and life-history traits (e.g., Futuyma & Slatkin 1983). The same type of approach has been used to depict the coevolutionary dynamics of predator-prey interactions, with one outcome of the reciprocal interaction being the notion of an evolutionary arms race of prey or host defenses and countermeasures by the parasite or predator (e.g., van Valen 1973). These coevolutionary arms races were also the driving force for Ehrlich and Raven's (1964) phylogenetic patterns of host plant and herbivore specialization. Mutualisms as coevolutionary interactions arrived on the scene with Janzen's studies (1966) of ants and acacias, where reciprocal changes in partners that benefit each other also came under the purview of coevolution.

More recent views of coevolution retain the concept of reciprocal changes in interacting species, be they antagonistic or mutualistic, but incorporate how the direction and magnitude of these interactions vary over time and space, and with the genetics of distinct populations of interacting species pairs within the surrounding milieu of other species (e.g., Bascompte et al. 2006, Thompson 1994, 2005). Thompson's (1994, 2005), geographic mosaic model of coevolution is based on the idea that coevolutionary dynamics are shaped by conflicting or complementary evolutionary forces. Interspecific interactions are influenced by local selection pressures acting on local populations, and interacting populations may evolve antagonistically, mutualistically, or not at all, such that characterizations of any given pair of interacting species as mutualists, parasite-host, or predator-prey may be fruitless over the geographic range of the species. Furthermore, migration between populations can thwart local selection and create populations maladapted for their environment (Thompson 1994, 2005). The interaction between selection and migration can create geographical mosaics, with some populations exhibiting coevolution ("hot spots") while others show maladaptation or no coevolution ("cold spots") because migration overwhelms local selection and populations are mismatched (Thompson 1994, 1999, 2005). Endophyte-host interactions are particularly intriguing interactions to consider within the various frameworks of coevolutionary theory because, at least for the systemic endophytes such as *Epichloë* and *Neotyphodium*, they are highly host specific and range in effects from antagonistic to mutualistic. Because of their specificity (Leuchtmann 1992, Schardl et al. 1997), they are also fairly easy to track phylogenetically and through evolutionary time, and transmission mode may provide some indication of the direction of the interaction (i.e., antagonistic or mutualistic) between the endophyte and its host.

The first paper to thoroughly examine endophyte-host interactions from a coevolutionary viewpoint was that of Schardl et al. (1997). They analyzed the phylogenetic relationships of *Epichloë* that varied by whether they were transmitted only sexually, and hence assumed as parasitic, or existed as pleiotrophic strains that can be transmitted either sexually or asexually. Sexually transmitted *Epichloë* form stromata and cause choke disease, which destroys the host grass inflorescence. They predicted that pleiotrophic *Epichloë* would more likely show long-term cospeciation and specialization than strictly sexual forms. Their results supported this prediction. Phylogenies based on beta-tubulin (*tub2*) and ribosomal DNA (rDNA) sequences showed concordance among host and symbiont for the pleiotrophic species and strains, at least at the tribe or genus level, while the strictly sexual species and a hybrid species did not. They reasoned that hybridization disrupts the long-term coadaptation of the host and *Epichloë* strain or species. Moreover, endophyte species that are pleiotrophic (or type II infections) (White 1988) are most evolutionarily stable because they can undergo both sexual and asexual reproduction. Schardl et al. (1997) assumed the direction of the interaction from the transmission type of the endophyte. In other words, sexual strains or species should be more parasitic or less mutualistic than asexual strains. At least for *Epichloë*, this assumption seems reasonable because vertically transmitted *Epichloë* appear to provide more protection, at least from insect herbivores, than sexually transmitted species (e.g., Brem & Leuchtmann 2001). However, this assumption, as we have seen for *Neotyphodium*, often does not hold based only on the endophyte's mode of transmission (chapters 4 and 6). Nor have there been thorough studies of the ecological interactions between different *Epichloë* species and strains and their various hosts. Nonetheless, Schardl et al. (1997) provided the first broad phylogenetic approach to understanding coevolutionary patterns of systemic endophytes and their host grasses, at least in the broad coevolutionary sense of cocladogenesis and cospeciation.

Tredway et al. (1999) followed Schardl et al.'s (1997) study using amplified fragment length polymorphisms (AFLPs) and rDNA sequences to more closely examine the phylogenetic relationships between seven *Epichloë festucae* strains in closely related fine fescue species and *Neotyphodium* endophytes. *E. festucae* is a common symbiont found in *Lolium*, *Festuca*, and *Koeleria* grass species and is ancestral or closely related to many *Neotyphodium* species (Schardl 2001). Tredway et al. (1999) found three major groups within the *E. festucae* group. One group from *Festuca ovina* (blue or sheep's fescue) was highly distinct from other grass species, the second from strong creeping red fescue (*F. rubra* ssp. *rubra*), and the third from hard fescue (*F. brevipila*) and Chewings fescue (*F. rubra* ssp. *fallax*). Their phylogenetic analyses suggested that the current taxonomic scheme of grouping fine fescues into two aggregates is incorrect, at least based upon endophyte relationships. More importantly, they concluded that the three clades indicate host specialization and coevolution with *E. festucae*. This is coevolution in the broad sense of concordant phylogenetic patterns. However, these approaches cannot show how interactions vary at the population level or how interactions change geographically within and among symbiont-host combinations.

5.3.1 Hybridization complicates the phylogenetic (and ecological) picture

More recent phylogenetic studies have established the intricate patterns of relationships between *Epichloë* and *Neotyphodium* species and strains and their grass hosts (Schardl 2001, Tsai et al. 1994). Many (about two-thirds) of known *Neotyphodium* species were formed by hybridization between two species of *Epichloë* or between a co-occurring *Epichloë* species and a *Neotyphodium* species (Moon et al. 2004, Schardl & Craven 2003). Moon et al. (2002) examined *Neotyphodium* endophytes from southern hemisphere native grasses using multiple gene phylogenies. They found three lineages of unique hybrid origin and one nonhybrid lineage within these grasses species. Moreover both hybrid and nonhybrid lineages occurred within one of the same host species, *Echinopogon ovatus* from Australia. In native grasses of Argentina, Gentile et al. (2005) found 27 *Neotyphodium* isolates from 10 different host species. Twenty-three of these were interspecific hybrids between *E. festucae* and *E. typhina*, with one hybrid of different origin, and three *Neotyphodium* were of nonhybird origin. Sullivan and Faeth (2004) also found multiple hybrids and nonhybrid *Neotyphodium* lineages not only within the same grass species, *Festuca arizonica*, but also within the same populations. Remarkably, in Moon et al.'s (2002) and Gentile's (2005) studies, *E. festucae* and *E. typhina* were often the same ancestors in different hybrid combinations across southern hemisphere continents as they are in the northern hemisphere. These results indicate a very cosmopolitan distribution and extensive across-continental gene flow of these two *Epichloë* species. It appears that interspecific hybridization of either two coinfecting *Epichloë* species or an *Epichloë* species with a *Neotyphodium* species occurs rampantly in nature and results in novel endophytes. These novel endophytes are thought to interact very differently with their hosts, not only in terms of transmission mode (e.g., vertically), but also in genetic and ecological consequences.

Hybridization is thought to arise from fusion of vegetative hyphae and then nuclear fusion to form heteroploids in hosts that are coinfected with either two *Epichloë* species or one *Epichloë* and one *Neotyphodium* species, rather than hybridization of sexual stromata of two *Epichloë* species (Selosse & Schardl 2007). Hybrids in nature are recognized by evidence of multiple copies of genes at various loci (e.g., Moon et al. 2004, Sullivan & Faeth 2004). Hybrids are apparently limited to asexual, vertical transmission and cannot revert to sexual forms by producing ascospores, presumably because heteroploidy interferes with chromosome pairing for meiosis (Selosse & Schardl 2007). Although hybrids produce conidia, there is no evidence to date that transmission can occur via asexual conidiospores. Similarly there is no evidence of interspecific hybrids that become haploidized (reduction of the heteroploid genome to the haploid state), although genome size is reduced somewhat after parasexual fusion (Kuldau et al. 1999, Schardl & Craven 2003). If these haploidized hybrids exist in nature, they would be difficult to distinguish from asexual, vertically transmitted haploid endophytes that evolved directly (i.e., without hybridization) from *Epichloë* (Selosse & Schardl 2007).

From an ecological and evolutionary standpoint, hybridization rapidly infuses genetic variation that should produce immediate changes in interactions with the host, as well as host interactions with local environmental factors, competitors, and consumers. Hybridization in the asexual endophytic fungus *Neotyphodium* is thought to infuse genetic variation that averts Muller's ratchet in the asexual fungus (Moon et al. 2004, Schardl & Craven 2003, Selosse & Schardl 2007) or the accumulation of deleterious mutations that may reduce fungal fitness. More importantly, however, hybridization has been postulated to rapidly increase host fitness via genetic variation that (1) immediately increases the diversity and types of alkaloids produced by the fungus, (2) may be adaptive in changing environments (Moon et al. 2004, Schardl & Craven 2003, Selosse & Schardl 2007), and (3) may be a way to achieve greater compatibility with a novel or unusual host plant (Selosse & Schardl 2007). It has been suggested that certain alkaloids, some of which are known for toxicity to vertebrate and invertebrate herbivores, are associated with hybrid endophytes (Clay & Schardl 2002, Moon et al. 2004, Schardl & Craven 2003, Selosse & Schardl 2007). Hybridization may also influence other well-known benefits of endophytes on host grasses, such as increased resistance to drought or increased nutrient uptake in poor soils, and generally enhance competitive abilities.

The sheer number of hybrid *Neotyphodium* endophytes in nature relative to nonhybrid *Neotyphodium* has been taken as de facto evidence that hybrids must be advantageous to the fitness of the host and endophyte (Selosse & Schardl 2007). For example, Moon et al.'s (2004) phylogenetic study indicated that the majority of hybridization events "followed rather than caused evolution of the strictly seedborne habit." Moon et al. (2004) and Selosse and Schardl (2007) suggested that the prevalence of hybrid species relative to nonhybrids implied a selective advantage of hybridization. However, much like mutualistic interactions were often presumed based on patterns of high infection frequencies (e.g., Clay 1998), which later proved to be unreliable (see section 5.2.2), care is necessary in extrapolating from frequency patterns to causation. Indeed, rigorous experimental tests of the hypothesis that hybridization increases host or endophyte fitness are still very rare.

There has, to our knowledge, been only one test of the hypothesis that hybridization alters host fitness and interactions within local environments and with other species, despite the prevalence of hybridization across *Neotyphodium*-infected grass species (Moon et al. 2004, Schardl & Craven 2003). In a reciprocal transplant experiment, Sullivan and Faeth (2008) showed that asexual and hybrid (H) *Neotyphodium* endophytes in native Arizona fescue hosts grew larger in terms of plant volume and had altered plant architecture (larger volume but less dense) relative to nonhybrid (NH) and E– hosts, but only in a relatively harsh environment with low soil moisture and nutrients. Plants with H *Neotyphodium* also had greater long-term survival than their NH counterparts. Because H, NH, and E– plants did not reproduce during the course of this experiment at the Clint's Well marginal site due to a prevailing drought, it is unclear if the change in architecture and life span provides a selective advantage in harsh environments, as has been proposed by others (e.g., Moon et al. 2004, Schardl & Craven 2003). Nevertheless, plants with hybrid *Neotyphodium* were common at two sites with low soil nitrogen and light, whereas hybrids were absent at a site with higher precipitation, soil nitrogen, and light, suggesting that variation resulting from hybrids may be advantageous in fringe or marginal habitats. Also, because hybrids did not occur in one of the populations in this reciprocal transplant experiment (Sullivan & Faeth 2008), it is uncertain if changes in survival and architecture in the nutrient-poor environment of hybrid-host symbiota were due to the hybrid endophyte per se or some other factor unique to the hybrid endophyte or host grass in this environment. Additional long-term experimental tests will be necessary to thoroughly test the hypothesis that hybrid *Neotyphodium* are generally more advantageous than nonhybrids in host grasses.

The prevalence of hybridization among *Neotyphodium* species indicates that gene flow may be augmented in *Neotyphodium* by the occasional influx of genetic variability by more widely dispersing spores of *Epichloë*. Hybrid *Neotyphodium* are thought to be more common because they have greater fitness and thus are more evolutionarily stable than nonhybrid *Neotyphodium* (Moon et al. 2004, Schardl & Craven 2003, Selosse & Schardl 2007, Tsai et al. 1994), although as stated above, there have been virtually no tests of this hypothesis. However, if hybrids increase host fitness and are more evolutionarily stable, then this sets up a very interesting, and counterintuitive, conclusion. The main way to get hybrid *Neotyphodium* is to occasionally have infections by the pathogenic, sexual form of *Epichloë*. This means that to maintain evolutionary stability of the mutualistic endophyte, the pathogenic form must be present in the population at some point. This is contrary to conventional wisdom that parasitic or pathogenic forms are thought to destabilize mutualisms (e.g., Bronstein et al. 2003, Pellmyr et al. 1996, Thompson 1994).

5.3.2 Coevolution and population-level interactions

The population level of biological organization is the arena where individual selection is played out, not the tribe or species level, although these population interactions may be reflected in phylogenetic patterns at these levels (Thompson 1997, 2005). Thompson's (1994, 1999, 2005) geographic mosaic model of coevolution is based on the idea that coevolutionary dynamics are influenced by conflicting or reinforcing evolutionary forces. Interspecific interactions are shaped by local selection pressures, with individual populations potentially evolving to different states. The geographic mosaic view of coevolution posits that interactions evolve in metapopulations, with the interaction between local selection pressures, migration, and drift creating a range of possible outcomes, including stable and dynamic antagonisms and mutualisms, and equilibrium points where both may exist (e.g., Burdon & Thrall 1999, Gandon 1998, 2002, Gandon et al. 1996, Nuismer et al. 1999, 2000). Migration between populations can overcome local selection and create populations maladapted for their environment (Thompson 1994, 2005). The interaction between selection and migration can create a geographical mosaic, with some populations exhibiting coevolution ("hot spots") while some populations may show maladaptation or no coevolution ("cold spots") because migration overwhelms local selection (Thompson 1994, 1999). When considering the community as a whole rather than just a pair of interacting species, asymmetries in the strength of interactions within a network of mutualistic interactions may promote coexistence and maintain biodiversity (Bascompte et al. 2006). Differences in gene flow rates between pairs of interacting species may also cause trait mismatching and inhibit coevolution. For symbiotic partners, differentials in gene flow change the interaction outcome because, in a given population, the genotype of one of the partners will be continually changing, thus making coevolution unlikely (Thompson 1999).

Most natural grass populations are spatially structured mosaics of E− and E+ plants, the latter varying in host and endophyte genotypic combinations (including hybrid and nonhybrid endophytes), and embedded in local biotic and abiotic environments (Bucheli & Leuchtmann 1996, Faeth 2002, Faeth & Fagan 2002, Sullivan & Faeth 2004). Thus the grass-endophyte symbiosis can be considered within the geographic mosaic view of coevolving interspecific interactions. The native grass-endophyte systems provide ideal systems for examining coevolutionary processes at the population level because grass endophytes span the spectrum of gene flow differentials between endophyte, modes of transmission, and host and endophyte interaction outcomes depending on endophyte strains, host genotypes, and environmental variation. Within *Epichloë* endophytes, for example, stromata (or disease-causing) and asymptomatic strains exist within the same host grass species (Bucheli & Leuchtmann 1996). Similarly, different asexual and seed-borne *Neotyphodium* haplotypes, some of hybrid and some of nonhybrid origin (e.g., Sullivan & Faeth 2004), often reside within the same host grass species. Moreover, the fitness of these different endophyte-host combinations changes depending on local environmental conditions within and across populations.

5.3.2.1 Coevolution and population-level interactions
in Epichloë

The pioneering research of Adrian Leuchtmann and associates (e.g., Brem & Leuchtmann 2003, Bucheli & Leuchtmann 1996) on European native woodland grasses shows a high degree of genetic differentiation between different populations of asymptomatic and disease-causing strains of *Epichloë* in European populations. The same genetic differentiation occurs in North American populations of A. *hypoxylon* that infects the pooid grass *Danthonia* (Leuchtmann & Clay 1996). Bucheli and Leuchtmann (1996) found remarkably high genetic differences among 173 asymptomatic and 93 disease-causing *Epichloë* strains from a single host plant species, *B. sylvaticum*, from 10 different geographic sites in Switzerland. The fungal strain or genotype largely controls whether *Epichloë* is sexual, and thus disease-causing, or asymptomatic (Meijer & Leuchtmann 2000, 2001). Bucheli and Leuchtmann (1996) concluded that the asymptomatic and disease-causing strains "do not belong to one panmictic population" and different patterns of disease or stromata formation in nature are "due to genetic differences among fungi in associations with their host plants." They suggested that selective pressures, such as drier environments, may favor vertical transmission, while moister environments may favor horizontal or disease-causing forms. Later research (Meijer & Leuchtmann 2000) on this same system indicated that contagious spread of the choke-forming strain was indeed more prevalent in shady, moister habitats.

In studies of another endophyte, *Epichloë bromicola*, which infects at least three different *Bromus* grass species, Brem and Leuchtmann (2003) found evidence for genetic differentiation of 26 isolates on the different *Bromus* hosts in Switzerland and France. *E. bromicola* apparently originated on *Bromus erectus* and then shifted to the other host species. More importantly, they suggested that the isolates infecting the more recent hosts, *Bromus benekenii* and *Bromus ramosus*, represent host shifts and genetic differentiation into host-adapted races or incipient species. Their reciprocal inoculations of the *Epichloë* strains showed that the ancestral sexual isolate from *B. erectus* retained broad capability of infecting the other host grass species, but asexual isolates from *B. benekenii* and *B. ramosus* could no longer infect *B. erectus* (see figure 4.8b). Thus after the host shifts to other *Bromus* species, the endophyte differentiated and specialized. The mechanism for this shift is host-mediated reproductive isolation after a shift to the asexual mode of reproduction in the new host grass species. Brem and Leuchtmann (2003) concluded that this represented a shift from pathogen to mutualist on the new hosts. An earlier experimental study showed the asexual *E. bromicola* in *B. benekenii* increased intraspecific competitive abilities (Brem & Leuchtmann 2002), supporting the view that shifts to asexuality increase the beneficial effects of *Epichloë* in host grasses. But as we have seen above, asexuality is certainly not a guarantee of mutualistic interactions. Furthermore, according to the geographic mosaic view of coevolution and evolutionary models, a differential in gene flow between host and symbiont caused by a shift to asexuality may actually disrupt coevolution, as discussed below.

5.3.2.2 Coevolution and population-level interactions in Neotyphodium

Unlike sexual *Epichloë*, the differential in gene flow rates may be great between a sexually reproducing host grass and the asexual and strictly seed-borne *Neotyphodium* endophyte (Sullivan & Faeth 2004). For *Neotyphodium* endophytes, gene flow is achieved only through seed dispersal, and dispersal may be limited to only a few meters from the maternal plant. In contrast, male gametes or pollen from the host have the ability to travel hundreds of meters in large quantities (Giddings et al. 1997, Nurminiemi et al. 1998, Warren et al. 1998). We have already seen that host genotype can significantly alter the outcome of grass-endophyte interactions (Cheplick et al. 2000, Faeth et al. 2002a, Faeth & Sullivan 2003) (chapters 2–4). Thus, at least for *Neotyphodium*, the host plant's genetic background in a population is changing over time due to gene flow while the local *Neotyphodium* genotype remains constant. The only published study of gene flow across *Neotyphodium*-infected wild grass

populations shows that gene flow is indeed very low among several populations of Arizona fescue (Sullivan & Faeth 2004). What are the coevolutionary
consequences of this gene flow differential between host and symbiont?

Gandon et al. (1996) and Kaltz et al. (1999) showed that when a parasite's
migration rate is lower than that of its host, it will not be locally adapted to
its host because the host's genetic background is changing too rapidly. The
effects of gene flow on more mutualistic interactions have not been thoroughly examined, but the notion that gene flow can disrupt favorable genotypic interactions between hosts and mutualistic symbionts should also hold
(Sullivan & Faeth 2004). These differing rates of flow might create coevolutionary "cold spots" because trait mismatching would occur and host-endophyte genotype combinations would not be stable over time (Saikkonen et al.
2004, Thompson 1999, 2005). Sullivan and Faeth (2004) suggested that for
asexual, vertically transmitted *Neotyphodium* to persist in populations against
the background of ever-changing host genotypic backgrounds, there may be
selection for endophytes providing few benefits but incurring little costs to
the host. In other words, *Neotyphodium* should become more neutral. For
example, there would be strong selection against *Neotyphodium* to produce
very high levels of alkaloids (a possible benefit, but only in environments with
intense and consistent herbivory and adequate nutrients) because they are also
costly in terms of host nitrogen and energetic demands for growth and reproduction (e.g., Faeth 2002). Indeed, very high alkaloid-producing *Neotyphodium*
appear the exception rather than the rule in natural populations where nutrients such as nitrogen are at a premium (Faeth 2002). Alternatively, there may
be selection for inducible or "on-demand" alkaloid production (e.g., Bultman &
Ganey 1995, Bultman et al. 1997, Gonthier et al. 2008) by infected plants
when herbivores are present to avert the costs of constitutive alkaloid production (Karban and Baldwin 1997). Strains or haplotypes of *Neotyphodium* that
incur low costs despite a host genetic background and local environment that
is different from the previous host generation may be the ones that usually
persist in populations.

5.3.2.3 Pathogen to mutualist?

In the larger context of sexual and asexual endophytes, the prevailing notion
is that sexual endophytes coevolve with their hosts from pathogenic interactions to mutualistic ones, coupled with the loss of sexuality (e.g., Carroll 1988,
Clay 1988a, Schardl & Clay 1997, Wilkinson & Schardl 1997). This view
derives from the general notion that parasitic or disease-causing interactions
attenuate in their antagonism over evolutionary time (e.g., Thompson 1994).
For some highly specialized symbioses where microbial symbiont genetic
diversity is low, this view seems to hold. For example, in *Buchnera*, an evolutionarily ancient obligate and vertically transmitted endosymbiont of various
insects that facilitates host nutrition, genome size has been greatly reduced
and genetic diversity across geographic areas is practically nil, such that a shift
from mutualistic to parasitic interactions is highly unlikely (e.g., Moran &

Wernegreen 2000). Many of these obligate and specialized symbionts also live intracellularly or in specialized tissues or organs within the host (Moran 2007). Indeed, endosymbiosis and the reduction of genome size and autonomy is thought to be a key process in the evolution of cellular organelles, such as mitochondria and chloroplasts, and thus eukaryotic organisms (e.g., Margulis 1991). However, for other symbioses where genetic variation is greater or when multiple symbionts are involved, the unidirectional pathway from pathogenic to mutualistic is not inevitable or even typical, and depends on locally inter-acting populations of species whose interaction outcomes depend on genet-ics and prevailing environments (Thompson 2005). For example, single gene mutations can convert a fungal pathogen to a mutualist and probably vice versa (e.g., Freeman & Rodriquez 1993, Kogel et al. 2006). In addition, asexu-ality and vertical transmission are not guarantees of mutualistic interactions, as evidenced by *Wolbachia* bacteria infecting many different species of arthro-pods (e.g., Werren & O'Neill 1997).

We have seen (section 5.3.1) that *Neotyphodium*, although somewhat spe-cialized and host specific, is by no means lacking in genetic diversity. Indeed, about two-thirds of all *Neotyphodium* species are interspecific hybrids, indi-cating occasional and rapid influx of genetic diversity into the genome. Furthermore, *Neotyphodium* genomes are not reduced in size relative to sexual *Epichloë*, and *Neotyphodium* grows intercellularly and does not occur in specialized tissues or organs. To the contrary, the genome size for non-hybrid *Neotyphodium* are equivalent to sexual *Epichloë*, and the genome of hybrid *Neotyphodium* is about twice the size of sexual *Epichloë* (Kuldau et al. 1999). Hybrids are well known for harboring additional copies of certain genes relative to *Epichloë* (e.g., Schardl & Craven 2003). At the population level, we now know that natural populations of host grasses harbor a diversity of hybrid and nonhybrid *Neotyphodium* species or haplotypes (e.g., Sullivan & Faeth 2004). In addition, coinfection with spore-dispersed strains of *Epichloë* or other systemic endophytes (e.g., *Atkinsonella, Balansia*) is always possible or even likely with the nonsystemic endophytes (e.g., Schulthess & Faeth 1998). Thompson (2005) predicted that with multiple symbionts present, the potential for geographic mosaics with a multiplicity of interaction outcomes increases. Variation in the direction of the interaction of *Neotyphodium* and its host, depending on genetics and environments, is now well documented (e.g., Faeth & Sullivan 2003, Saikkonen et al. 2006). Thus the notion that pathogenic *Epichloë* inevitably evolve into mutualistic *Neotyphodium* appears inconsistent with accumulating empirical evidence and recent conceptual developments of coevolution.

An ecological consequence of the evolutionary view that pathogenic, sex-ual endophytes inexorably evolve to mutualistic, asexual ones is that grasses harboring asexual *Neotyphodium* or *Epichloë* should have greater fitness than E− hosts or those infected by sexual *Epichloë*. For example, Brem and Leuchtmann (2003) argued that the change from sexuality to asexuality in *Epichloë* leads to local adaptation, increased fitness, and host race formation, and conforms to the view that host pathogens coevolve with their hosts to become more mutualistic over time (e.g., Schardl & Clay 1997). However, within the context of the geographic mosaic view of coevolution developed by Thompson (1994, 1999, 2005), Saikkonen et al. (2004) argued that maintenance of sexual reproduction in the endophyte or other ways that increase fungal genetic diversity (e.g., hybridization) or suppression of host genetic diversity (promotion of clonal growth) should be favored to avoid genetic mismatches between the host and endophyte. This view of coevolution between host and endophyte at the population level that may be subverted due to differentials in gene flow and genetic mismatches is very different from that of Brem and Leuchtmann (2003). More importantly, this view, in concert with geographic mosaic theory, argues that sexual and presumably more pathogenic endophytes do not inexorably evolve to specialized mutualisms. Lipsitch et al. (1996) theoretically argued along similar lines against the "the notion of a virulence-avirulence continuum between horizontal and vertical transmission." Growing evidence from natural communities indicates that endophyte-grass hosts do indeed exist as a collection of spatially structured metapopulations, where genetics of specific endophyte-host combinations, differentials in gene flow, other interacting species, and local selective pressures create multiple interaction outcomes (e.g., Saikkonen et al. 2004, Sullivan & Faeth 2004). Clearly additional studies in natural populations will be necessary to discriminate between these two views. Both make specific predictions that can be tested (see chapter 7).

Phylogenetic studies of endophytes and their host grasses have revealed interesting patterns of cospeciation and cocladogenesis at the host tribe or species level. However, evolutionary ecological studies at the population level, where processes of selection at the interspecific level of interaction determine these patterns, are very scarce and greatly needed. It seems logical that such studies are best conducted across wild populations of infected grass species where evolutionary forces have resulted in structured and distinct population differences. Moreover, because endophyte-host grass interactions operate in natural communities within a web of other interacting species (e.g., Bascompte et al. 2006), some of which may also interact mutualistically (e.g., mycorrhizae) (Novas et al. 2005, Vanderkoornhuyse et al. 2003), understanding coevolution of endophyte and host grasses should include these other interactions.

6

Community and Ecosystem Consequences of Grass Endophytes

6.1 ENDOPHYTES WITHIN THE CONTEXT OF COMMUNITY GENETICS

Neuhauser et al. (2003) and Whitham et al. (2003) argued that heritable variation within individual species has profound consequences at the community and ecosystem level of organization. In effect, genetic variation among individuals within species or "extended phenotypes" at one organizational level (e.g., the primary producers) alters properties and processes at other ecological levels (e.g., detritivores, herbivores, and natural enemies). Also, the repercussions of genetic variation are often amplified across trophic levels. For example, phytochemistry of individual cottonwood trees, determined partly by tree genotype, alters herbivorous insect communities and the birds foraging on these insects, and ecosystem functions, such as decomposition rates (Bailey et al. 2006). In a recent study of arthropod diversity in old fields, Crutsinger et al. (2006) showed that plant genotypic diversity of goldenrods (*Solidago altissima*) dictated community diversity of associated arthropods and affected ecosystem-level aboveground net primary productivity. Whereas there is considerable debate about whether such community-level traits are heritable and under natural selection (e.g., Collins 2003, Morin 2003), the idea that genetic and phenotypic variation at one trophic level influences ecological patterns and processes at other trophic levels and ecosystem properties has particular relevancy to endophytes inhabiting grasses and their consequences for communities and ecosystems.

Seed-borne endophytes and their host plants are ideal systems to consider within the context of "community genetics" (see Neuhauser et al. 2003

145

for the origin of this phrase), where the imprint of genetic variation at one trophic level reverberates across others. First, seed-borne endophytes are maternally inherited components of cool-season grasses, similar to mitochondria. Infections can be lost through hyphae inviability (Siegel et al. 1984b) or occasional failure of hyphae in culms to infect developing seeds (do Valle Ribeiro 1993, imperfect transmission *sensu* Ravel et al. 1997b), but are not gained through horizontal transmission (Clay 1990b, Saikkonen et al. 1998, Schardl et al. 2004, Wilkinson & Schardl 1997, but see Moy et al. 2000). Thus endophyte infection is heritable and variable due to different endophyte haplotypes and strains that persist at various frequencies (see chapter 4), at least in natural populations (e.g., Bony et al. 2001, Piano et al. 2005, Sullivan & Faeth 2004). Second, the effects of the endophyte are traceable empirically through food webs and often have distinct footprints on community and ecosystem properties and processes (e.g, Faeth & Hammon 1992, Rudgers & Clay 2005, Rudgers et al. 2004). Compounding the genetic variation in the endophyte is that of the host grass, and although incorporating both endophyte and host grass genetics is challenging, endophyte-host grass systems provide unique opportunities to test the concept of community genetics and extended phenotype because both the endophyte strain and host grass genotype can be manipulated (chapter 2). In this chapter we examine the effects of endophytes in grasses on community- and ecosystem-level patterns and processes.

6.2 COMMUNITY EFFECTS

6.2.1 Plant diversity

As maternally inherited components of infected grasses, endophytes may have profound effects on community patterns and processes, despite their miniscule biomass relative to other species in biological communities (Clay & Holah 1999). In a pioneering, long-term experimental test of this hypothesis, Clay and Holah (1999) established eight experimental plots in an Indiana old field (a former agricultural field that had undergone plant succession), four with high densities of E+ and four with E− tall fescue. They then monitored species richness of other plant species over the next 4 years. Species richness declined by an average of 2.2 plant species in the E+ relative to E− plots, and E+ dominance increased at the expense of other species, especially other grasses. E+ fescue became increasingly dominant in their plots, while the biomass of other plants species concomitantly declined. They concluded that E+ tall fescue, widely planted in North America, may reduce plant diversity in native plant communities, although they note that neither tall fescue nor many of the plant species in their old field communities were native species.

In a later study using another experimental site, an old field in a bottomland in Indiana, Clay et al. (2005) showed that endophyte frequency in Kentucky 31 tall fescue increased 30% when insect and small mammalian herbivores were present over a 4.5-year period. This shift in endophyte frequency in agronomic tall fescue is well known in pastures under heavy grazing pressure from livestock (Clay 1998: table 10.5). Furthermore, species composition of the plant communities changed in the herbivore and herbivore-exclusion plots, although biomass remained the same. Thus the presence of infected tall fescue has community-level effects by altering plant community composition. Tracy and Renne (2005) renovated nine pastures originally containing high frequencies of E+ tall fescue and planted them with mixtures of E– tall fescue (Barcel) and a combination of seven other grass and forb species to test whether the pastures would become infected with E+ over time. In contrast to the findings of Clay et al. (2005), they found that the percentage of E+ tall fescue did not increase relative to E– tall fescue over 3 years in pastures, even though pastures were subjected to moderate grazing pressure from livestock. Tracy and Renne (2005) concluded that relatively wet growing seasons contribute to the lack of competitive advantage of E+ versus E– tall fescue. Their study implies that it is resistance to drought stress rather than resistance to herbivores that mainly dictates the competitiveness of E+ tall fescue.

Also in contrast to Clay and Holah's (1999) findings, Spyreas et al. (2001) found a positive relationship between the frequency of endophyte infection and plant diversity in a 3-year-old successional field in Illinois. In this study, endophyte infection was not experimentally manipulated, but relied upon established fields. However, different plots were either mowed or unmowed and fertilized or unfertilized within the fields. They found the highest plant diversity in mowed plots with the highest endophyte frequencies. They surmised that mowing decreased soil moisture, which reduced the interspecific competitive abilities of tall fescue, and thus promoted increased diversity. Nonetheless, infection frequency throughout all the plots was positively correlated with plant species richness. Spyreas et al. (2001) concluded that "the previously reported negative relationship between endophyte infection and community diversity is probably overly simplistic in complex ecological settings." Like Tracy and Renne (2005), Spyreas et al. (2001) concluded that it was mainly resistance to abiotic factors rather than herbivory that provides E+ tall fescue its advantage over E– tall fescue, although neither of these studies manipulated or controlled herbivory. As we explore below, this complexity becomes even more pronounced when one considers multiple endophyte haplotypes residing within many host grass genotypes situated amid highly diverse and variable natural communities.

We know virtually nothing about the effects of endophyte infection on plant diversity in natural communities. The studies of both Clay and Holah (1999) and Spyreas et al. (2001) involved a single cultivar of tall fescue (Kentucky 31) harboring a single *Neotyphodium* endophyte in successional old fields where introduced Eurasian weeds comprise a significant component of the plant community. The only study of endophyte infection and the relationship to plant diversity in natural communities of which we are aware involves Arizona fescue in semiarid Ponderosa pine-grassland communities. In 1996,

Faeth and Rambo surveyed 34 populations across high-elevation communities for understory plant diversity and *Neotyphodium* infection frequency in a correlational study. Plant species richness of forbs, grasses, shrubs, and trees was recorded in ten 1 m^2 quadrats at 10 m intervals along a randomly placed transect. Details of the plant species richness survey can be found in Rambo and Faeth (1999). Infection frequency was determined by randomly sampling at least 50 Arizona fescue plants in each population near the transects. Plant species richness was significantly ($F = 5.31$, df = 1,33, $P = 0.03$) and negatively related to frequency of endophyte infection (figure 6.1). However, the regression explained only 14% of the variation, suggesting factors other than endophyte frequency may better explain variation in plant species richness. Also, there was no difference between grazed ($N = 5$) and ungrazed ($N = 29$) populations (by large vertebrates grazers) in terms of the relationship between plant species diversity and endophyte frequency, although grazed populations had higher plant species richness (Rambo & Faeth 1999). Clearly, other studies of the relationship between endophyte infection frequency and plant species richness are needed, especially in natural plant communities, before any firm conclusions can be drawn.

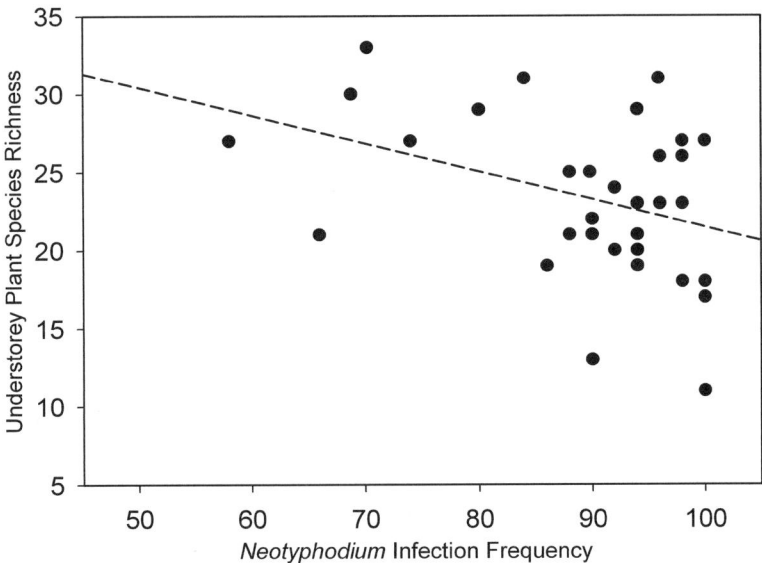

Figure 6.1 Regression of *Neoyphodium* infection frequency in Arizona fescue and understory plant species richness in high-elevation, semiarid Ponderosa pine-grassland communities in Arizona. Plant species richness data from Rambo and Faeth (1999). Infection frequency data partially from Faeth et al. (2002b). Figure reprinted from Faeth and Saikkonen (2007) with permission of the New Zealand Grassland Association.

6.2.2 Herbivore and natural enemy abundance and diversity

Because they alter the physiological and morphological properties (chapter 2), as well as the allelochemistry (via alkaloid production) (chapter 3) of their hosts, seed-borne endophyte infections, as maternally inherited components, potentially influence herbivores and the natural enemies of herbivores. Infections certainly alter herbivore abundance, often negatively, but sometimes positively, depending on the endophyte haplotype, plant genotype, herbivore species, and environmental background (chapter 3). Changes in the relative abundance of one herbivore species may, in turn, influence abundances and presence/absence of other herbivorous species, or the predators and natural enemies that feed upon them. Indeed, endophytic fungi and their alkaloids may promote coexistence of herbivore species. Härri (2007) recently conducted greenhouse experiments involving two species of aphids, *Rhopalosiphum padi* and *Metopolohium festucae*, feeding on E+ and E− perennial ryegrass. The former, *R. padi*, is highly susceptible to endophyte alkaloids, while *M. festucae* is less so. Although *R. padi* has higher growth rates and higher population densities in pure populations than on both E+ and E− perennial ryegrass, this advantage is offset when both aphid species are grown together on E+ ryegrass. On E+ plants, *M. festucae* outcompeted *R. padi*. Thus infection permits coexistence of *M. festucae* with *R. padi*, although it is an inferior competitor when feeding on E− plants.

There is little information on how endophyte infections influence diversity at higher trophic levels. Omacini et al. (2001) experimentally examined food webs associated with E+ and E− agronomic Italian ryegrass (*Lolium multiflorum*) grown in large containers. The relative abundances of some herbivore and parasitoid species differed between E+ and E− plots; however, this shift in relative abundances did not lead to radical changes in herbivore or natural enemy diversity. The only change in herbivore or natural enemy species richness was one additional generalist hyperparasitoid in E− plots, such that E+ plots harbored eight total herbivores, primary parasitoids, and secondary (hyper-) parasitoids, while E− plots had nine species (table 6.1). However, addition of this hyperparasitoid effectively increased the length of the food chain relative to E+ Italian ryegrass plots. Other effects of infection on trophic interactions in this experiment are detailed below (section 6.2.3). In a recent experimental study, Finkes et al. (2006) found that the richness of spider families and morphospecies was greater in E− compared to E+ agronomic tall fescue plots in Indiana, but total spider abundance was not different. Moreover, declines in relative abundances of two spider families, Linyphiidae and Thomisidae, occurred in the infected tall fescue plots. Thus infection in agronomic tall fescue affects not only herbivores, but also species richness and familial composition of the spider communities that prey upon the herbivores.

How does endophyte infection in primary producers influence herbivore and natural enemy diversity in natural communities, where genetic and

Table 6.1 Summary of results from Omacini et al. (2001) and Chaneton and Omacini (2007) on the effect of *Neotyphodium* infection in plots of introduced Italian ryegrass (*Lolium multiflorum*) on trophic and community-level properties relative to E– plots.

Trophic/Community Property	E+ *L. multiflorum* plots (relative to E–)
Species richness	Slightly decreased (from nine to eight species)
Overall herbivore (aphid) abundances	3× greater
Herbivore (aphid) relative abundances	Dominance of *Rhopalosiphum padi* declined; *Metopolohium festucae* increased
Parasitoid relative abundances	Shifted relative to E– plots
Primary parasitism rate on aphids	Decreased significantly
Primary parasitism success on aphids	Increased success
Secondary parasitism rates	Decreased significantly (presumably due to reduced food quality/alkaloid transfer)
Connectance (number of trophic links)	Decreased
Median strength of interactions	Significantly decreased
Evenness of interactions	Significantly increased

environmental effects complicate the effects of inherited symbionts in host grasses on herbivores and natural enemies? Faeth (2008) conducted a field experiment with multiple plant genotypes of Arizona fescue with and without their endophytes, and subjected to treatments of varying limited resources (water) and herbivory. This long-term experiment used replicates of four naturally infected maternal genotypes of Arizona fescue, half from which the endophyte had been removed experimentally. Plants were subjected to three water treatments (reduced, ambient, and supplemented) and two herbivore treatments (ambient herbivory from invertebrates and small mammals and greatly reduced herbivory via hardware cloth cages and periodic insecticide treatments). Arthropods were collected from all plants over two field seasons. E+ plants had greater abundances of total arthropods, herbivores, predators, and omnivores, but equivalent abundances of detritivores and parasites relative to E– plants. Rarefaction analyses (Gotelli & Entsminger 2007) showed that species richness was greater for predators, but less for parasites on E+ than E– plants. Evenness, an important component of diversity that measures how individuals are distributed among species, was greater for herbivores, predators, and detritivores on E+ than E– plants. Overall, Shannon-Weiner diversity (an index that combines richness and evenness) was greater on E+ plants for all arthropods, herbivores, and predators, but less for parasites on E+ grasses. However, herbivore:natural enemy ratios (and herbivore:all other feeding guilds ratios) were equivalent on E+ and E– plants, indicating that, unlike Omacini et al.'s (2001) results, natural enemies were likely responding to increases in herbivore abundances rather than because of alkaloid-based reductions in food quality in E+ plants. Of interest, E+ experimental plants averaged 11.05 ppm peramine, which are levels far beyond those known to negatively affect insects (e.g., 3.0 ppm) (Siegel & Bush 1996). At least for this native grass, inherited endophytic symbionts alter herbivore and natural enemy diversity, but apparently through increased productivity rather than reduced food quality due to alkaloids (e.g., Omacini et al. 2001). Instead, the increased productivity of E+ plants cascades upward and increases abundances of herbivores, predators, and omnivores.

The only other study of grass endophytes and their effects on herbivores and natural enemies in natural communities involved *Achnatherum robustum* (sleepygrass). Jani (2005) examine the diversity of herbivores, predators, parasites, and detritivores collected from 87 infected individuals of *A. robustum* that varied in alkaloid levels from a natural community near Cloudcroft, New Mexico. Sleepygrass is a notoriously toxic grass, at least to livestock, often with extraordinarily high levels of alkaloids, but these alkaloid levels vary widely within and across populations (Faeth et al. 2006). Insects were vacuumed from infected plants in 2002 and alkaloid levels determined. Because few plants were uninfected, only infected plants were used in the analysis and plants were grouped by alkaloid levels (no alkaloids, low alkaloids [<68 ppm], and high alkaloids [>68 ppm]). Interestingly, Jani (2005) found significantly higher abundances of herbivores on low- and high-alkaloid grasses relative to infected grasses without alkaloids. These results contradict expectations that alkaloids should reduce abundances of herbivores in communities (see chapter 3). In terms of diversity, rarefaction analyses showed that total species richness did not differ between no alkaloid and low- and high-alkaloid plants, but herbivore communities on high-alkaloid plants were significantly more diverse than those on low-alkaloid plants. The same pattern held for herbivore species richness: herbivore richness on E+ plants without alkaloids was not different from either low- or high-alkaloid plants, but high-alkaloid plants had more herbivore diversity than low-alkaloid plants. When the richness of natural enemies was examined, this pattern was reversed: high-alkaloid plants supported a lower richness of natural enemies than low-alkaloid plants (no-alkaloid plants had too few natural enemies for meaningful comparisons). This difference is due to higher richness of predators on low-alkaloid plants; parasitoid richness did not differ significantly between low- and high-alkaloid plants. Evenness also differed for arthropod communities on sleepygrass plants with varying levels of alkaloids. For example, evenness of herbivore species was greater on high- versus low-alkaloid plants, but the reverse was true for natural enemies, especially predators (once again, no-alkaloid plants did not differ from either low- or high-alkaloid plants).

The study of Jani (2005) was correlational and therefore higher herbivore levels could be related to factors other than alkaloids, such as plant genotype or local abiotic factors. Therefore Jani and Faeth conducted a controlled field experiment where four different sleepygrass plant genotypes, three of which were infected but produced different levels of ergot alkaloids, were planted and herbivory and water (the main limiting resource in semiarid grasslands) were manipulated. The four plant genotypes were E−, E+ with no alkaloids, E+ with moderate alkaloids, and E+ with high alkaloids. Although this experiment is ongoing, results from the first year support those from the correlational study. Abundances of some herbivore families, such as Cicadellidae (leafhoppers) were significantly higher on infected plants with high alkaloids relative to E− plants or E+ plants with no alkaloids.

These results suggest complex changes in arthropod and trophic structure due to variations in alkaloid levels among endophyte-infected plants. Alkaloids in sleepygrass appear to have more negative effects on the third trophic level, especially predators, similar to the studies of Omacini et al. (2001) and de Sassi et al. (2006). Furthermore, alkaloid variation in sleepygrass, likely due to genetic variation in strains of *Neotyphodium* within the same community (Faeth et al. 2006), effects changes in relative abundances and evenness of individuals within various trophic levels, often in unexpected ways. Jani (2005) suggested that herbivores may be able to escape in "chemical" space from predators on infected plants with high alkaloids. This notion that alkaloids in infected plants may affect natural enemies of herbivores more so than the herbivores themselves challenges the long-held notion that alkaloids are acquired defenses of the infected grass (e.g., Clay 1990b, Cheplick & Clay 1988, Clay & Schardl 2002). The complication of alkaloid "defenses" used by natural enemies of herbivores against their own natural enemies is explored further in section 6.3 and section 6.4.

Given that there has been only one study of herbivore and natural enemy diversity and abundances in agronomic grasses and two in native grasses, with results that differ, it seems premature to make any conclusion about the effects of endophytes on diversity and abundances of herbivores, predators, omnivores, and detritivores. Further long-term studies, particularly in natural communities where species interactions are intricate and multifaceted, are needed to understand the relative effects of endophytes and their alkaloids. Natural communities are composed of mosaics of uninfected and infected plants (Faeth 2002, Faeth & Fagan 2002), with the latter varying in endophyte and host genotype which dramatically alter phenotypic properties, such as alkaloid levels, of the endophyte-host symbiota. Thus understanding how endophytes in grasses influence plant, animal, and microbial diversity in natural communities is a challenging as well as exciting and essential task.

6.2.3 Trophic interactions

Endophyte-associated trait alterations in the host grass may cascade upward in communities and ecosystems to alter trophic structure and feeding relationships (Chaneton & Omacini 2007, Faeth & Bultman 2002). One of the longstanding ecological paradigms is that energy and resources from the bottom of food webs, in concert with relative efficiencies of consumers, limits the biomass of each trophic level and the number of trophic levels in communities (Elton 1926, Post 2002, Schoener 1989). Given that infected grasses often produce more biomass and can be more fecund (e.g., Clay 1993, and many examples in chapter 2), one simple prediction from this view is that communities with a greater frequency of infected grasses are more productive in terms of plant, herbivore, predator, and parasite biomass. Clay and Holah (1999: figure 3A) found that community plant biomass was only marginally greater for six of seven sampling periods in heavily infected tall fescue plots than in E− plots. However, plants species composition changed in E+ plots, with a declining proportional biomass of other grasses relative to E− plots.

Despite the purported increase in primary productivity of infected grasses, this additional biomass may not be transferred to higher trophic levels if endophyte-associated alkaloids reduce herbivory (e.g., Chaneton & Omacini 2007, Clay 1994b, Omacini et al. 2001). Alkaloids can certainly deter or increase the mortality of some herbivores (see chapter 3), and thus reduce plant quality and transfer to higher trophic levels. Alternatively, alkaloids can be transferred to higher trophic levels and thus protect herbivores from their natural enemies such as predators or parasitoids, and even parasitoids from secondary parasitoids (e.g., Bultman et al. 1997) (see section 6.4). Given that natural grass communities typically harbor not only uninfected and infected grasses, but infected grasses that vary enormously in the alkaloids they produce due to host and endophyte genetics and local environments, then we can expect an exceedingly complex network of community interactions that may be influenced by fungal endophytes. A summary of the results of studies of the complex effects of endophytes on the third trophic level, parasites and predators, is shown in table 6.2.

In a novel experimental test to simplify the effect of grass endophytes on these complex interactions, Omacini et al. (2001) examined trophic interactions in E+ and E− monocultures of Italian ryegrass (L. multiflorum), an introduced agronomic grass, in Argentina. They found that overall density of two aphid species, R. padi and M. festucae, increased threefold in E− relative to E+ plants. Of interest, only R. padi increased in E− plots, and as noted in chapter 3, this aphid species is likely hypersensitive to endophyte infection and alkaloids. The relative abundance of M. festucae actually decreased in density in E− containers. The overall increase in aphid density translated into an eightfold increase in parasitism, the next trophic level, in E− containers. However, in the E− containers, primary parasitoids were disproportionately less successful because of increased attack by secondary parasitoids. Apparently the effects of alterations in food quality, presumably reduced by alkaloids in E+ plants, were transferred through two trophic levels, and reduced the growth of secondary parasitoids. Overall, Omacini et al. (2001) showed that herbivores and the top parasites (secondary parasitoids) responded positively in terms of density and biomass to removal of endophytes and the absence of their alkaloids, while primary parasitoids showed no change in reproductive success, even though they increased their attack rate. Furthermore, connectance (the number of trophic linkages) and the strength of interactions declined in E+ relative to E− plots (table 6.1). These results indicate shifts in community trophic interactions due to a microbial symbiont that is an inherited component among the primary producers in this system. A schematic of the complex interactions in this relatively simple community on introduced Italian ryegrass is summarized in figure 6.2.

Table 6.2 Complex and contradictory effects of endophyte infection on the third trophic level, parasites and predators of herbivores.

Natural enemy	Community richness or composition	Searching efficiency	Growth	Developmental time	Survival
Arthropod predators	Altered[1]		Reduced[11,12]	Slowed[11,12]	Decreased[11,12]
Vertebrate predators		Increased[4]			
Nematode predators/parasites		Reduced[8,12,13,14]	Reduced[8,12,13,14]		Decreased[12,13,14]
Primary parasitoids	Increased richness[3]	Reduced[7] Increased[9,10]	Reduced[10]	Slowed[5,10]	None[5] Decreased[6,10]
Hyperparasitoids	Reduced richness[2]		Reduced[2]		

[1]Finkes et al. 2006, [2]Omacini et al. 2001, [3]Jani 2005, [4]Huitu et al. 2005, [5]Krauss et al. 2006, [6]Bultman et al. 2006, [7]Goldson et al. 2000, [8]Grewal et al. 1995, [9]Barker & Addison 1997, [10]Bultman et al. 1997, [11]de Sassi et al. 2006, [12]Richmond et al. 2004b, [13]Kunkel & Grewal 2003, [14]Kunkel et al. 2004.

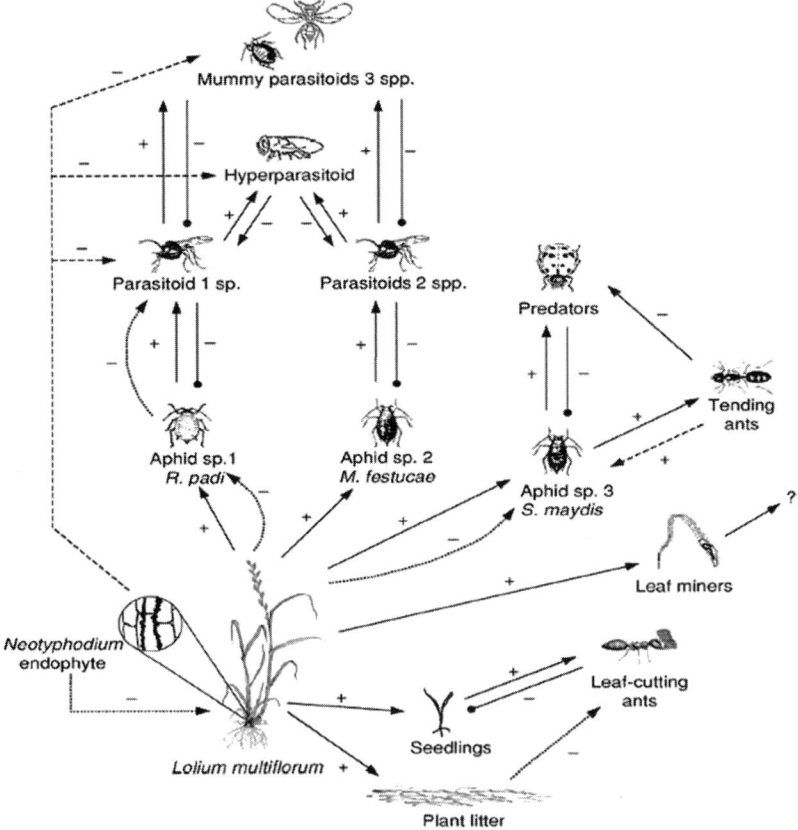

Figure 6.2 Complex interactions for insect consumers in a relatively simple food web on endophyte-infected Italian ryegrass (*Lolium multilforum*). Direct and indirect interactions are shown by different line styles. Direct interactions are shown by solid lines indicating the direction of energy flow (arrowhead) and negative effects from consumption (bullet-head). Indirect interactions comprise density-mediated (dashed lines) and trait-mediated (dotted lines) effects. Reprinted with the permission of Cambridge University Press from Chaneton and Omacini (2006).

Omacini et al. (2001) clearly showed that endophytes and their alkaloids potentially influence both bottom-up (resource-based) and top-down (natural enemy-based) forces in communities. However, it should be kept in mind that this experiment, despite its complex interactions, involved monocultures of an introduced agronomic grass in containers, with one endophyte haplotype, limited plant genetic diversity, only two herbivore species, and relatively uniform environments. We should expect even more unpredictable and complex interactions in speciose natural communities with mosaics of endophyte-host combinations. Furthermore, diversity and abundances are typically measured as the number of species and individuals. However, food web models are based on biomass or energy transfer and not individuals (e.g., Chase et al. 2000, Power 1992). Using individuals in measuring energy transfer, trophic dynamics, and food web structure can be erroneous because individuals of different species within the same trophic level are not equivalent in terms of biomass. For example, a single herbivorous beetle may weigh 100 times that of a single herbivorous thrip.

The only studies of which we are aware that measure biomass differences of herbivores, omnivores, parasites, predators, and detritivores are those by Faeth (2008) and Jani (2005), described in section 6.2.2. In addition to determining the number of individuals, Faeth (2008) measured the length of individuals of each species and then converted length to biomass via well-established equations relating insect length to biomass (Sabo et al. 2002). Infected plants had significantly greater herbivore biomass, reflecting increases in abundance as noted above (section 6.2.2). Thrips and mirid bugs comprised the bulk of the herbivore biomass, and like total biomass, were significantly greater on E+ compared to E− plants. Similar to individuals, these increases were translated into greater biomass of other trophic levels—biomass of predators and omnivores were also greater on E+ plants. Jani (2005) found that the mean biomass of herbivores was twofold greater on low- and high-alkaloid plants relative to plants without alkaloids, although not significantly so after alpha levels were adjusted for multiple comparisons. Thus in both of these studies, endophyte effect reverberated through the herbivores to predators and omnivores in terms of both arthropod abundances and biomass.

6.2.4 Spatial heterogeneity of E+ and E− plants, herbivory, and trophic interactions

It is now well established that the spatial arrangement of plants that vary in phenotypic traits, such as allelo- or secondary chemistry, can profoundly alter the distribution of arthropod herbivores and their natural enemies, predators and parasites (e.g., Karban 1997). Focal plants surrounded by other plant species may experience reduced herbivory via associational resistance because neighboring plants produce allelochemicals that deter herbivores or attract natural enemies of the herbivores (e.g., Root 1973, Tahvanainen & Root 1972). Alternatively, surrounding plants may also increase the vulnerability of plants to herbivores (associational susceptibility), either via attraction of herbivores or a reduction in natural enemies (e.g., Karban 1997). Whereas these changes in higher trophic levels based upon neighboring plants are typically viewed from the plant species level, they may also occur when conspecific plants that vary in phenotypic traits surround a focal plant. For example, Rodriquez-Saona and Thaler (2005) tested whether the spatial arrangement of conspecific tomato plants that varied in the ability to induce chemical defenses altered arthropod herbivores, omnivores, and predators. They found that plants in homogeneously induced patches (clusters of plants) contained fewer herbivores, but more predators relative to homogeneously noninduced patches. At the neighborhood scale of individual plants that varied phenotypically, induced plants in heterogeneous patches contained more herbivorous arthropods and omnivores compared with induced plants in homogeneous patches. Thus they concluded "the abundance of some herbivores and omnivores on induced plants varied depending on the phenotype of the other plants within the patch" (Rodriquez-Saona & Thaler 2005).

Endophytes in grasses obviously alter host phenotypic expression in terms of physiology, morphology (chapter 2), and allelochemistry, notably alkaloids (chapter 3). Therefore the spatial configuration at the neighborhood and patch level of E− and E+ grasses in populations should profoundly affect insect herbivores, omnivores, predators, parasites, and parasitoids. The constitutive types and levels of alkaloids in neighboring plants may influence the outcome of third trophic level interactions on E− or other E+ grasses that differ in alkaloid levels. Furthermore, because herbivory may induce increases in alkaloids in infected agronomic grasses (e.g., Bultman & Ganey 1995) and also in infected native grasses such as *Glyceria striata* (T. J. Sullivan, personal communication), changes in herbivore attacks and natural enemies are expected based on the proximity of other infected plants that show induced alkaloid changes. It is well known that induced defenses in plants in general alter the abundances and behavior of herbivores, omnivores, and natural enemies (e.g., Dicke & Vet 1999, Karban & Baldwin 1997, Kessler & Baldwin 2001).

Despite the importance of the spatial configuration of E− and E+ plants relative to each other and other plant species, we know of no studies of how host spatial relationships affect the herbivores, omnivores, and natural enemies associated with a particular grass-endophyte system. Spatial scale and arrangement should be particularly important in natural populations and communities where surrounding plant diversity is typically greater than planted agronomic pastures or lawns, and where greater plant genotypic diversity and multiple endophyte haplotypes results in a plethora of host-endophyte phenotypic outcomes. Furthermore, theoretical work has shown that the inclusion of spatial effects in biological models can contribute substantially to understanding the evolutionary processes that result in variable outcomes in mutualistic symbioses (Gomulkiewicz et al. 2003, Yamamura et al. 2004). There is clearly an urgent need to begin to incorporate spatial scale and arrangement into studies of the effects of endophytes on grass-herbivore-natural enemy interactions.

This absence of studies also applies to vertebrate herbivores. We should also expect that endophytes and their associated alkaloids might affect vertebrate patterns of foraging behavior and ultimately abundances and diversity as well. Generalist vertebrate grazers and browsers are well known for mixing diets to increase nutritional balance (Westoby 1978) and to reduce the detrimental effects of plant allelochemicals (Freeland & Janzen 1974, Wiggins et al. 2006). The spatial distribution of E− and E+ plants in natural communities where E+ plants vary widely in alkaloid types and levels due to endophyte or plant genetics and environmental factors (e.g., Faeth et al. 2006) may thus determine foraging patterns and efficiencies of vertebrate herbivores. For example, Wiggins et al. (2006) showed that the spatial arrangement of two *Eucalyptus* species that differed in plant allelochemicals alter the foraging behavior and efficiency of brushtail possums and depended on small- or large-scale heterogeneity of the two plant species. Because alkaloids are essentially plant allelochemicals or secondary metabolites, but produced by the endophytic symbiont rather than by the plant, we should expect similar alterations in vertebrate foraging behavior and efficiencies, depending upon the spatial distribution of E– and E+ grasses in the population or community, the latter of which vary in allelochemical alkaloids. Unfortunately, little is currently known about the spatial arrangement of E– and E+ plants with varying alkaloids in natural populations and communities, either at a geographic or local scale (Faeth et al. 2006), and how spatial configurations affect vertebrate grazing or browsing activities.

6.3 THE PARADOX OF ENDOPHYTE-ASSOCIATED PLANT DEFENSES

Traditionally endophytes in grasses have been viewed as defensive mutualists (e.g., Clay 1988b, 1990b, Clay & Schardl 2002, Cheplick & Clay 1988) because of the negative effects of endophytic alkaloids on herbivores (chapter 3). Empirical support for this hypothesis has, until recently, been derived largely from studies of two agronomic grasses, tall fescue and perennial ryegrass, originating in Europe and Asia, and imported and selectively bred as cultivars that are now planted worldwide (Saikkonen et al. 1998, 2006). Agronomic varieties of these grasses are high in alkaloids (at least until recent manipulations have reduced toxicity) and most species that have been assayed for resistance tend to be agronomic generalist pests, at least one of which (R. padi) appears hypersensitive to alkaloids. We have seen, however, in chapter 3, that more recent evidence indicates wide variability in response to endophyte infection in general and alkaloids specifically, especially in native grasses (Saikkonen et al. 2006). At least part of the explanation for this variability may reside in the effects of endophyte-associated alkaloids on the third trophic level—parasites and predators of herbivores.

Alkaloids not only influence herbivores, but also either directly or indirectly affect their natural enemies, just like constitutive (e.g., Barbosa et al. 1991) or induced (e.g., Faeth 1994) plant-based allelochemicals that alter the efficacy of natural enemies. Any direct or indirect mechanism that reduces location, attack rate, survival, or performance of natural enemies of herbivores can potentially nullify endophyte benefits to the host via increased deterrence and toxicity to herbivores.

There are several ways that endophytic alkaloids can either increase or decrease natural enemy attack (Faeth & Bultman 2002).

- First, alkaloids may slow growth and prolong development such that herbivores are exposed to natural enemies for longer periods. This is essentially the slow-growth, high-mortality hypothesis of Clancy and Price (1987) that states the reduced food quality or allelochemicals found in plants can act to increase the efficiency of predators and parasitoids. Support for this hypothesis is mixed, and appears to depend on the type of herbivore and plant chemistry (see references in Faeth & Bultman 2002). For endophyte-based alkaloids, insects feeding on E+ grasses may show delayed development (Breen 1994, Clay et al. 1985, Hardy et al. 1985, 1986, Popay & Rowan 1994). However, there have been few empirical tests of this hypothesis, especially under field conditions. Goldson et al. (2000) found that rates of parasitism by *Micronotus hyperodae* on the Argentine stem weevil were negatively related to the density of *Neotyphodium lolii* and peramine alkaloids in perennial ryegrass. If endophytes and their alkaloids increased attack by parasitoids, we should expect the opposite. Furthermore, Rowan et al. (1990) showed that development time of the Argentine stem weevil fed diets containing peramine did not differ from controls.

- Second, alkaloids, similar to plant-based allelochemicals, may lower herbivore resistance to pathogens and parasites. Grewal et al. (1995) found that Japanese beetles feeding on roots of E+ tall fescue were more susceptible to an entomopathogenic nematode than beetles feeding on E− tall fescue roots. Grewal et al. (1995) suggested that the weakened condition of the beetles from feeding on poorer quality food increased their vulnerability to the parasite. Alkaloids may also alter search efficiencies of natural enemies of herbivores on infected grasses. For example, Barker and Addison (1997) found that prior experience with Argentine stem weevil feeding on E+ and E− perennial ryegrass can influence efficiency of the parasite *M. hyperodae*. Naïve parasites showed no preference for weevils on either E+ or E− plants. However, parasites with prior experience with weevils on E+ or E− diets were more efficient at parasitizing the respective hosts. However, this increased search efficiency is apparently outweighed by decreased survival of the parasite on weevil hosts reared on E+ perennial ryegrass (Bultman et al. 2003).

- Endophytes may also increase the resistance of herbivores to their predators in unexpected and unusual ways. Huitu et al. (2008) found that field voles (*Microtus agrestis*) feeding on infected meadow fescue (*Lolium pratense*) in Finland lost body mass relative to voles feeding on E− meadow fescue. More intriguing was that the urine of voles feeding on E+ meadow fescue shifted in peak ultraviolet (UV) fluorescence from >380 nm to 370 nm. The latter is the maximum sensitivity of UV pigments in the eyes of raptor predators that feed on voles and locate their prey by UV visibility of urine (Huitu et al. 2008). Thus endophyte-associated alkaloids may alter the susceptibility of mammalian herbivores to their predators via changes in UV properties of their urine. They have now established E+ and E− meadow fescue plots to test if infection alters vole survival and predation efficiency of raptors.

- Third, alkaloids may directly increase the resistance of herbivores to predators and parasites as they move through the food web and are sequestered by the herbivores as part of their own defensive repertoire (Chaneton & Omacini 2007, de Sassi et al. 2006, Faeth & Bultman 2002). In contrast to the first two indirect mechanisms, which enhance the benefits of endophyte-produced alkaloids to the grass host, sequestration of alkaloids by herbivores would short-circuit the purported defensive mutualism of endophytes and grasses. Sequestration of plant allelochemicals by herbivores is well known in the plant-herbivore literature (e.g., Bowers 1990). For example, there is the classic case of monarch butterfly larvae sequestering cardiac glycosides from milkweed host plants which then render them toxic to avian predators

(Moranz & Brower 1998). Plant-based alkaloids are also used by herbivores to reduce the growth and survival of larval parasitoids (e.g., Barbosa et al. 1991). Therefore we might expect endophyte-based alkaloids to be incorporated into the defenses of herbivores. Bultman et al. (1997) tested the effect of alkaloids in tall fescue on the growth and survival of two *Euplectrus* parasitoids of the fall armyworm, a generalist herbivore, feeding on E+ tall fescue. Both parasitoid species had reduced pupal mass. When armyworms were fed artificial diets with loline alkaloids, survival of the parasitoids was reduced relative to control diets. Likewise, they found that *Aphenilus asychis* parasitoids of R. *padi* aphids had lower mass when aphids were reared on E+ tall fescue relative to E− tall fescue. In later experiments, Bultman et al. (2003) showed that the survival and development time of the parasitoid M. *hyperodae* decreased and slowed, respectively, when its host, the Argentine stem weevil, another generalist herbivore, was reared on two cultivars of perennial ryegrass. A third infected cultivar had no effect on parasitoid development or survival. This is the same system as Barker and Addison (1996, 1997) (above), who found that the parasitoid can learn to search more efficiently for E+ and E− hosts. As noted above, Omacini et al. (2001) showed similar results in a field experiment with Italian ryegrass. Infection in Italian ryegrass reduced the rate of parasitism by secondary parasitoids (table 3.1). Similarly Jani (2005) showed that high-alkaloid, infected sleepygrass apparently provide enemy-free space (*sensu* Jeffries & Lawton 1984) for herbivores, and relative abundances of herbivores increase on these grass hosts.

- Alkaloids may also be transmitted via food webs to generalist predators. de Sassi et al. (2006) found that development time of generalist ladybird beetles (*Coccinella septempunctata*) increased, while fecundity and survival decreased, when feeding on generalist cereal aphids (R. *padi*) feeding on E+ perennial ryegrass.
- All of the above studies involved herbivores that are generalists. We should expect the effects of alkaloids should be even more pronounced for natural enemies of specialized insect herbivores. Specialists often evolve not only to detoxify plant toxins, but to require them in their diets, use allelochemicals for host location and oviposition cues, and sequester them as defenses against natural enemies (e.g., Barbosa and Letourneau 1988, Bowers 1990). We know of no studies of the effects of endophytic alkaloids on natural enemies of specialized herbivores of grasses.

The picture gets even more complicated when other interacting species are involved. Entomopathogenic nematodes have their own symbiotic bacteria, which when injected into the host insect, multiply and serve as food for the offspring of the nematodes. Richmond et al. (2004b) experimentally examined survival of fall armyworm larvae reared on E+ and E− perennial ryegrass and exposed to the nematode *Steinernema carpocapsae* or injected with just its symbiotic bacteria, *Xenorhabdus nematophila*. Larvae fed on E+ grass survived better than those fed E− in both instances. Thus Richmond et al. (2004b) concluded that alkaloids used by the herbivore "effectively turns the tables on both plant and natural enemy." Endophytes and their alkaloids also negatively affected the effectiveness of the same nematode on another host, the black cutworm (*Agrotis ipsilon*), by also interfering with the symbiotic bacteria of the nematode (Kunkel & Grewal 2003, Kunkel et al. 2004). Once again, this interaction would thwart the defensive mutualisms of the endophytes with their host grasses, at least as related to defense against herbivores.

Do endophytes and their alkaloids act as defensive mutualists of plants, given that herbivores can "turn the tables" and perhaps consume their infected hosts with relative impunity? The simple answer is that there is not a simple answer. Most of the studies to date have involved relatively simple laboratory assays of a single herbivore species with one or two natural enemy species, and fed E+ and E− agronomic grasses, with limited genetic (and thus alkaloid) variability. The question has yet to be tested rigorously in natural communities, where multiple herbivore and natural enemy species, including microbial ones, form complex food webs on a mosaic of grass hosts that vary enormously in endophyte and host genotype, and thus in traits such as alkaloid types and levels. Omacini et al.'s (2001) elegant study serves as a caveat that considering even relatively simple food webs on one strain of endophyte in a single grass cultivar produces enormously complex effects of infection for some species and trophic levels (table 6.1, figure 6.2). In natural communities, the increased complexity of food webs, amplified genotypic variation in the endophyte and host, and environmental fluctuations will magnify the possible outcomes of endophyte infection in grasses at the community level (figure 6.3). Moreover, as discussed in chapter 3, herbivory may increase seed output and thus increase the fitness of the endophyte, perhaps at the expense of the host's fitness. Perhaps a better question is, given the variability in endophyte, host, and the abiotic and biotic (e.g., plant competition) environment in natural communities, does endophyte infection make a difference in the long-term fitness of the host? This question will be a challenging one for ecologists to address since it involves long-term studies and experiments in complex communities. However, as we have seen above, researchers studying endophyte-plant-herbivore-natural enemy interactions are already beginning to incorporate this natural complexity into long-term observational and experimental studies.

6.3.1 More interacting plant species, more variation in endophyte-host outcomes

We have already seen that endophytes often have complex and unexpected outcomes on herbivory when other interacting species are considered (e.g., Richmond et al. 2004b). In natural communities, infected grass hosts do not exist in monocultures of other infected and uninfected grass hosts, but instead are typically surrounded by other interacting plant species. These other plant species also modulate and may reverse the expected outcome of grass-endophyte interactions and effects at higher trophic levels. For example,

Agronomic Grass Populations   **Wild Grass Populations**

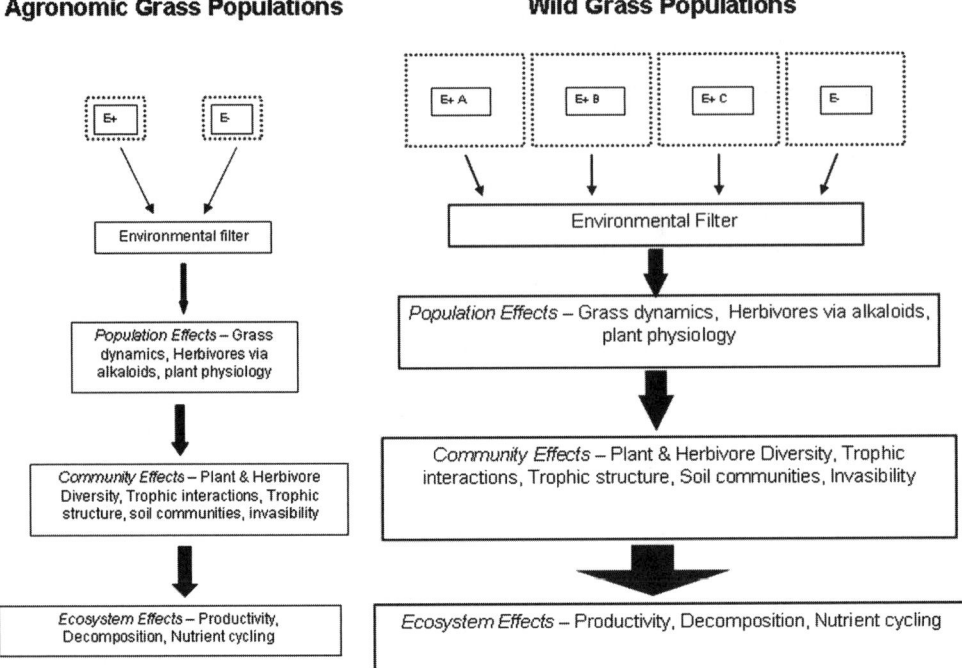

Figure 6.3 Relative variability in endophyte effects on population, community, and ecosystem patterns and processes in agronomic and wild (natural) grass populations. Variation in endophyte haplotypes (A, B, C) is represented by multiple solid boxes embedded within hosts that differ genetically (the dashed boxes, with the size of the dashed boxes representing host genetic variability). For the wild grass population, three different endophyte haplotypes, plus the E– plants are shown, while agronomic grasses typically harbor only one endophyte haplotype. Because wild grass populations are mosaics of endophyte-infected (with multiple host and endophyte combinations) and uninfected grasses in a matrix of other plant species, and environments are inherently more variable than agroecosystems, variation becomes magnified at each higher organizational level (increasingly thicker arrows).

Lehtonen et al. (2005b) found that endophyte-produced alkaloids can also be sequestered by hemiparasitic plants. In experimental studies of *L. pratense* infected with *Neotyphodium uncinatum*, they showed that a root hemiparasitic plant (*Rhinanthus serotinus*) acquires alkaloids from the host grass. The purloined alkaloids increase resistance of the hemiparasitic plant to the generalist aphid *Aulacorthum solani*. The mutualistic endophyte became parasitic in the presence of the hemiparasite, as indicated by reduced growth of the infected grass. Thus the presence of another plant species, in this case a root hemiparasite, can reverse the interaction of the endophyte with its host grass. Had the grass-endophyte interaction been studied in isolation, very different conclusions about the direction of interaction between the endophyte and host grass would have ensued.

Although not yet well studied, we should also expect that other types of plant-plant interactions alter the effects of endophytes at the community level. Intra- and interspecific plant competition alters the nutrition, growth, and allelochemistry in plants, and thus resistance to herbivores (Agrawal 2004, Karban et al. 1989). Neighboring plants can also cause associational resistance, where neighboring plants affect resistant traits of a focal plant or attract natural enemies that reduce herbivory (Agrawal 2004, *sensu* Root 1973). Plant competition may also alter induced and constitutive chemical defenses (e.g., Cipollini 2004, Karban & Baldwin 1997). It is yet untested how plant competition may influence the ability of infected plants to produce constitutive levels of alkaloids or those induced by herbivore damage (e.g., Bultman et al. 2004, Bultman & Ganey 1995). These effects of neighboring and competing plants in natural communities, well known from the plant-herbivore literature, should produce highly variable and conditional outcomes of grass-endophyte interactions.

Complicating species interactions are just not limited to other plants, herbivores, and consumers of herbivores. As we saw in chapter 3, other plant microbial partners, such as mycorrhizae, can influence the effects of endophytes on their hosts. In a recent paper, Márquez et al. (2007) showed that a virus is a necessary partner for the endophyte *Curvularia protuberata* to increase tolerance to high soil temperatures in *Dichanthelium lanuginosum*. When infected grasses were cured of the virus experimentally, heat tolerance to geothermal soils was lost. However, heat resistance could be restored by experimentally reintroducing the virus, providing strong support that the mutualistic effects of the endophyte were dependent on the presence of the virus. *Curvularia* is thought to be horizontally transmitted, although Márquez et al. (2007) and Redman et al. (2002) indicate it is still tightly associated with host grasses, so it is yet unclear how viruses may interact with systemic, seed-borne endophytes in cool-season grasses. Nevertheless, Zabalgogeazcoa et al. (2001) found that two viruses in the families Totitviridae and Narnaviridae often infect isolates of *Epichloë festucae* from wild populations of *Festuca rubra* from dehesa habitats (semiarid warm savannahs) in Spain. This poses the intriguing possibility that viruses may be involved in the drought- and heat-resistance properties conferred to some grasses by their systemic endophytes.

6.4 COMMUNITY AND ECOSYSTEM EFFECTS

6.4.1 Invasibility and productivity

Diversity in communities has long been thought to increase stability in terms of ecosystem properties such as productivity and nutrient retention (Tilman et al. 1997) and cycling (Cardinale et al. 2002, Heemsbergen et al. 2004, Rudgers et al. 2004). Another consequence of increased diversity in communities may also be resistance to invasion by alien species (Levine & D'Antonio 1999). In theory, more diverse communities harbor more redundant and complementary species that fill available resource bases and niches, thus inhibiting invading species from gaining footholds (e.g., Heemsbergen et al. 2004, Loreau et al. 2001). However, empirical tests of this concept are conflicting, with diverse communities showing either less (e.g., Stohlgren et al. 2003) or more resistance to invasion (Levine & D'Antonio 1999).

Rudgers et al. (2004) hypothesized that fungal symbionts influence invasibility via changing relationships between diversity and ecosystem functions. They first developed a graphical model relating endophyte infection to diversity and ecosystem function. From this model they predicted that endophytic infections reduce the strength of the correlation between diversity and ecosystem function, either in terms of productivity or invasibility. Basically, in communities where endophyte-infected grasses are present, other plant species achieve lower biomass because of the enhanced competitive properties of E+ relative to E− grasses. However, Renne et al. (2004) showed that, contrary to Rudgers et al. (2004) where Kentucky 31 tall fescue is considered invasive to native plant communities due to competitive (Clay & Holah 1999) or allelopathic effects (Sutherland et al. 1999, Matthews & Clay 2001), infected Kentucky 31 tall fescue had negative allelopathic effects on germination and seedling growth of only 1 of 13 native tallgrass prairie species. They concluded that "inhibitory effects are limited." Likewise, Spyreas et al. (2001), in a correlational study, found no negative effects of Kentucky 31 tall fescue on plant diversity.

Rudgers et al. (2004) used long-term experimental plots sown with E+ and E− agronomic (Kentucky 31) tall fescue in old fields in Indiana that had been used in other studies (e.g., Clay & Holah 1999, Clay et al. 2005, Finkes et al. 2006, Lemons et al. 2005, Matthews & Clay 2001). The presence of the endophyte altered the relationship between diversity and productivity. Diversity and biomass were negatively correlated in endophyte-free plots, but not correlated when in E+ tall fescue plots. The endophyte also altered invasibility of tall fescue, at least as measured by aboveground biomass. The biomass of tall fescue was more negatively correlated with diversity when the endophyte was absent than when it was present. Rudgers et al. (2004) concluded that endophytic fungi "may contribute more to the dynamics of communities than previously thought" because many of the studies relating diversity to productivity involve grasses that are known to be infected by systemic endophytes (Rudgers et al. 2004: table 2).

Rudgers et al. (2004) noted that "few experimental studies have been conducted on non-agronomic grasses." We still know very little about how infection in native grasses alters plant, animal, and microbial diversity (see section 6.2), and more relevant here, the diversity-productivity relationships in communities or their invasibility. Even studies of agronomic grasses dispute the invasiveness of infected cultivars such as Kentucky 31 tall fescue and their effects on plant diversity. Clay et al. (2005) found that biomass of the infected cultivar (Kentucky 31) was enhanced by herbivores and concluded that herbivore pressure should make Kentucky 31 tall fescue a more potent invader of novel habitats. However, Spyreas et al. (2001), for example, showed that infected tall fescue was positively related to plant species diversity and did not decrease local diversity to a greater extent than E– tall fescue. Spyreas et al. (2001) concluded that E+ tall fescue is not a "supergrass, unencumbered by herbivory [and] outcompeting all other palatable neighbors," and is mutualistic in a narrow ecological range. Thus it is unclear even for Kentucky 31 tall fescue, with limited genetic variability in the host and endophyte strain relative to tall fescue in its native habitats (Clay 1993, Piano et al. 2005, Saikkonen 2000), whether infection alters community diversity and invasibility. We should therefore expect that in natural communities, underlying variation in plant and endophyte coupled with fluctuations in the environment will lead to many possible outcomes involving endophyte effects on diversity, productivity, and invasibility of communities (figure 6.3).

6.4.2 Community change—succession

We know of only one direct and experimental test of whether endophytes in grasses alter community succession—the changes in species composition over time. Rudgers et al. (2007), using the same experimental plots set up by Clay and Holah (1999) and an additional set of bottomland plots seeded with E+ and E– tall fescue (Kentucky 31) in Indiana old fields, measured tree colonization, tree growth, and species turnover over a 12-year period. Woody plant succession is a typical outcome of old field succession in the midwestern United States. Tree abundances and sizes were generally reduced in E+ plots relative to E– ones. Furthermore, tree composition was different in E+ and E– plots, with relative abundances of certain tree species such as box elder, silver maple, and red osier dogwood more negatively affected in E+ plots than other species. In E+ plots, voles increased predation on seedlings of trees because E+ tall fescue is unpalatable to voles (Clay et al. 2005) and therefore tree seedlings are targeted. Rudgers et al. (2007) concluded that endophyte infection could "alter future forest composition" by selectively feeding on certain tree species and thus inhibiting their establishment and growth and reducing species turnover. Notably, voles also preferred several native species over the nonnative white mulberry, and thus may also promote dominance of nonnative species at the expense of native trees.

Whether systemic endophytes have similar effects on successional processes in other agroecosystems and natural communities is uncertain. We have seen (section 6.2.1) that even for agronomic tall fescue, plant diversity in old fields infected with *Neotyphodium* may increase rather than decrease (Spyreas et al. 2001), and infected tall fescue may (e.g., Orr et al. 2005) or may not (e.g., Renne et al. 2004) inhibit the growth of other species. Furthermore, the negative effects of agronomic grasses on native plant species may be direct through allelopathic effects (Orr et al. 2005) or indirect by altering soil properties (Matthews and Clay 2001), rendering other plants more susceptible targets for herbivory (Rudgers et al. 2007) or altering mycorrhizal associations of neighboring plants (Omacini et al. 2006). Therefore it is still unclear whether agronomic tall fescue has consistently negative effects on plant succession in old fields in areas of introduction and what the mechanisms are by which it suppresses or alters succession.

For native grass populations and their associated community, there have been no tests of how infection by endophytes may alter the course of plant succession. We have seen that the frequency of endophyte infection in natural populations of Arizona fescue is negatively correlated with plant diversity (section 6.2.1, figure 6.1), but the causality underlying this relationship is not clear. It is also not known if similar patterns might occur in other wild populations and communities containing endophyte-infected grasses. Clearly this is an area for future research endeavors.

6.4.3 Decomposition and nutrient cycling

Although endophytes in grasses inhabit the aboveground tissues, they have the potential to influence belowground ecosystem processes, such as rates of decomposition and nutrient cycling (Bernard et al. 1997, Franzluebbers et al. 1999, Matthews & Clay 2001, Mayer et al. 2005, Omacini et al. 2004). As inherited components of some grasses, endophytes may be akin to genetic and phenotypic variation in plants that ripples through communities and trophic levels to affect ecosystem-level processes (e.g., Whitham et al. 2003). In terms of nutrient cycling, endophytes may increase phosphorous and nitrogen uptake in nutrient-poor soils (e.g., Malinowski et al. 1999b, 2000) (see section 2.2.2).

Alternatively, fungal endophytes, at least in agronomic grasses, may inhibit mycorrhizae (Bernard et al. 1997, Chu-Chou et al. 1992, Guo et al. 1992, Müller 2003, Omacini et al. 2006) and endophytes may thus reduce nutrient uptake via mycorrhizae. Alternatively, a recent study of a native grass, *Bromus setifolius*, in Argentina showed that grasses infected by *Neotyphodium* had increased root colonization by arbuscular mycorrhizae and increased growth relative to E– grasses (Novas et al. 2005). Also, E+ agronomic tall fescue indirectly alters soil properties, which in turn modifies the success of other plant species either through allelopathic mechanisms (e.g., Sutherland et al. 1999), altered soil pathogens, or changes in available resources (Matthews & Clay 2001).

In terms of decomposition, infection in agronomic E+ grasses often retards decomposition rates, similar to ecosystem-level consequences of genetic variation among cottonwood trees (LeRoy et al. 2006). Omacini et al. (2004) demonstrated with litter bag experiments that E+ *L. multiflorum* (Italian ryegrass) decomposed slower than a co-occurring and uninfected grass, *Bromus unioloides*, under a range of soil moisture conditions. Furthermore, E+ Italian ryegrass changed the decomposition environment such that *Bromus* decomposed more slowly under a mat of E+ ryegrass compared to E– ryegrass. Similarly, Lemons et al. (2005) found that decomposition was slower in E+ agronomic tall fescue plots relative to E– plots in an experimental old field in Indiana. The composition of the soil invertebrate community shifted, especially in the Collembola (springtail communities). In contrast to these studies, Mayer et al. (2005) found that decomposition rates of old fields in Oklahoma were significantly and positively related to the relative proportion of infected tall fescue. They concluded that macroinvertebrate decomposers were attracted to the better litter quality of tall fescue, a C_3 species, relative to other forbs and C_4 grasses in the old field, despite tall fescue's toxicity. It seems, therefore, that the relative decomposition rates and effects on soil invertebrate communities may depend upon local plant species composition and soil environments.

We are unaware of any studies of decomposition rates of infected native grasses in natural communities. However, given the conflicting results even from agronomic tall fescue studies, it would not be surprising that the effects of endophytes in natural communities on nutrient cycling and decomposition rates are multifarious and context dependent. Certainly such studies are needed in natural communities before any broad conclusions of the effects of endophytes on decomposition and nutrient cycling can be made.

6.5 CONCLUSIONS

Endophytes can be viewed within the context of community genetics, where relatively little genetic variation at one trophic level may ripple through the community to affect community structure and diversity, trophic dynamics, and ecosystem functions and processes. However, the effects of variation due to just maternally inherited endophyte infection at the primary producer level may be particularly difficult to trace to other community trophic levels and to ecosystem function in natural communities due to their inherent complexity. So far, nearly all studies of endophyte effects at the community and ecosystem level have used agronomic tall fescue, especially the Kentucky 31 variety, and agronomic perennial ryegrass, and to lesser extent, annual ryegrass. The former two may make relatively poor model systems because they fail to capture the remarkable variation found in natural communities and ecosystems (Saikkonen et al. 2006). Even within these systems, nearly every study of the effects of endophytes on plant and consumer diversity, trophic structure and dynamics, productivity, invasibility, decomposition, and nutrient cycling has a counterexample with opposite conclusions. These agronomic grasses are basically crop plants, much like rice, corn, or wheat, except they have been specifically selected or bred for lawns and pastures, with limited genetic diversity in host and endophyte that may not accurately reflect the diversity in endophyte-host symbioses in complex natural communities (e.g., Clay 1993, Saikkonen et al. 2006). Studies in these systems have provided an essential foundation and critical insights into the interactions of endophytes, host grasses, herbivores, and natural enemies, as well as how infected grasses, when naturalized or invasive, have dramatic effects on the surrounding community structure, diversity, and ecosystem processes. However, just as ecologists and evolutionary biologists would hesitate to extrapolate from agricultural communities and ecosystems involving selectively bred crops to natural communities, it may be dangerous to do likewise for endophyte-host grass systems.

We argue that attention should now turn to natural communities to address whether repercussions of infections and phenotypic variation in host and endophyte at the primary producer level can be detected in community and ecosystem structure and function (see chapter 7). Natural communities are mosaics of E+ and E− plants, both of which vary in host genotypes, embedded in a matrix of other interacting plant species (Faeth 2002, Faeth & Fagan 2002). Moreover, E+ plants within a community often vary in endophyte haplotype, and some endophyte genomes are of hybrid origin (Sullivan & Faeth 2004) (see chapter 4). Endophyte and plant genotype alter phenotypic traits of the host, ranging from morphology to alkaloid chemistry, which in turn may alter within- (intra- and interspecific competition) and between- (herbivory, predation, and parasitism) trophic level patterns and processes. The answers derived from natural communities are likely to be much more divergent and less predictable than those from agronomic grass systems (figure 6.3), but this is the nature of complex communities and ecosystems. The community genetics concept may provide a good starting point because many systemic endophytes are an inherited, maternal component of endophyte-grass symbioses and the phenotypic traits that endophytes affect, such as alkaloid variation within the host population, can be traced through communities and ecosystems.

7

Future Directions

Throughout this book we have emphasized how systemic fungal endophytes and their host grasses provide useful symbiotic systems to test contemporary ecological and evolutionary theory. Cool-season grasses form the productivity base for many ecosystems and agroecosystems, and together with their endophytes, can impact the surrounding community of competing plants, herbivores, and their consumers. Because these grasses often provide forage for grazing animals, the grass-endophyte symbiosis has tremendous agronomic importance, as evidenced by the plethora of studies on endophyte-infected tall fescue and perennial ryegrass. Knowledge of the basic ecology and evolution of endophytes and their hosts is needed to improve our understanding of fundamental symbiotic interactions and the coevolutionary process. In addition, the continued application of the knowledge gained from endophyte research to agronomy and other applied fields will depend on the data generated from a diversity of scientific approaches to investigating ecological and evolutionary questions. In this chapter we hope to explore some of the many questions that remain, suggest future avenues for research, and encourage the application of endophyte research in a diverse set of biological subdisciplines.

7.1 NEW INFORMATION AND FUTURE RESEARCH

7.1.1 Extensions to other grass-endophyte systems

Our current understanding of the biology of the grass-endophyte symbiosis has been greatly aided by the numerous studies of the two agronomically important forage species, *Lolium arundinaceum* (tall fescue) and *Lolium perenne* (perennial ryegrass) and their *Neotyphodium* endophytes. However, much of the conceptual framework for thinking about grass-endophyte interactions has predominantly been based on the behavior of these two host species in agricultural or former agricultural settings (e.g., old fields), or in controlled environments (Malinowski & Belesky 2006). More recently, new insights on the invasiveness of infected agronomic grasses into surrounding plant communities have emerged (e.g., Rudgers et al. 2007). Due to the inherent difficulties of extrapolating from just a few model systems, some of the "conventional wisdom" about the nature of the symbiosis may need to be reexamined as new information becomes available from inherently more complex natural systems (Saikkonen et al. 2006).

Although research on grass endophytes has greatly accelerated in the past several decades, many studies continue to use agronomic grasses as models and test systems. There is still a paucity of grass-endophyte studies in wild grass populations and natural communities, and these studies continue to lag far behind those involving agronomic grass cultivars (figure 7.1). Yet, some key questions concerning coevolutionary processes, multitrophic interactions, and community diversity (see below) can only be adequately addressed in natural grass populations and communities.

In addition to tall fescue and perennial ryegrass being forage species important in many ecosystems, other similarities exist, including the tight phylogenetic relationship of these two species, which are now included in the same

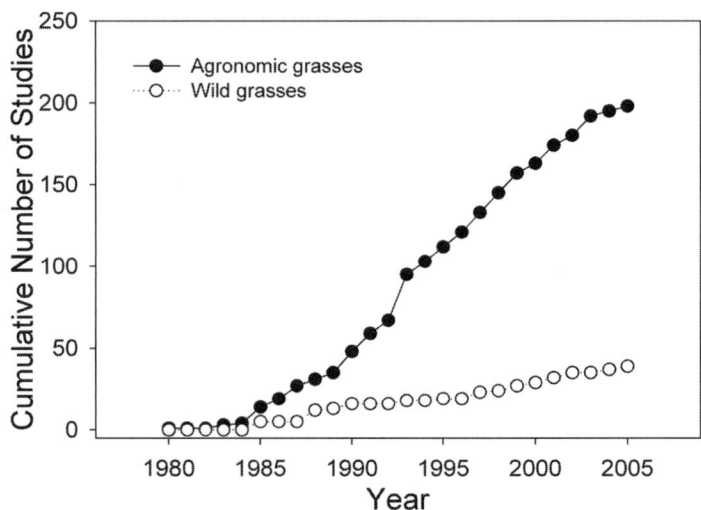

Figure 7.1 Cumulative number of endophyte studies by year involving either introduced agronomic grasses or wild (native, nonagronomic) grasses through 2005. Figure reprinted from Faeth and Saikkonen (2007) with permission of the New Zealand Grassland Association.

genus (*Lolium*), and the fact that both are infected by related endophytes in the same genus (*Neotyphodium*) that are asexual, vertically transmitted fungi that do not occlude host sexual reproduction. Thus these two grass-endophyte systems constitute a "type 3 [III] association" (White 1988), which may represent only one point along the coevolutionary axis from antagonism to mutualism. Even within type III associations, traditionally considered as highly mutualistic, accumulating evidence indicates wide variability in the interaction contingent upon host and endophyte genotype and environment.

Clearly it would be desirable to perform more field studies on other grass-endophyte systems in a more diverse set of habitats. Studies of native grasses in their natural environments may shed more light on the relative effects of endophytic fungi on their associated community. Research efforts devoted to the native grasses *Festuca arizonica* and *Achnatherum robustum* and their genetically complex *Neotyphodium* endophytes (Faeth et al. 2002b, 2004, 2006, Faeth & Hamilton 2006, Faeth & Sullivan 2003, Sullivan & Faeth 2004) exemplify the type of ecological research needed on other grass-endophyte symbioses.

Perhaps because research in the past has largely focused on less complex agronomic grasses and their endophytes, there has also been a tendency to consider grass endophytes, especially type III associations, as special cases and within an ecological conceptual framework separate from other microbial-plant interactions (Saikkonen et al. 1998, 2004). For example, type III endophyte-grass interactions have sometimes been deemed as predominantly mutualistic. As noted above and throughout this book, we now know that endophyte-grass interactions are highly variable and contingent upon specific host genotype and endophyte haplotype combinations and environmental factors, just like other symbiotic interactions. As we have emphasized throughout this book, endophyte-grass interactions are best viewed within existing ecological and evolutionary concepts of microbial-plant interactions. While this might tarnish slightly the often cited uniqueness of the endophyte-grass symbiosis, we believe this broader context provides a more realistic view and greatly expands research opportunities.

It is also clear from the foregoing paragraphs that most endophyte research has been devoted to *Neotyphodium* spp. and their associated host grasses. Other grasses with endophytic fungi, such as *Atkinsonella hypoxylon* (Clay 1984, 1994a, Kover & Clay 1998, McCormick et al. 2001) and various *Epichloë* spp. (Brem & Leuchtmann 2002, 2003, Bucheli & Leuchtmann 1996, Groppe et al. 1999, Meijer & Leuchtmann 2000, 2001, Pan & Clay 2002, 2003, Zabalgogeazcoa et al. 2006a,b), have been examined, but by a smaller pool of researchers. A glance at table 1.2 shows a fair variety of endophyte species that can infect a wide variety of grass genera, not all of which were included in this table. Some of these fungi, such as *Nigrocornus*, are taxonomically very widespread (it has been recorded on 29 grass genera) (Ryley 2003), yet little is known about its potential effects on the ecology and evolution of its many hosts. New grass-endophyte systems are still being described (e.g., Wei et al. 2006) and undoubtedly many more remain to be discovered.

7.1.2 Do endophytic fungi really matter?

Meta-analysis of the primary literature has revealed that, while endophytes do tend to enhance host resistance to herbivores, their overall effects on plant performance and competitive ability are markedly inconsistent (Saikkonen et al. 2006). Even in well-studied agronomic grasses, endophyte effects depend greatly on host genotype and environmental conditions, as discussed earlier in chapters 2 and 4. The environmental and genotypic contingency of many endophyte-mediated effects may make it especially difficult to adequately characterize the significance of endophytes to the evolutionary ecology of host populations. Nevertheless, some real, community-level consequences to the presence of endophyte-infected grasses have been described for some ecosystems, as discussed in chapter 6.

As it has become increasingly clear that a plant in nature is more accurately described as a "merger of fungal cells with plant tissues" (Wilson 1993), with many cellular and tissue components of fungal origin (Astatt 1988), it has become important for ecologists to recognize the ubiquity of fungal symbionts (Carroll 1988) and their potential to impact biological properties at all levels of the ecological hierarchy from individual through ecosystem. Wilson (1993) maintained that, while fungal endophytes are often clandestine and out of sight, they should not be out of mind. But how important are endophytes to their host populations and the surrounding community? Do endophytic fungi really matter?

These two important questions are not likely to ever be satisfactorily answered by any single study. Although specific endophyte-mediated effects on host growth or reproduction may be readily documented in a controlled environment such as a greenhouse, it is quite another matter to extrapolate the results to the complex field situation. Likewise, it is notoriously difficult to extrapolate from the results of greenhouse studies of competition to predict performance in a natural community of competing plants.

To better show that fungal endophytes really do matter to ecology and evolution in nature, it will be necessary to carefully monitor the dynamics of infected and uninfected segments of host populations growing in a diverse set of habitats. If the habitats differ in particular abiotic factors or in the competitive milieu they present, the resulting variation in the ecological outcome of the symbiotic interaction could be expected to supply part of the raw material for coevolution (Thompson 1999, 2005). The ecological outcome may be measurable as endophyte-mediated changes in specific organismal features (e.g. leaf area, tiller, or panicle production) that translate into changes in the vital rates of the host population. These vital rates should be metrics important to population growth, fecundity, or mortality (Gibson 2002). One important goal for future research is to establish the link between organismal features of the host species that are known to be affected by endophyte infection and the vital rates of the host population over broad geographic ranges.

To extend the question of whether or not fungal endophytes really matter, more investigations at the community and ecosystem levels will be required. Several examples of endophyte effects at these levels were presented in chapter 6. Nevertheless, the potential impact of endophytes on biodiversity, productivity, decomposition and nutrient cycles, pathogens and other fungi, and herbivores and their predators have only been studied in a few ecosystems, and the majority of these are in agroecosystems, former agroecosystems (old fields), or for agronomic species that have invaded surrounding plant communities (Faeth & Bultman 2002, Finkes et al. 2006, Lemons et al. 2005, Omacini et al. 2001, Orr et al. 2005, Rudgers & Clay 2005, Rudgers et al. 2004). These studies have provided the basis for understanding how endophytes may have community- and ecosystem-wide effects. However, it is still too early to say if broad-scale impacts of endophytes are common to ecological communities in managed, invasive, or natural ecological communities. We have seen in chapter 6 that limited studies to date with native grasses in natural communities suggest that community and trophic effects of endophytes may vary from those of agronomic grasses in managed agroecosystems, old fields, and invasive situations in natural communities.

One other important, but often overlooked consideration relevant to the question of whether endophytes really matter is that systemic endophytes in grasses constitute only a small fraction of endophytes in grass hosts overall. Cool-season grasses resemble woody plants (e.g., Arnold et al. 2000) in terms of the diversity and abundance of horizontally transmitted endophytes that form local infections in leaves, stems, culms, and even seeds (Hamilton & Faeth 2005, Nan & Li 2001, Schulthess & Faeth 1998). The ecological role of most of these more weedy and opportunistic fungi is unknown, but some may act as plant mutualists (e.g., Arnold et al. 2003, Márquez et al. 2007, Redman et al. 2002), pathogens (Hamilton & Faeth 2005), or saprophytes (Faeth & Hammon 1997) during some life cycle stages. Yet studies focusing on systemic endophytes in grasses usually ignore these other diverse and abundant endophytes and how systemic endophytes may interact with their locally infective counterparts, except in the few cases where they become systemic pathogens of grasses (Wäli et al. 2006). If we are to understand whether and how endophytes matter, then it seems prudent to begin including the nonsystemic endophytes in grasses as well as the systemic ones.

7.1.3 Host biology

In chapter 2, many documented effects of endophytic fungi on growth, photosynthetic physiology, and reproduction of their hosts were presented. The sections on abiotic stress tolerance (2.2) and competitive ability (2.3) revealed that endophyte-mediated effects are highly contingent on the prevailing abiotic and biotic environment. Multiple inconsistencies among the results of various studies, including those on the same grass-endophyte system (e.g., tall fescue with *Neotyphodium coenophialum*), made it especially difficult to generalize about the ecological impact of endophytes on hosts growing under various stresses such as drought, low mineral nutrients, and competition.

More studies that employ a cost-benefit analysis to endophyte-infected hosts growing under a variety of realistic conditions in the field would improve our basic understanding of the ecology of the grass-endophyte symbiosis. Fungal endophytes need carbohydrates to supply substrates to meet their energy demands during general metabolism (e.g., aerobic respiration). The only sources of these carbohydrates are those manufactured by the host via photosynthesis. Thus at least some fraction of the available photoassimilates must be utilized by the endophyte in infected hosts. This loss to the infected host should be manifested as a cost, perhaps shown at the organismal level by reduced growth or reproduction relative to uninfected hosts under unfavorable conditions (Cheplick 2007, Marks & Clay 2007). Likewise, endophyte production of secondary compounds such as nitrogen-rich alkaloids (Petrini et al. 1992, Schardl & Panaccione 2005) requires elements that can be limiting to other plant functions, such as nitrogen for growth and reproduction. The production of alkaloids, which may provide benefits by deterring herbivores, may exact costs on the host plant in terms of limited nutrients allocated for growth and reproduction (Faeth 2002). This is not different from the assumed cost of plant-based secondary chemical defenses relative to allocation to growth and reproduction (e.g., Herms & Mattson 1992). However, to date, formal cost-benefit analyses have not been performed on most grass-endophyte systems. Furthermore, although possible costs of infection have sometimes been suggested during the interpretation of experimental results (Ahlholm et al. 2002, Cheplick 2004a, 2007, Cheplick et al. 1989, Faeth & Sullivan 2003, Olejniczak & Lembicz 2007), some have maintained that the cost of the endophyte's "service" to the infected host (in a mutualistic symbiosis) is "essentially nil" (Bacon & Hill 1996). Cost-benefit analyses represent an open area for future research on grass-endophyte systems used for a variety of ecologically meaningful experiments. Calculation of the relative interaction intensities (Armas et al. 2004), as described in section 1.3 and presented in figure 1.6 for a diverse set of grass species and their endophytes in different environments, may facilitate comparisons among grass-endophyte systems.

Many basic questions remain regarding the possible effects of endophytes on host development, physiology, and reproductive biology. The mechanisms responsible for the documented effects of endophytes on the growth, physiology, and reproduction of host grasses (described in chapter 2) remain elusive. This is probably because the descriptive work of simply documenting these effects is much easier than determining their mechanisms. Growth-regulating substances such as hormones produced by the fungi (e.g., de Battista et al. 1990) are likely to be involved, as they regulate many developmental processes in plants. Because many studies have reported the results of cumulative growth processes at one point in time, little is known about potential developmental changes induced by endophytic fungi as hosts grow and mature. Researchers should attempt to obtain multiple measurements of the variables of interest in E− and E+ individuals as they develop over time. Recent studies using real-time polymerase chain reaction (PCR) to examine up- and down-regulation of key genes and gene clusters, such as genes for alkaloid synthesis (e.g., Schardl et al. 2006) and photosynthetic pathways, in response to environmental factors such as nutrient availability or herbivore damage offer great promise. For example, Sullivan et al. (2007) showed that Kentucky 31 tall fescue (*L. arundinaceum*) infected by *N. coenophialum* and damaged by herbivores responded after 10 days with increases in messenger RNA (mRNA) for *lolC*, a gene required for loline biosynthesis. The use of real-time PCR, DNA microarrays, and mRNA differential display reverse transcription PCR techniques will likely become increasingly important in unraveling the complex genetic interactions of endophytes with their host grasses over developmental time and as biotic and abiotic environmental factors vary. These techniques are already widely used to elucidate changes in transcription of many genes for specific induced chemical defenses in response to herbivores in other plants (e.g., Schmidt et al. 2005).

Growth of the grass plant entails the reiteration of modules known as phytomers along developing tillers (Nelson 1996). It is presently not known how endophytes might impact the developmental process as phytomers are added and matured on the elongating tiller. Modularity is especially relevant to the evolution of development and life histories, as it can produce new subunits of genetic expression and natural selection (West-Eberhard 2003). Modular growth in the grass plant can permit phenotypic adjustment to changing environmental conditions via the partitioning of resources (Cheplick 2006, Preston & Ackerly 2004). Thompson (2005) speculated that modular organisms "may be able to control their symbiont assemblages by partitioning the symbionts" among body parts or modules. This hypothesis of "symbiont partitioning" could apply to endophyte-infected grasses whereby the fungal hyphae are distributed unequally among phytomers within a tiller and among whole tillers of the same host plant. Newly formed tillers of infected individuals can sometimes be endophyte-free; it is unknown whether or not phytomers within the same tiller can sometimes be endophyte-free. The induction of resistance to further endophyte infection of new tillers or phytomers by a developing E+ host may be an adaptive plastic response if endophytes reduce reproductive fitness (e.g., choke disease). Such types of adjustment, if they occur, would have to be accomplished at the modular level.

Another aspect of basic host biology that has been neglected in much grass-endophyte research is the potential effect of the symbiotic fungi on the ability of the host to respond phenotypically to environmental changes (i.e., phenotypic plasticity) (Schlichting & Pigliucci 1998). If endophytes can change the expression of phenotypic variation among host genotypes, then microevolutionary processes such as natural selection could differentially affect the infected and uninfected segments of the host population (section 4.2). In theoretical models, allowing for phenotypic plasticity in host-pathogen interactions results in the evolution of reduced pathogen virulence (Taylor et al. 2006). The role of phenotypic plasticity of both host and endophyte in population microevolution has not been similarly analyzed with theoretical models. Nonetheless, in an experimental study of 12 genotypes of perennial ryegrass (cv. Yorktown III), the plastic responses of host morphology (leaf area, tiller numbers) to changes in soil nutrient levels were found to depend on endophyte infection (Cheplick 1997a). It is presently not known how the impact of endophytes on phenotypic variation in a genotypically variable host might influence ecological and coevolutionary processes.

Closely allied to the developmental concept of phenotypic plasticity is phenotypic integration, which Pigliucci (2004) described as a "network of multivariate relationships among the phenotypic traits that define the morphology and life history of a living organism." Covariation among a set of traits may be an adaptive feature of the life history if it improves growth, survival, or reproduction under some conditions, while also acting as a constraint on further phenotypic evolution (Merilä & Björklund 2004). The potential for endophytic symbionts to alter phenotypic integration among morphological or physiological traits certainly exists, but few studies have yet addressed this issue. If endophytes can generally affect carbon translocation within the plant body, as shown in infected *Glyceria striata* (Pan & Clay 2004), then physiological integration and patterns of resource allocation to diverse functions such as clonal growth, tiller production, and sexual reproduction might be impacted. The possible effects of endophytes on the partitioning of photoassimilates and biomass are not well known (Cheplick 2004b), especially in underinvestigated grass-endophyte systems.

7.1.3.1 Host reproductive biology

As outlined in section 2.1.4, a variety of effects of endophytic fungi on the reproductive biology of host grasses have been reported, ranging from parasitic castration (e.g., choke disease in *Bromus* spp. caused by *Epichloë* spp.) (Groppe et al. 1999) to increased seed production (e.g., tall fescue infected by *Neotyphodium*) (Rice et al. 1990).

For the many grass species infected by fungal endophytes (e.g., *Neotyphodium* spp.) that do not prevent the production of spikelet-bearing inflorescences, endophytic hyphae can be detected within individual flowers, especially within the ovary wall (Sugawara et al. 2004). These hyphae will later be located underneath the seed coat following seed maturation (White et al. 1993) and permit vertical transmission of the endophyte, which will grow into the seedling as the seed germinates.

We do not yet have a good picture of how any endophyte impacts evolutionary fitness within a host population. In part, this is due to the inherent difficulties in forming a reasonable conception of what fitness is in long-lived perennial plants. Genet survival and tillering ability might be considered as indirect components of fitness because both are related to the probability of flowering and to the number of flowering culms, and perhaps spikelets and seeds, that will be produced (section 5.1.1, figure 5.3).

Seed production, an important component of sexual reproduction, can be viewed as one manifestation of female fitness. For grasses infected by asymptomatic endophytes (e.g., *Neotyphodium* spp.) that do not cause inflorescence abortion, infection can sometimes be correlated with increased seed production (Hesse et al. 2003, 2004, Rice et al. 1990). However, it is presently not clear if improved seed maturation would necessarily result in greater seedling establishment in natural communities.

The astute reader may have noticed that in the consideration of the potential impacts of endophyte infection on the sexual reproduction of their hosts (section 2.1.4), no reference was made to the male component of sex allocation. This is because virtually nothing is known about this particular aspect of the grass-endophyte symbiosis. Although experimental tests of sex allocation theory (Charnov 1982) have been conducted on male and female functions in some plants (Campbell 2000, de Jong & Klinkhamer 2005), potential endophyte-related effects on male fitness components in grasses have not been extensively explored. Male fitness gains are expected to be related to increases in allocation to male function (Campbell 2000). In grasses, pollen production is typically correlated with anther length, which can vary among plants and among floret positions within spikelets (McKone 1989). As predicted by sex allocation theory, in wind-pollinated species such as grasses, sex allocation may be significantly male-biased (McKone et al. 1998). When endophytic hyphae are exclusively vertically transmitted via seeds (but not pollen), one would predict that selection on the endophyte would favor a reduction in allocation to male function and a greater allocation to female function. This prediction appears to hold for the asexual, maternally transmitted *Wolbachia* symbiont that infects many invertebrates (e.g., Werren & O'Neill 1997). Infection by *Wolbachia* reduces male function, increases feminization, and reduces male:female sex ratios of the host (e.g., Werren & O'Neill 1997). Note that this contention assumes a trade-off exists in allocation to male versus female function, which is not necessarily true in all plants (Campbell 2000). Nevertheless, it should not prove especially difficult to quantify the relevant parameters in the inflorescences of endophyte-infected grasses (for comparison to uninfected hosts). Given the modular construction of flowers, it may be best to approach the study of allocation (and integration) of floral functions from a hierarchical perspective (see Armbruster et al. 2004, Tomimatsu & Ohara 2006).

We know of only one test of allocation to male and female reproduction in infected and uninfected grasses. Hamilton (2002) compared flower, pollen, and seed allocation in four E+ and five E− Arizona fescue (*F. arizonica*) plants from the same wild population in northern Arizona. Anther length did not differ between E+ and E− plants. However, mean pollen counts were significantly greater for E+ relative to E− plants, contrary to the assumption that anther length alone is correlated with pollen production. More importantly, E+ plants did not allocate more to female function as predicted. Neither mean flower production nor the ratio of seeds to flowers differed between E+ and E− plants. Given the small sample sizes (four E+ and five E− plants) and that the male and female function were determined in only one growing season of a long-lived perennial grass, these results should be interpreted cautiously. Nonetheless, Hamilton's (2002) study suggests at the very least that the effects of endophyte infection on male and female reproduction allocation may be very challenging to address in wild grasses and may not always support conventional expectations of reproductive allocation in plants.

Because fungal endophytes can affect plant size and architecture (Sullivan & Faeth 2007), it is important to examine possible size- or shape-dependent effects on reproductive allocation (Cheplick 2005, de Jong & Klinkhamer 2005) in endophyte-infected grasses. To date, the allometric scaling of allocation to male and female functions has not been adequately described for any endophyte-infected grass species. It is also unclear if endophytes contribute to the future short- and long-term costs of sexual reproduction (section 5.1.2, figure 5.4).

7.1.4 Coevolutionary outcomes

We saw in chapter 5 that the processes of coevolution—natural selection acting on various symbiotic combinations that are genetically structured because of differential gene flow, genetic drift, hybridization, or random mutations— occur at the population level. These genetic differences, combined with varying local environments, result in a geographic selection mosaic where certain endophyte-host combinations may be more or less fit than others, and coevolutionary hot spots and cold spots appear (Thompson 1994, 1999, 2005). This view contrasts with the long-held notion that asexual endophyte forms, especially type III associations, are typically mutualistic derivatives of type I or II ancestors that are more pathogenic (Clay 1988a, 1990b, Clay & Schardl 2002, White 1988, Wilkinson & Schardl 1997). Coupled with this view is that type II (facultatively asexual) endophytes that become isolated and shift to type III endophytes also become more mutualistic and increase host fitness (Brem & Leuchtmann 2003). Geographic mosaic theory (GMT), in contrast, suggests a much more spatially and temporally dynamic suite of endophyte-host interactions.

Thus, in terms of coevolution, one of the key questions is how dynamic and changeable are type III interactions with their hosts? Are they largely mutualistic, as the traditional view holds, or does the interaction depend on local environments, genetic matches or mismatches between host and endophyte, differential gene flow and drift, and genotype × haplotype × environment interactions within the framework of the GMT of species interactions? This question seems best addressed in natural population and communities where the processes of coevolution have played out, and continue to do so, over evolutionary and ecological time and across geographic space. Agronomic grasses in managed agroecosystems may not be suitable for addressing broad questions about long-term coevolutionary processes and patterns. Communities where agronomic grasses have become naturalized provide ideal opportunities to examine shorter term coevolutionary processes, such as the dynamic interactions of E+ and E− grasses and herbivores and natural enemies of herbivores. However, many agronomic grass communities are relatively recent in origin. For example, Kentucky 31 tall fescue, the object of many ecological studies, has become widespread in North America only since the late 1940s and 1950s, and naturalized thereafter (Buchner & Bush 1979). Moreover, even naturalized agronomic grasses typically lack genetic variation in the endophyte, so questions about coevolution from the perspective of the endophyte may not be possible.

These two contrasting views produce different predictions that are testable. According to the traditional view (e.g., Clay 1990, Clay & Schardl 2002, Schardl & Clay 1997), asexual endophytes should generally be more mutualistic than their ancestral type I or II ancestors. Furthermore, type III asexual endophytes represent local adaptations to their hosts and increase host fitness relative to type I or type II endophytes. GMT predicts more variable outcomes of type III interactions with their hosts, depending on genetic matches or mismatches, gene flow differences, and local environmental conditions. Local adaptation may occur, but local nonadaptation or maladaptation may also occur if genes or traits are mismatched or favorable combinations are disrupted by differential gene flow due to outcrossing among host plants. Traditional views posit that the presence of sexually reproducing (type I or II) endophytes disrupt local adaptation and thus mutualistic interactions. A GMT or metapopulation view suggests that the presence of type I or type II infections may infuse genetic variation and counteract genetic mismatches between hosts and type III infections because asexual endophytes are continuously confronted by a sexually reproducing and thus constantly changing host genotype background (Saikkonen et al. 2004). Indeed, the GMT view suggests that the occasional presence of sexually reproducing and typically disease-causing endophytes may actually be necessary to maintain mutualisms between asexual, seed-borne endophytes and their grass hosts (Sullivan & Faeth 2004, 2007).

Hybridization, which rapidly infuses genetic variation into endophyte genomes, appears commonly in grass endophyte lineages (e.g., Moon et al. 2004, Schardl & Craven 2003). The traditional view of prevalence of hybrids is (1) avoidance of Muller's ratchet or the accumulation of deleterious mutations and (2) infusion of additional genes that increase fitness for hosts of mutualistic asexual endophytes, such as increased alkaloid types or levels that deter herbivores (Schardl & Craven 2003). We saw in chapter 5 that there have been virtually no tests of this hypothesis, and the one of which we are aware (Sullivan & Faeth 2008) has equivocal results (no increases in alkaloids, altered architecture, and slight increases in longevity in hybrid infected grasses relative to nonhybrid infected grasses). GMT provides another explanation. Hybridization may be another mechanism to overcome differential gene flow and the ensuing genetic mismatches between sexually reproducing hosts and asexual, seed-borne endophytes (chapter 5).

Saikkonen et al. (2004) postulated that genetic mismatches may explain the loss of infection and thus the observation that many natural populations are not 100% infected. With each generation of the host there is an increasing probability that an asexual endophyte that once increased host, and thus endophyte fitness, no longer does so because the endophyte-host combination no longer "matches" well. They predicted that genetic mismatching may also explain differences in infection frequencies between (1) pioneering and established older populations of perennial grasses, (2) agronomic and natural grass populations, and (3) perennial and annual grasses. They predicted that for grasses colonizing a new habitat, the seeds produced by outcrossing should have a high frequency of genetic mismatches and incompatibilities, and thus lower infection frequencies, than in older, established populations. In older, established populations, successful host-endophyte combinations persist and can be propagated by asexual cloning. Note that this prediction is opposite to the traditional view that hosts with asexual endophytes should have a consistent advantage over uninfected plants because of their mutualistic attributes. Similarly, high infection frequencies in agronomic pastures may be the result of, first, artificial selection of well-matched host-endophyte combinations throughout selective breeding and, second, maintenance of relatively constant environmental conditions in terms of soil nutrients and moisture where these favorable combinations thrive. Finally, genetic mismatching may explain why relatively few asexual, seed-borne endophytes occur in annual grasses. More than 80% of type III infections occur in perennial caespitose grasses, 16% occur in annuals, and 3% in perennial rhizomatous grasses (Clay 1988a, 1998). Annual grasses often reproduce each generation via outcrossing and can show a higher proportion of polymorphic loci and more allelic diversity than perennial grasses (Godt & Hamrick 1998). High rates of recombination and outcrossing and no vegetative reproduction or inbreeding may prohibit long-term stability of compatible endophyte-host genetic combinations. A corollary of these predictions is that asexual endophytes, if they retain control over host growth and reproductive functions, should shift reproduction to clonal growth or vivipary such that the current favorable combination is maintained. These predictions are fairly easy to test, yet we know of no observational or experimental tests so far, except research showing that at least one endophyte species promotes clonal growth (e.g., Pan & Clay 2003). For example, the prediction of lower infection frequencies due to endophyte-host mismatching in pioneering relative to established populations could be easily tested by examining endophyte frequencies in established core and colonizing fringe populations of the same host grass. It seems imperative that such tests be conducted in the future.

7.1.5 Herbivory

The traditional mainstay of mutualistic interactions of endophytes and their hosts is deterrence of herbivory via the production of endophytic alkaloids. Historically the effects of endophytic alkaloids on herbivores has been determined through bioassay-type experiments, typically using one or two generalist aphid species (e.g., Siegel et al. 1990) or other agronomic generalist pests, such as the fall armyworm (e.g., Clay & Cheplick 1989), or toxicity effects observed in domesticated livestock (e.g., Eerens et al. 1998c). However, as we saw in chapters 3 and 6, alkaloids, like other constitutive or induced plant allelochemicals, can have widely varying effects on invertebrate and vertebrate herbivores depending on whether herbivores are specialists or generalists and the presence of other interacting species. For the former, specialist herbivores often have the capability of detoxifying alkaloid compounds, using them to locate host plants and requiring them for growth and reproduction. For the latter, some herbivores or natural enemies of herbivores apparently sequester alkaloids from their food resources and use them as defense against their own consumers (e.g., Omacini et al. 2001). Alkaloids may also alter competitive hierarchies of multiple herbivore species and promote herbivore coexistence, which may or may not increase the fitness of the host plant (e.g., Härri 2007). The presence of parasitic plant species that steal alkaloids for their own defense could render the grass host less protected against herbivores (Lehtonen et al. 2005b).

Simple bioassays involving choice or no choice tests can provide valuable information on the deterrent and toxic effects of endophytic alkaloid types and levels on a species by species basis. Many alkaloids have strong deterrent or toxic effects on various herbivores, depending on the alkaloid type and level (e.g., Bush et al. 1997, Siegel & Bush 1996). However, these effects are not always translatable to field situations or natural communities where food web complexity prevails, and other interacting species alter the outcome of host grass and herbivore interactions. Furthermore, there is the implicit assumption in studies of herbivory in endophyte-grass interactions that herbivory reduces the fitness of the host and endophyte. Herbivory does not necessarily reduce fitness for cool-season grasses because many grasses are adapted to tolerate herbivory and grazing (Cullen et al. 2006, Faeth 2002) and typically show compensatory responses to defoliation (Ferraro & Oesterheld 2002). Likewise, from the endophyte's perspective, increased herbivory could sometimes increase allocation to seed production, and hence increase the fitness of seed-borne endophytes (chapter 6).

The overarching question for endophyte researchers is not whether alkaloids increase resistance to herbivores, but rather, how the cost and benefits of alkaloids translate into fitness changes for the host and endophyte within the context of other interacting species and variable environments. It is important to note that these fitness changes need not be equivalent, or even in the same direction, for both host and endophyte, even for type III endophytes. Addressing this question will require much more complex and ingenious experiments and observations than simple bioassays due to the inherent complexity of food webs and environments, even in agroecosystems. There is already solid footing for conducting these more complex experiments for both agronomic (e.g., Finkes et al. 2005, Härri 2007, Omacini et al. 2001) and wild grasses (e.g., Faeth & Saikkonen 2007).

Within this main question is an important subset of more specific questions: (1) Are specialist and generalist herbivores differentially susceptible to endophytic alkaloids? (2) What are the metabolic costs and ecological benefits of variable alkaloid types and levels so often found in natural populations and among host grass species? Is there any predictability as to when low or high alkaloid production should be favored by natural selection?

(3) How do alkaloids vary as a function of host and endophyte genotype, environment, and seasonality? How does this variation affect the invertebrate and vertebrate communities and behavior of specific herbivore species? (4) Are alkaloids sequestered by herbivore and natural enemy species and, if so, what are the consequences for herbivore and natural enemy populations and species coexistence? Do endophyte-host combinations that produce high levels of alkaloids provide enemy-free space to herbivores because natural enemies avoid or are deterred on these plants? (5) How do plant competition and plant parasites or pathogens influence the outcome of endophyte-host-herbivore interactions? These questions are not unique to endophyte-host interactions, but are similar to those asked by researchers studying plant-herbivore interactions in general. Once again, we emphasize that it is important to consider endophyte-grass-herbivore interactions as part and parcel of existing concepts of plant-herbivore interactions rather than as a separate and unique biological interactions.

7.1.6 Other interacting species and multitrophic interactions

It is becoming increasingly clear that studying pairwise species interactions in isolation often provides a restricted and sometimes erroneous view of the interaction (e.g., Morin 1999). This appears especially true of endophyte-host interactions, where incorporating other interacting species and higher trophic levels often leads to very different outcomes. For example, the outcome of performance tests of E+ and E− grasses can be reversed when aboveground and belowground herbivores and their natural enemies are added to the mix. The conventional wisdom that higher alkaloid levels lead to reduced herbivory, and thus increased fitness, of the host-endophyte combination may not necessarily hold when predators and parasites of the herbivores or plant parasites are considered (chapters 3 and 6), especially when specialized herbivores that can sequester alkaloids for their own defense are present. The bottom line is that host-endophyte interactions in real-world situations are enormously complex, just like other plant-herbivore or plant-microbial symbiont interactions, and predictions about the direction and magnitude of the host-endophyte interaction become more and more uncertain as other species in the community are incorporated.

Nonetheless, for progress to be made in understanding the ecology and evolution of endophyte-host interactions, it seems imperative to begin studies that include at least part of the complexity found in natural communities. Some of these have been successfully initiated in agroecosystems or former agroecosystems (e.g., old fields), where other interacting species and other trophic levels are easier to manipulate or study due to reduced complexity in these systems (e.g., Finkes et al. 2006, Omacini et al. 2001). These studies have provided a data- and empirical-rich backbone and generate important insights for endophyte-host interactions and their community and ecosystem consequences. Just as studies in agroecosystems provided the foundation for ecological concepts of insect population dynamics (e.g., Price 1984) and food web control (e.g., Mills & Getz 1996), these pioneering studies have done the same for endophyte-host interactions. But the drawbacks are that these systems can be genetically simplistic and the other interacting species are usually a limited subset of those occurring in natural communities. For example, many of the invertebrate herbivores are generalist pests and the vertebrate herbivores are usually domesticated livestock. Interactions in natural communities may be quite different (chapter 6) and typically more complex because of multiple phenotypes of grass-host combinations that exist in metapopulations. The great challenge is initiation of comprehensive studies of endophyte-host interactions that are embedded within multifarious interspecific interactions, patchy mosaics of endophyte-host combinations, and varying environments of natural communities.

7.1.7 The fungal perspective

From the perspective of the endophytic fungus, very little is currently known about microevolutionary processes such as natural selection and gene flow. A thorough understanding of endophyte evolution necessitates investigation of the role of genetic variation in determining fungal fitness and natural selection. To date, there are no standardized measures of fungal fitness (Gilchrist et al. 2006, Pringle & Taylor 2002). Genetic markers, when available, can be used to track the ecological success of different fungal genotypes (Pringle & Taylor 2002) and to estimate gene flow (Sullivan & Faeth 2004).

For many filamentous fungi, an appropriate measure of fitness is the production of asexual spores (e.g., conidia) or sexual spores (e.g., ascospores) (Pringle & Taylor 2002). However, for fungal endophytes, a relationship would need to exist between high spore production and increased likelihood of horizontal transmission, because spores that fail to infect a new host would not survive to produce new hyphae. Ascospores of *Epichloë typhina* are known to mediate horizontal transmission, following ejection from the ascostromata on a host plant, by infecting the florets of neighboring plants (Chung & Schardl 1997a, Schardl & Leuchtmann 2005). If greater ascospore release corresponds to a greater rate of successful horizontal transmission, then fungal fitness in sexual endophytes like *E. typhina* could be estimated by ascostromata size and ascospore production. However, like seed production in plants, spore production is still an indirect measure of fitness and may or may not correspond to contributions to future generations.

Fungal fitness would be quite different for asexual *Neotyphodium* spp. endophytes that are exclusively vertically transmitted as hyphae within host seeds. Here the fitness of the endophyte would invariably be tied to the reproductive fitness of its host. The ability of the endophyte to successfully invade developing ovules, increase seed production or quality (e.g., seed mass), enhance seed germination, and infect the emerging seedling will ultimately determine fungal fitness. To date, we know of no studies that have collectively examined all these aspects of the grass-*Neotyphodium* association.

7.1.8 Molecular ecology and evolution

Great strides have been made in elucidating the phylogenetic and cospeciation patterns of endophytes and grass host species (e.g. Schardl et al. 1997). One of the interesting features of this work is that *Epichloë* endophytes such as *E. festucae* and *E. typhina* and *Neotyphodium* endophytes such as *N. coenophialum* and *N. trembladerae* appear to have an intercontinental distribution either as individual endophytes or as part of hybrid genomes. The distribution patterns and the mechanisms for dispersal across continents (before intentional human introductions) are questions that need to be resolved.

Similarly, phylogenetic patterns of endophyte hybridization across host species are well established (Moon et al. 2004, Schardl & Craven 2003, Schardl & Wilkinson 2000). More recently, the structure and function of individual genes within the endophyte genome that dictate key traits such as alkaloid production (e.g., Schardl & Panaccione 2005, Spiering et al. 2006b) and fungal growth and morphology (Zhang et al. 2006) have also been elucidated. However, there is a still a large gap in our knowledge of endophyte and host genetic variation within and across populations. Since the population level is where species interactions occur and natural selection plays out, it seems imperative to document distributional patterns of endophyte haplotypes within and across populations (e.g., see Wäli et al. 2007). Likewise, there have been very few attempts to correlate endophyte haplotype and genotypes of nonagronomic grasses, except at the species or higher levels. Knowledge of endophyte and grass genotypes at the population level is essential to addressing questions of genetic matches or mismatches that determine geographic variation in the outcome and coevolution of species interactions (e.g., Thompson 2005). This knowledge could also be instrumental in deciding the debate of whether asexual endophytes like *Neotyphodium* are truly type III (only maternally transmitted in the seeds) or may also occasionally disperse horizontally by spores (e.g., Moy et al. 2000). If the same plant maternal genotype harbors different *Neotyphodium* endophytes within or across populations, then this would indicate that *Neotyphodium* is transmitted in other ways besides vertical transmission.

There is clearly a great need for more studies of molecular genetic variation at the population level in wild grasses if we are to understand the coevolution of endophytes and their hosts. The process of sorting host genotypes by natural selection, described in section 5.1.3, is predicated on the ability to distinguish among host and endophyte genotypes. The description of related processes important to microevolution, such as gene flow and genetic drift, also requires information on the underlying genetic structure of infected grass populations. Additional data are needed on the way endophyte and host molecular genetics interact to influence phenotypic variation in traits that are subject to selection in natural ecosystems. Each host genotype with its particular endophytic variant should probably be viewed as a unique entity. The process of genotypic sorting should be expected to differentially impact infected and uninfected segments (or subpopulations) of the host population. Furthermore, not all molecular genetic variation will necessarily be relevant to selection among phenotypic variants unless it has fitness consequences among host genotypes. Precise measurement of the distribution of genetic variants within and among host populations coupled with in situ investigation of basic population ecology under field conditions will be difficult, but important, to future research on the evolutionary ecology of grass-endophyte symbioses.

7.2 APPLICATION OF GRASS-ENDOPHYTE RESEARCH

7.2.1 Global environmental change

Rising levels of greenhouse gases such as atmospheric carbon dioxide (CO_2) and increasing habitat fragmentation are primary components of global environmental change likely to impact biotic interactions in natural communities (Jump & Penuelas 2005, see the collection of papers in Karieva et al. 1993). For example, for a number of grass species, the benefits of mycorrhizal fungi were reduced by CO_2 enrichment (Johnson et al. 2005). The studies that have explored the effect of elevated CO_2 on the growth and physiology of endophyte-infected grasses were reviewed in section 2.2.3. To date, the experimental approaches have been organismal, focusing on the infected host (Hunt et al. 2005, Marks & Clay 1990, 1996 Newman et al. 2003), and the lack of consistency of both CO_2 and endophyte-related effects have made it difficult to extrapolate to natural communities and ecosystems. However, a better understanding of how plant traits such as size and growth rate affect ecosystem processes will be necessary to predict the consequences of global environmental change (Chapin 2003). Reconstruction of historical plant communities (e.g., Clark et al. 2001) suggests that a warmer, drier climate brought about by rising levels of CO_2 would expand the range of C_4 grasses and concomitantly shrink the range of cool-season, C_3 grasses and lead to an increase in endophyte infection frequency, if endophytes impart greater drought resistance to their hosts (section 2.2.1). Stachowicz (2001) predicted that the importance of "habitat-ameliorating positive interactions" would be likely to increase as global environmental stress increases. It remains to be determined whether endophytic fungi will be key biotic factors in the response of grass-dominated communities to future climate change.

The consequences of elevated CO_2 might be expected to extend to other trophic levels. For example, Marks and Lincoln (1996) showed that the relative consumption of tall fescue by the fall armyworm was greater under increased CO_2 and consumption was reduced by infection; however, there was no significant interaction between CO_2 enrichment and infection status. In a later study, Newman et al. (2003) found that elevated CO_2 generally reduced crude protein content in tall fescue, but this reduction was less in infected plants than uninfected plants. They concluded that under scenarios of increased CO_2 and global warming, such changes in plant quality are "likely to have large significant effects on herbivore populations." Unfortunately, in both studies, only one genotype of cultivar Kentucky 31 was used in the experiment, and extrapolation to natural communities with infected grasses, their herbivores, and the consumers of herbivores is not yet possible.

Habitat fragmentation along with global climate change (Quinn & Karr 1993) has critical effects on the dynamics and persistence of plant populations (e.g., Holsinger 2000, Wagenius 2006). For *Bromus erectus*, small-scale habitat fragmentation has been shown to significantly increase the incidence of choke disease caused by *Epichloë bromicola* (Groppe et al. 2001). This effect was probably due to changes in abiotic conditions (light, temperature, soil nutrient availability) along the edges of habitat fragments. In the future, the dynamics of coevolutionary change are likely to be further reshaped by anthropogenic alteration of the earth's many diverse ecosystems (Thompson 2005).

7.2.2 Ecology of invasive species

A mutualistic symbiosis could benefit the ecological success of a plant species as it invades a new habitat (Callaway et al. 2004, Richardson et al. 2000). Microbial symbionts such as mycorrhizal and endophytic fungi that can be beneficial to their host in stressful environments are easily overlooked or ignored during investigations into the ecology and evolution of invasive species (Desprez-Loustau et al. 2007). The competitiveness of infected tall fescue in plant communities of the United States where it is not native has been well established (Clay & Holah 1999, Clay et al. 2005, Hill et al. 1998) and its endophyte has been implicated in its ability to invade such communities (Orr et al. 2005, Rudgers et al. 2005, 2007). The endophyte-grass symbiosis represents a useful system in which to examine how a microbial symbiont could alter invasive properties, as many of the infected hosts (e.g., *Lolium* and *Festuca* spp.) are widely distributed species that have colonized and spread to geographic areas far from their original centers of evolutionary origin.

7.2.3 Ecological restoration

More than 40% of the terrestrial habitats worldwide are grasslands and these provide important ecosystem services such as forage for livestock, reservoirs of biodiversity, and soil nutrient cycles and stabilization (e.g., Hawkes et al. 2005, 2006, Mack & D'Antonio 2003). Yet many of these native grasslands have been degraded by livestock overgrazing, soil erosion from grazing and agriculture, suppression of fire, and invasion of woody and exotic species. Native grasslands are now considered the rarest habitat worldwide (e.g., Noss et al. 1995, Samson & Knopf 1994) and governmental and private organizations have intensified efforts to restore native or seminative grasslands.

It is now widely recognized that consideration of plant microbes are essential in efforts to restore native grasslands that are either degraded or suffer from invasion from exotic species. Nearly all these efforts, however, have been focused on plant pathogens (e.g., Mitchell & Power 2003) or soil microbial communities, notably mycorrhizal associations (Callaway et al. 2004, Hawkes et al. 2005). In the case of pathogens, exotic plant species may gain footholds and spread because they are released from fungal and viral pathogens (e.g., Mitchell & Power 2003). In the case of mycorrhizae, exotic species may alter the soil microbial community or soil processes such as decomposition and nitrogen cycling such that establishment and growth of the exotic species is enhanced at the expense of native grass species (e.g., Callaway et al. 2004).

Little attention has been paid to how grass endophytes influence the success or failure of restoration efforts. Many of these restoration endeavors involve cool-season grasses, which are often highly palatable forage grasses for livestock and wildlife, and constitute major components of many grasslands. Indeed, some studies of the effects of exotic species on mycorrhizae-mediated effects on natives species have used native C_3 grasses that are highly infected with *Neotyphodium* endophytes as target species in experiments, but without regard to whether the experimental native grass was infected or not. Failure to consider endophytes is short-sighted, given that aboveground endophytes can dramatically alter belowground mycorrhizal colonization (negatively or positively), soil processes, susceptibility to plant pathogens, root structure (chapters 3 and 6), below- and aboveground herbivory, and general establishment, performance, growth, and reproduction of host grasses (chapter 2).

Restoration projects involving cool-season pooid grasses must begin to include grass endophytes in the restoration effort or otherwise they may be doomed to failure. Beyond agronomic cultivars, there are few native pooid grasses that are commercially available for restoration projects. Those that are, do not indicate infection status. For example, Redondo, a commercial variety of native Arizona fescue (*F. arizonica*), was released in the 1970s by agricultural stations of New Mexico State University and Colorado State University and the USDA. However, all the seeds tested from this variety were endophyte-free, even though the native source populations from whence the seed stock was collected, near Los Alamos, New Mexico, were nearly 100% infected (S. H. Faeth, unpublished data). Furthermore, because the GMT of coevolution predicts, and empirical studies show, that endophyte haplotype and plant genotype greatly affect herbivore resistance, growth, reproduction, and competitive abilities in different environments (chapters 3 and 4), specific endophyte-grass combinations may be suitable for restoration in given localities but not others. Therefore, successful restoration attempts involving pooid grasses may need to include detailed consideration of not just infection status, but the endophyte-grass genetic combination that is adapted to the specific biotic and abiotic environment of the targeted restoration area.

7.2.4 Emerging infectious disease

Emerging infectious diseases (EIDs) are diseases that have recently evolved or shifted hosts or have rapidly increased in virulence, geographic range, or frequency (e.g., Dashak et al. 2000). EIDs such as the West Nile virus have attracted heightened research interest because of practical concerns about their effects on human health and economic losses to livestock. However, from a conservation biology standpoint, EIDs are usually considered detrimental to species diversity and a disruptor of community structure and function (e.g., Woolhouse et al. 2005). EIDs are also a concern in crop plants, where infectious fungal, bacterial, and viral diseases can potentially devastate agricultural crops, causing great economic loss.

Since cultivation of cool-season turf and pasture grasses such as tall fescue, perennial ryegrass, and Italian ryegrass for sod or for seeds is a large and substantial agronomic industry, any EID that infects these grasses could potentially cause great economic losses. Type II, and especially type I, *Epichloë* endophytes which cause choke disease, involving destruction of inflorescences and no seed yield, are obvious candidates as EIDs. For example, exotic *E. typhina* appeared in orchardgrass (*Dactylis glomerata*) seed farms in the Willamette Valley of the Pacific Northwest, in 1996 and spread rapidly (Rao et al. 2005, Rao & Bauman 2004). Rao and Bauman (2004) found that distribution and abundance was not limited by the presence of the fungal-"pollinating" and endemic fly *Phorbia phrenione*, even though the fly larvae used the exotic stroma as a food source. Rather, the fungus was dispersed from plant to plant in the absence of the flies. At least in this study, the introduced *Epichloë* no longer depended on specific fly vectors (e.g., Bultman & Leuchtmann 2003), and thus the probability of widespread transmission increased. This study suggests that *Epichloë* endophytes, like other diseasing-causing systemic endophytes, have the potential to become EIDs. With increasing human travel, trade, and commerce, the probability that seeds and plants infected with exotic *Epichloë* will be introduced into cool-season cultivated and natural grasslands concomitantly increases, as it does for other pathogens (e.g., Cunningham et al. 2003). The consequences of introduction of these endophytes are potentially devastating economically to seed farms, pastures, commercial and recreational turfgrass systems, and natural grasslands.

Certainly introduction of exotic endophytes that reduce the abundance or competitiveness of native host grasses may have repercussions for local biodiversity. Just as the introduced Asian chestnut blight fungus caused the effective extinction of American chestnuts and radically altered the structure of both plant and animal communities in the eastern deciduous forests of North America (e.g., Anagnostakis 2001), introduction of nonnative and virulent *Epichloë* endophytes could also alter biodiversity and community structure in native grasslands. However, in the case of *Neotyphodium*-plant interactions, the infectious disease component, *Epichloë*, may be necessary to occasionally input genetic variation in order to maintain the mutualistic nature of the interaction between *Neotyphodium* and its native host grasses (Sullivan & Faeth 2004) (see chapter 5). Whereas *Epichloë* in the short term might be considered an emerging infectious plant pathogen, in the longer term, the same *Epichloë* may infuse genetic variation via hybridization with the *Neotyphodium* fungus persisting in the same host grass, and thus increase fitness of the hybrid symbiota in certain environments.

Global climate change (section 7.2.1) may also influence the emergence of new infectious diseases as species ranges contract or expand and environmental changes alter the susceptibility of hosts to disease or increase transmission frequency. We know little about how climate change might affect contagious diseases, such as choke disease, caused by certain *Epichloë* species. Meijer and Leuchtmann (2000) found that elevated CO_2 levels had negligible effects on disease expression of *E. sylvatica* in the woodland grass *Brachypodium sylvaticum*, and instead the endophyte-host symbiota genotype mostly determined disease expression. However, higher nitrogen levels, increased shading, and the absence of preexisting infection all promote contagious disease spread and expression. Thus anthropogenic changes in climate (increased or decreased cooling) and soil nutrients (nitrogen inputs via fertilizer use and airborne emissions) may increase the probability of contagious disease spread in cool-season grasses. Global climate change and other anthropogenic alterations to the environment, such as habitat fragmentation (Groppe et al. 2001), should be a fruitful arena for studying EIDs in cool-season grasses.

7.2.5 Agronomy and agroecosystems

Although this text has focused on the ecology and evolution of the grass-endophyte symbiosis predominantly from a basic perspective, the contribution of agronomic research to understanding the intricacies of the interaction has been enormous. A perusal of the reference list reveals many citations of applied researchers who regularly publish in journals such as *Agronomy, Crop Science,* or *Grass and Forage Science.* The importance of applied biologists to our present understanding of *Neotyphodium* spp. endophytes in forage grasses cannot be overlooked.

One important goal of agronomic research is to improve forage cultivars and endophytic fungi by maintaining the beneficial aspects of infection while reducing toxicity to livestock (Bouton & Easton 2005). This work involves the introduction of novel endophyte haplotypes into specific cultivars (or genotypes) of forage crops followed by selection for desirable agronomic traits. The selection of new strains of endophytes that show reduced toxicity to grazing animals, yet still retain resistance to environmental stresses has been referred to as the "domestication" of endophytes (Malinowski & Belesky 2006). Similar to forage crops, endophytes have also proven to be useful in the development of new turfgrass cultivars of *Lolium* and *Festuca* species, especially in regard to enhanced resistance to insect pests and abiotic stresses (Bacon et al. 1997, Brilman 2005, Clement et al. 1994).

In the future, conserved endophyte-infected seeds in germplasm repositories should prove valuable in supplying genetically diverse *Neotyphodium* species and haplotypes for grass improvement programs (Clement 2001, Clement et al. 1994, Wilson 1996). Furthermore, to better understand the evolution of forage-endophyte associations, it would be helpful to have more information on hosts from the native part of their geographic range. Global repositories of grass accessions contain seeds collected from many areas, including the Mediterranean regions, thought to be the center of origin for both tall fescue

(Sleper & West 1996) and perennial ryegrass (Beddows 1967). In a survey of *Lolium* germplasm, the greatest infection frequency occurred in the Middle East, "suggesting that this region may be a center of origin of clavicipitaceous endophyte-grass associations" (Wilson 1996). Also, more than 85% of the accessions of tall fescue collected from Morocco, Tunisia, and Sardinia (Italy) were endophyte-infected (Clement et al. 2001).

Endophytic fungi may also have a significant role in advancing our current understanding of the domestication and evolution of certain economically important cereals. For example, *Neotyphodium* endophytes have been found in the wild relatives of wheat (*Triticum* spp.) from Turkey (Marshall et al. 1999). Although endophytes were not found in accessions of cultivated wheat (*Triticum aestivum*), the researchers speculated that such investigations into the endophytes of related *Triticum* species could "lead to new methods and strategies of controlling pests and subsequently increasing yields in cultivated wheat" (Marshall et al. 1999). Endophytic fungi have also been reported in germplasm of various *Hordeum* species (Wilson et al. 1991a), wild relatives of the cultivated barley (*Hordeum vulgare*). These infected accessions of "wild barley" have been useful in studies of endophyte-mediated insect resistance (Clement et al. 1997, 2005).

Endophyte-infected seeds are available for some *Lolium* and *Festuca* accessions from the National Plant Germplasm System of the U.S. Department of Agriculture (http://www.ars-grin.gov/npgs). Plants in these particular collections had previously been examined for the presence of endophytic fungi. Nevertheless, Clement (2001) noted that the majority of grass accessions held in global repositories have never been examined for endophyte presence. This implies the unfortunate circumstance that some researchers who utilize grass germplasm collections for experimental work not directed at endophytes may be unaware that their plants are infected. Furthermore, endophytic fungi in collected and stored seeds may lose viability over time and with improper storage (e.g., Siegel et al. 1984b), such that native seeds in repositories may not accurately reflect actual endophyte infections in natural populations. Future comparisons between commercially available forage cultivars or introduced populations and indigenous populations collected from the natural communities in which they evolved should yield new insights into the ecology and evolution of grass-endophyte associations.

Although much has been learned about the nature of the symbiotic interaction of *Neotyphodium* endophytes, their host grasses, and foraging animals (see the collected papers in Bacon & Hill 1997, Popay & Thom 2007, Roberts et al. 2005), many questions remain regarding the establishment, maintenance, and persistence of endophytes in populations of agronomic grasses. In their review of the ecological importance of *Neotyphodium* endophytes in agroecosystems, Malinowski and Belesky (2006) stated that the "ecological consequences of novel endophyte introduction…are not well understood." Future release of agronomic grasses containing genetically modified endophytes will require knowledge of the possible impact of novel host genotype-haplotype combinations on the agroecosystems they will inhabit.

7.3 CONCLUSIONS

Microscopic fungal endosymbionts are pervasive in most vascular plant species, but there still is much to be learned regarding their potential effects on the population biology of their hosts and the community matrix in which they naturally occur. In this book we have described the documented effects of endophytic fungi on the morphological, physiological, and reproductive traits of their graminoid hosts under a variety of environmental conditions (chapter 2). From an ecological genetics perspective, we have demonstrated the potential for endophytes to impact the dynamics of grass populations and described the coevolutionary dynamics of endophytes and their hosts (chapter 5). The evidence for and against endophytes acting as defensive mutualists by protecting grasses from herbivory was explored in chapter 3. Finally, we considered the possible ramifications of endophyte infection on the associated plant and animal community and on ecosystem-level processes (chapter 6).

The pronounced contingency of the grass-endophyte symbiosis on environmental (chapter 2) and genetic factors (chapter 4) has been emphasized repeatedly throughout this book. Like most symbiotic relationships between fungi and plants, grass-endophyte relationships have been shown to range from highly parasitic to commensalistic to mutualistic. We have argued that endophyte-grass interactions are conceptually not different from other microbial-plant interactions nor are endophyte-host-herbivore-natural enemy interactions conceptually different from other multitrophic interactions. We believe that viewing endophyte-host interactions as a subset of other species interactions in communities will serve to expand ideas and research involving grass endophytes.

As is noticeable from the many examples given in this book, a major portion of what is presently known about grass-endophyte symbioses comes from investigations of tall fescue and perennial ryegrass and their *Neotyphodium* endophytes. Furthermore, most studies of these two agronomically important forage crops have involved commercially available cultivars, especially Kentucky 31 tall fescue (Saikkonen et al. 2006). There has been a slow, gradual increase in endophyte research directed at wild grass populations and communities over the past few decades, but a proportionate research effort still lags behind studies involving agronomic grass cultivars (section 7.1.1, figure 7.1). Nevertheless much additional research on other species involved in grass-endophyte symbioses within natural settings will be required to reveal how much fungal endophytes really matter to ecology and evolution in nature.

References

Aarssen, L. W. 2005. On size, fecundity, and fitness in competing plants. In *Reproductive allocation in plants*, eds. E. G. Reekie and F. A. Bazzaz, pp. 215–244. Burlington, MA: Elsevier Academic Press.

Aarssen, L. W., and T. Keogh. 2002. Conundrums of competitive ability in plants: what to measure? *Oikos* 96:531–542.

Agee, C. S., and N. S. Hill. 1994. Ergovaline variability in *Acremonium*-infected tall fescue due to environment and plant genotype. *Crop Science* 34:221–226.

Agrawal, A. A. 2000. Overcompensation of plants in response to herbivory and the by-product benefits of mutualism. *Trends in Plant Science* 5:309–313.

Agrawal, A. A. 2004. Resistance and susceptibility of milkweed: competition, root herbivory, and plant genetic variation. *Ecology* 85:2118–2133.

Ahlholm, J. U., M. Helander, S. Lehtimäki, P. Wäli, and K. Saikkonen. 2002. Vertically transmitted fungal endophytes: different responses of host-parasite systems to environmental conditions. *Oikos* 99:173–183.

Amalric, C., H. Sallanon, F. Monnet, A. Hitmi, and A. Coudret. 1999. Gas exchange and chlorophyll fluorescence in symbiotic and non-symbiotic ryegrass under water stress. *Photosynthetica* 37:107–112.

Anagnostakis, S. L. 2001. American chestnut sprout survival with biological control of the chestnut-blight fungus population. *Forest Ecology and Management* 152:225–233.

Anderson, L. L., and K. N. Paige. 2003. Multiple herbivores and coevolutionary interactions in an *Ipomopsis* hybrid swarm. *Evolutionary Ecology* 17:139–156.

Anzhi, R., G. Yubao, W. Wei, and W. Jinlong. 2006. Photosynthetic pigments and photosynthetic products of endophyte-infected and endophyte-free *Lolium perenne* L. under drought stress conditions. *Frontiers of Biology in China* 1:168–173.

Arachevaleta, M., C. W. Bacon, C. S. Hoveland, and D. E. Radcliffe. 1989. Effect of the tall fescue endophyte on plant response to environmental stress. *Agronomy Journal* 81:83–90.

Arachevaleta, M., C. W. Bacon, R. D. Plattner, C. S. Hoveland, and D. E. Radcliffe. 1992. Accumulation of ergopeptide alkaloids in symbiotic tall fescue grown under deficits of soil water and nitrogen fertilizer. *Applied Environmental Microbiology* 58:857–861.

Armas, C., R. Ordiales, and F. I. Pugnaire. 2004. Measuring plant interactions: a new comparative index. *Ecology* 85:2682–2686.

Armbruster, W. S., C. Pélabon, T. F. Hansen, and C. P. H. Mulder. 2004. Floral integration, modularity, and accuracy: distinguishing complex adaptations from genetic constraints. In *Phenotypic integration: studying the ecology and evolution of complex phenotypes*, eds. M. Pigliucci and K. Preston, pp. 23–49. Oxford: Oxford University Press.

Arnold, A. E., and F. Lutzoni. 2007. Diversity and host range of foliar fungal endophytes: Are tropical leaves biodiversity hotspots? *Ecology* 88:541–549.

Arnold, A. E., Z. Maynard, G. S. Gilbert, P. D. Coley, and T. A. Kursar. 2000. Are tropical endophytes hyperdiverse? *Ecology Letters* 3:267–274.

Arnold, A. E., L. C. Mejia, D. Kyllo, E. I. Rojas, Z. Maynard, N. Robbins, and E. A. Herre. 2003. Fungal endophytes limit pathogen damage in a tropical tree. *Proceedings of the National Academy of Sciences USA* 100:15649–15654.

Arroyo García, R., J. M. Martínez Zapater, B. García Criado, and I. Zabalgogeazcoa. 2002. Genetic structure of natural populations of the grass endophyte *Epichloë festucae* in semiarid grasslands. *Molecular Ecology* 11:355–364.

Asay, K. H., C. J. Nelson, and G. L. Horst. 1974. Genetic variability and net photosynthesis in tall fescue. *Crop Science* 14:571–574.

Assuero, S. G., C. Matthew, P. D. Kemp, G. C. M. Latch, D. J. Barker, and S. J. Haslett. 2000. Morphological and physiological effects of water deficit and endophyte infection on contrasting tall fescue cultivars. *New Zealand Journal of Agricultural Research* 43:49–61.

Atsatt, P. R. 1988. Are vascular plants "inside-out" lichens? *Ecology* 69:17–23.

Bacon, C. W. 1993. Abiotic stress tolerances (moisture, nutrients) and photosynthesis in endophyte-infected tall fescue. *Agriculture, Ecosystems and Environment* 44:123–141.

Bacon, C. W., and N. S. Hill. 1996. Symptomless grass endophytes: products of coevolutionary symbioses and their roles in the ecological adaptations of grasses. In *Endophytic fungi in grasses and woody plants*, eds. S. C. Redlin and L. M. Carris, pp. 155–178. St. Paul, MN: American Pathological Society Press.

Bacon, C. W., and N. S. Hill, eds. 1997. *Neotyphodium/grass interactions*. New York: Plenum Press.

Bacon, C. W., J. K. Porter, J. D. Robbins, and E. S. Luttrell. 1977. *Epichloë typhina* from toxic tall fescue grasses. *Applied and Environmental Microbiology* 34:76–81.

Bacon, C. W., M. D. Richardson, and J. F. White, Jr. 1997. Modification and uses of endophyte-enhanced turfgrasses: a role for molecular technology. *Crop Science* 37:1415–1425.

Bacon, C. W., and J. F. White, Jr. 1994. Stains, media, and procedures for analyzing endophytes. In *Biotechnology of endophytic fungi of grasses*, eds. C. W. Bacon and J. F. White, Jr., pp. 47–56. Boca Raton, FL: CRC Press.

Bailey, J. K., S. C. Wooley, R. L. Lindroth, and T. G. Whitham. 2006. Importance of species interactions to community heritability: a genetic basis to trophic-level interactions. *Ecology Letters* 9:78–85.

Ball, D. M., J. F. Pedersen, and G. D. Lacefield. 1993. The tall-fescue endophyte. *American Scientist* 81:370–379.

Ball, O. J.-P., E. C. Bernard, and K. D. Gwinn. 1997. Effect of selected *Neotyphodium lolii* isolates on root-knot nematode (*Meloidogyne marylandi*) numbers in perennial ryegrass. In *Proceedings of the 50th New Zealand Plant Protection Conference*, ed. M. O'Callaghan, pp. 65–68. Hastings, New Zealand: New Zealand Plant Protection Society.

Ball, O. J.-P., T. A. Coudron, B. A. Tapper, E. Davies, D. Trently, L. P. Bush, K. D. Gwinn, and A. J. Popay. 2006. Importance of host plant species, *Neotyphodium* endophyte isolate, and alkaloids on feeding by *Spodoptera frugiperda* (Lepidoptera: Noctuidae) larvae. *Journal of Economic Entomology* 99:1462–1473.

Barbosa, P., P. Gross, and J. Kemper. 1991. Influence of plant allelochemicals on the tobacco hornworm and its parasitoid, *Cotesia congregata*. *Ecology* 72:1567–1575.

Barbosa, P., and D. K. Letourneau, eds. 1988. *Novel aspects of insect-plant interactions*. New York: John Wiley & Sons.

Barker, G. M. 1987. Mycorrhizal infection influences *Acremonium*-induced resistance to Argentine stem weevil. *Proceedings of the New Zealand Weed and Pest Control Conference* 40:199–203.

Barker, G. M., and P. J. Addison. 1996. Influence of clavipitaceous endophyte infection in ryegrass on development of the parasitoid *Microctonus hyperodae* Loan (Hymenoptera: Braconidae) in *Listronotus bonariensis* (Kuschel) (Coleoptera: Curculionidae). *Biological Control* 7:281–287.

Barker, G. M., and P. J. Addison. 1997. Clavicipitaceous endophytic infection in ryegrass influences attack rate of the parasitoid *Microctonus hyperodae* (Hymenoptera: Braconidae, Euphorinae) in *Listronotus bonariensis* (Coleoptera: Curculionidae). *Environmental Entomology* 26:416–420.

Barrow, J., M. Lucero, I. Reyes-Vera, and K. Havstad. 2007. Endophytic fungi structurally integrated with leaves reveals a lichenous condition of C_4 grasses. *In Vitro Cellular and Developmental Biology – Plant* 43:65–70.

Bascompte, J., P. Jordano, and J. M. Olsen. 2006. Asymmetric coevolutionary networks facilitate biodiversity maintenance. *Science* 312:431–433.

Baskin, C. C., and J. M. Baskin. 1998. Ecology of seed dormancy and germination in grasses. In Population biology of grasses, ed. G. P. Cheplick, pp. 30–83. Cambridge: Cambridge University Press.

Bazely, D. R., J. P. Ball, M. Vicari, A. T. Tanentzap, M. Berenger, T. Radocevic, and S. Koh. 2007. Broad-scale geographic patterns in the distribution of vertically-transmitted, asexual endophytes in four naturally-occurring grasses in Sweden. *Ecography* 30:367–374.

Bazely, D. R., M. Vicari, S. Emmerich, L. Filip, D. Lin, and A. Inman. 1997. Interactions between herbivores and endophyte-infected *Festuca rubra* from the Scottish islands of St. Kilda, Benecula Rum. *Journal of Ecology* 34:847–860.

Bazzaz, F. A. 1996. *Plants in changing environments*. Cambridge: Cambridge University Press.

Beddows, A. R. 1967. Biological flora of the British Isles: *Lolium perenne* L. *Journal of Ecology* 55:567–587.

Begon, M., C. R. Townsend, and J. L. Harper. 2006. *Ecology: from individuals to ecosystems*, 4th ed. Malden, MA: Blackwell Publishing.

Belanger, F. C. 1996. A rapid seedling screening method for determination of fungal endophyte viability. *Crop Science* 36:460–462.

Belesky, D. P., O. J. Devine, J. E. Pallas, Jr., and W. C. Stringer. 1987. Photosynthetic activity of tall fescue as influenced by fungal endophyte. *Photosynthetica* 21:82–87.

Belesky, D. P., and J. M. Fedders. 1996. Does endophyte influence regrowth of tall fescue? *Annals of Botany* 78:499–505.

Belesky, D. P., and D. P. Malinowski. 2000. Abiotic stresses and morphological plasticity and chemical adaptations of *Neotyphodium*-infected tall fescue plants. In *Microbial endophytes*, eds. C. W. Bacon and J. F. White, Jr., pp. 455–485. New York: Marcel Dekker.

Belesky, D. P., W. C. Stringer, and N. S. Hill. 1989. Influence of endophyte and water regime upon tall fescue accessions. I. Growth characteristics. *Annals of Botany* 63:495–503.

Bernard, E. C., K. D. Gwinn, C. D. Pless, and C. D. Williver. 1997. Soil invertebrate species diversity and abundance in endophyte-infected tall fescue pastures. In *Neotyphodium/grass interactions*, eds. C. W. Bacon and N. S. Hill, pp. 125–135. New York: Plenum Press.

Bertoni, M. D., D. Cabral, N. Romero, and J. Dubcovsky. 1993. Endofitos fungicos en especies SudAmericanas de *Festuca* (Poaceae). *Boletin de la Sociedad Argentina de Botánica* 39:25–34.

Bhusari, S., L. B. Hearne, D. E. Spiers, W. R. Lamberson, and E. Antoniou. 2006. Effect of fescue toxicosis on hepatic gene expression in mice. *Journal of Animal Science* 84:1600–1612.

Boege, K., and R. J. Marquis. 2005. Facing herbivory as you grow up: the ontogeny of resistance in plants. *Trends in Ecology and Evolution* 20:441–446.

Boning, R. A., and T. L. Bultman. 1996. A test for constitutive and induced resistance by tall fescue (*Festuca arundinacea*) to an insect herbivore: impact of the fungal endophyte, *Acremonium coenophialum*. *American Midland Naturalist* 136:328–335.

Bony, S., N. Pichon, C. Ravel, A. Durix, C. Balfourier, and J.-J. Guillaumin. 2001. The relationship between myotoxin synthesis in fungal endophytes of *Lolium perenne*. *New Phytologist* 152:125–137.

Boucher, D. H., S. James, and K. H. Keeler. 1982. The ecology of mutualism. *Annual Review of Ecology and Systematics* 13:315–347.

Bouton, J., and S. Easton. 2005. Endophytes in forage cultivars. In *Neotyphodium in cool-season grasses*, eds. C. A. Roberts, C. P. West, and D. E. Spiers, pp. 327–340. Ames, IA: Blackwell Publications.

Bowers, M. D. 1990. Recycling plant natural products for insect defense. In *Novel aspects of insect-plant relations*, eds. P. Barbosa and D. Letourneau, pp. 273–311. New York: John Wiley & Sons.

Bradshaw, A. D. 1959. Population differentiation in *Agrostis tenuis* Sibth. II. The incidence and significance of infection by *Epichloe typhina*. *New Phytologist* 58:310–315.

Braun, K., J. Romero, C. Liddell, and R. Creamer. 2003. Production of swainsonine by fungal endophytes of locoweed. *Mycological Research* 107:980–988.

Breen, J. P. 1994. *Acremonium* endophyte interactions with enhanced plant resistance to insects. *Annual Review of Entomology* 39:401–423.

Brem, D., and A. Leuchtmann. 2001. *Epichloë* grass endophytes increase herbivore resistance in the woodland grass *Brachypodium sylvaticum*. *Oecologia* 126:522–530.

Brem, D., and A. Leuchtmann. 2002. Intraspecific competition of endophyte infected vs uninfected plants of two woodland grass species. *Oikos* 96:281–290.

Brem, D., and A. Leuchtmann. 2003. Molecular evidence for host-adapted races of the fungal endophyte *Epichloë bromicola* after presumed host shifts. *Evolution* 57:37–51.

Brilman, L. A. 2005. Endophytes in turfgrass cultivars. In *Neotyphodium in cool-season grasses*, eds. C. A. Roberts, C. P. West, and D. E. Spiers, pp. 341–349. Ames. IA: Blackwell Publications.

Briske, D. D., and J. D. Derner. 1998. Clonal biology of caespitose grasses. In *Population biology of grasses*, ed. G. P. Cheplick, pp. 106–135. Cambridge: Cambridge University Press.

Bronstein, J. L. 1994a. Conditional outcomes in mutualistic interactions. *Trends in Ecology and Evolution* 9:214–217.

Bronstein, J. L. 1994b. Our current understanding of mutualism. *Quarterly Review of Biology* 69:31–51.

Bronstein, J. L. 2001. The costs of mutualism. *American Zoologist* 41:825–839.

Bronstein, J. L., W. G. Wilson, and W. F. Morris. 2003. Ecological dynamics of mutualist/antagonist communities. *American Naturalist* 162:S24–S39.

Brundrett, M. C. 2002. Coevolution of roots and mycorrhizas of land plants. *New Phytologist* 154:275–304.

Bucheli, E., and A. Leuchtmann. 1996. Evidence for genetic differentiation between choke-inducing and asymptomatic strains of the *Epichloë* grass endophyte from *Brachypodium sylvaticum*. *Evolution* 50:1879–1887.

Buchner, R. C., and L. P. Bush, eds. 1979. *Tall fescue*. Agronomy Series 20. Madison, WI: American Society of Agronomy.

Bultman, T. L., and G. D. Bell. 2003. Interaction between fungal endophytes and environmental stressors influences plant resistance to insects. *Oikos* 103:182–190.

Bultman, T. L., G. Bell, and W. D. Martin. 2004. A fungal endophyte mediates reversal of wound-induced resistance and constrains tolerance in a grass. *Ecology* 85:679–685.

Bultman T. L., K. L. Borowicz, R. M. Schneble, T. A. Coudron, R. J. Crowder, and L. P. Bush. 1997. Effect of a fungal endophyte and loline alkaloids on the growth and survival of two *Euplectrus* parasitoids. *Oikos* 78:170–176.

Bultman, T. L., and D. T. Ganey. 1995. Induced resistance to fall armyworm (Lepidoptera: Noctuidae) mediated by a fungal endophyte. *Environmental Entomology* 24:1196–1200.

Bultman, T. L., and A. Leuchtmann. 2003. A test of host specialization by insect vectors as a mechanism for reproductive isolation among entomophilous fungal species. *Oikos* 103:681–687.

Bultman T. L., M. R. McNeill, and S. L. Goldson. 2003. Isolate-dependent impacts of fungal endophytes in a multitrophic interaction. *Oikos* 102:491–496.

Bultman, T. L., J. F. White, Jr., T. I. Bowdish, A. M. Welch, and J. Johnston. 1995. Mutualistic transfer of *Epichloë* spermatia by *Phorbia* flies. *Mycologia* 87:182–189.

Bultman, T. L., C. Pulas, L. Grant, T. Bell, and T. J. Sullivan. 2006. Effects of fungal endophyte isolate on performance and preference of bird cherry-oat aphid *Environmental Entomology* 35:690–1695.

Burdon, J. J. 1987. Diseases and plant population biology. Cambridge: Cambridge University Press.

Burdon, J. J., and P. H. Thrall. 1999. Spatial and temporal patterns in coevolving plant and pathogen associations. *American Naturalist* 153:S15–S33.

Burdon, J. J., and P. H. Thrall. 2001. The demography and genetics of host-pathogen interactions. In *Integrating ecology and evolution in a spatial context*, eds. J. Silvertown and J. Antonovics, pp. 197–217. Oxford: Blackwell Science.

Burpee, L. L., and J. H. Bouton. 1993. Effect of eradication of the endophyte *Acremonium coenophialum* on epidemics of *Rhizoctonia* blight in tall fescue. *Plant Disease* 77:157–159.

Busch, J. W., M. Neiman, and J. M. Koslow. 2004. Evidence for maintenance of sex by pathogens in plants. *Evolution* 58:2584–2590.

Bush, L. P., H. H. Wilkinson, and C. L. Schardl. 1997. Bioprotective alkaloids of grass-fungal endophyte symbioses. *Plant Physiology* 114:1–7.

Buwalda, J. G., and K. M. Goh. 1982. Host-fungus competition for carbon as a cause of growth depressions in vesicular-arbuscular perennial ryegrass. *Soil Biology and Biochemistry* 14:103–106.

Cabral, D., M. J. Cafaro, B. Saidman, M. Lugo, P. V. Reddy, and J. F. White, Jr. 1999. Evidence supporting the occurrence of a new species of endophyte in some South American grasses. *Mycologia* 91:315–325.

Callaway, R. M., G. C. Thelen, A. Rodriquez, and W. E. Holben. 2004. Soil biota and exotic invasion. *Nature* 427:731.

Campbell, D. R. 2000. Experimental tests of sex-allocation theory in plants. *Trends in Ecology and Evolution* 15:227–232.

Cardinale, B. J., M. A. Palmer, and S. L. Collins. 2002. Species diversity enhances ecosystem functioning through interspecific facilitation. *Nature* 415:426–429.

Carlile, M. J., S. C. Watkinson, and G. W. Gooday. 2001. *The fungi*, 2nd ed. San Diego: Academic Press.

Carrière, Y., A. Bouchard, S. Bourassa, and J. Brodeur. 1998. Effect of endophyte incidence in perennial ryegrass on distribution, host-choice, and performance of the hairy chinch bug (Hemiptera: Lygaeidae). *Journal of Economic Entomology* 91:324–328.

Carroll, G. 1988. Fungal endophytes in stems and leaves: from latent pathogen to mutualistic symbiont. *Ecology* 69:2–9.

Carroll, G. C. 1991. Beyond pest deterrence. Alternative strategies and hidden costs of endophytic mutualisms in vascular plants. In *Microbial ecology of leaves*, eds. J. H. Andrews and S. S. Monano, pp. 358–375. New York: Springer-Verlag.

Chaneton, E. J., and M. Omacini. 2007. Bottom-up cascades induced by fungal endophytes in multitrophic systems. In *Ecological communities: plant mediation in indirect interaction webs*, eds. T. Ohgushi, T. P. Craig, and P. W. Price, pp. 164–187. Cambridge: Cambridge University Press.

Chapin, F. S., III. 2003. Effects of plant traits on ecosystem and regional processes: a conceptual framework for predicting the consequences of global change. *Annals of Botany* 91:455–463.

Charnov, E. L. 1982. *The theory of sex allocation*. Princeton, NJ: Princeton University Press.

Chase, J. M., M. A. Leibold, A. L. Downing, and J. B. Shurin. 2000. The effects of productivity, herbivory, and plant species in grassland food webs. *Ecology* 81:2483–2497.

Cheplick, G. P. 1993a. Effect of simulated acid rain on the mutualism between tall fescue (*Festuca arundinacea*) and an endophytic fungus (*Acremonium coenophialum*). *International Journal of Plant Sciences* 154:134–143.

Cheplick, G. P. 1993b. Reproductive systems and sibling competition in plants. *Plant Species Biology* 8:131–139.

Cheplick, G. P. 1997a. Effects of endophytic fungi on the phenotypic plasticity of *Lolium perenne* (Poaceae). *American Journal of Botany* 84:34–40.

Cheplick, G. P. 1997b. Responses to severe competitive stress in a clonal plant: differences between genotypes. *Oikos* 79:581–591.

Cheplick, G. P. 1998a. Genotypic variation in the regrowth of *Lolium perenne* following clipping: effects of nutrients and endophytic fungi. *Functional Ecology* 12:176–184.

Cheplick, G. P. 1998b. Seed dispersal and seedling establishment in grass populations. In *Population biology of grasses*, ed. G. P. Cheplick, pp. 84–105. Cambridge: Cambridge University Press.

Cheplick, G. P. 2003. Evolutionary significance of genotypic variation in developmental reaction norms for a perennial grass under competitive stress. *Evolutionary Ecology* 17:175–196.

Cheplick, G. P. 2004a. Recovery from drought stress in *Lolium perenne* (Poaceae): Are fungal endophytes detrimental? *American Journal of Botany* 91:1960–1968.

Cheplick, G. P. 2004b. Symbiotic fungi and clonal plant physiology. *New Phytologist* 164: 413–415.

Cheplick, G. P. 2005. The allometry of reproductive allocation. In *Reproductive allocation in plants*, eds. E. G. Reekie and F. A. Bazzaz, pp. 97–128. Burlington, MA: Elsevier Academic Press.

Cheplick, G. P. 2006. A modular approach to biomass allocation in an invasive annual (*Microstegium vimineum*; Poaceae). *American Journal of Botany* 93:539–545.

Cheplick, G. P. 2007. Costs of fungal endophyte infection in *Lolium perenne* genotypes from Eurasia and North Africa under extreme resource limitation. *Environmental and Experimental Botany* 60:202–210.

Cheplick, G. P. 2008. Host genotype overrides fungal endophyte infection in influencing tiller and spike production of *Lolium perenne* (Poaceae) in a common garden experiment. *American Journal of Botany* 95:1063–1071.

Cheplick, G. P., and R. Cho. 2003. Interactive effects of fungal endophyte infection and host genotype on growth and storage in *Lolium perenne*. *New Phytologist* 158:183–191.

Cheplick, G. P., and T. Chui. 2001. Effects of competitive stress on vegetative growth, storage, and regrowth after defoliation in *Phleum pratense*. *Oikos* 95:291–299.

Cheplick, G. P., and K. Clay. 1988. Acquired chemical defenses in grasses: the role of fungal endophytes. *Oikos* 52:309–318.

Cheplick, G. P., K. Clay, and S. Marks. 1989. Interactions between infection by endophytic fungi and nutrient limitation in the grasses *Lolium perenne* and *Festuca arundinacea*. *New Phytologist* 111:89–97.

Cheplick, G. P., A. Perera, and K. Koulouris. 2000. Effect of drought on the growth of *Lolium perenne* genotypes with and without fungal endophytes. *Functional Ecology* 14:657–667.

Cheplick, G. P., and T. P. White. 2002. Saltwater spray as an agent of natural selection: no evidence of local adaptation within a coastal population of *Triplasis purpurea* (Poaceae). *American Journal of Botany* 89:623–631.

Christensen, M. J. 1995. Variation in the ability of *Acremonium* endophytes of *Lolium perenne*, *Festuca arundinacea*, and *F. pratensis* to form compatible associations in the three grasses. *Mycological Research* 99:466–470.

Christensen, M. J., O. J.-P. Ball, R. J. Bennett, and C. L. Schardl. 1997. Fungal and host genotype effects on compatibility and vascular colonization by *Epichloë festucae*. *Mycological Research* 101:493–501.

Christensen, M. J., R. J. Bennett, and J. Schmid. 2001. Vascular bundle colonisation by *Neotyphodium* endophytes in natural and novel associations with grasses. *Mycological Research* 105:1239–1245.

Christensen, M. J., R. J. Bennett, and J. Schmid. 2002. Growth of *Epichloe/Neotyphodium* and p-endophytes in leaves of *Lolium* and *Festuca* grasses. *Mycological Research* 106:93–106.

Christensen, M. J., and G. C. M. Latch. 1991. Variation among isolates of *Acremonium* endophytes (*A. coenophialum* and possibly *A. typhinum*) from tall fescue (*Festuca arundinacea*). *Mycological Research* 95:1123–1126.

Christensen, M. J., A. Leuchtmann, D. D. Rowan, and B. A. Tapper. 1993. Taxonomy of *Acremonium* endophytes of tall fescue (*Festuca arundinacea*), meadow fescue (*Festuca pratensis*) and perennial ryegrass (*Lolium perenne*). *Mycological Research* 97:1083–1092.

Christensen, M. J., W. R. Simpson, and T. Al Samarrai. 2000. Infection of tall fescue and perennial ryegrass plants by combinations of different *Neotyphodium* endophytes. *Mycological Research* 104:974–978.

Christensen, M. J., and C. R. Voisey. 2007. The biology of the endophyte/grass partnership. In *Proceedings of the 6th International Symposium on Fungal Endophytes of Grasses*, eds. A. J. Popay and E. R. Thom, pp. 123–133. Christchurch, New Zealand: New Zealand Grassland Association.

Chu-Chou, M., B. Guo, Z.-Q. An, J. Hendrix, R. Ferriss, M. Siegel, C. Dougherty, and P. Burrus. 1992. Suppression of mycorrhizal fungi in fescue by the *Acremonium coenophialum* endophyte. *Soil Biology and Biochemistry* 24:633–637.

Chung, K.-R., W. Hollin, M. R. Seigel, and C. L. Schardl. 1997. Genetics of host specificity in *Epichloë typhina*. *Phytopathology* 87:599–605.

Chung, K.-R., and C. L. Schardl. 1997a. Sexual cycle and horizontal transmission of the grass symbiont, *Epichloë typhina*. *Mycological Research* 101:295–301.

Chung, K.-R., and C. L. Schardl. 1997b. Vegetative compatibility between and within *Epichloë* species. *Mycologia* 89:558–565.

Cipollini, D. 2004. Stretching the limits of plasticity: Can a plant defend against both competitors and herbivores? *Ecology* 85:28–37.

Clancy, K. M., and P. W. Price. 1987. Rapid herbivore growth enhances enemy attack: sublethal plant defenses remain a paradox. *Ecology* 68:733–737.

Clark, E. M., J. F. White, and R. M. Patterson. 1983. Improved histochemical techniques for the detection of *Acremonium coenophialum* in tall fescue and methods of in vitro culture of the fungus. *Journal of Microbiological Methods* 1:149–155.

Clark, J. S., E. C. Grime, J. Lynch, and P. G. Mueller. 2001. Effects of Holocene climate change on the C-4 grassland/woodland boundary in the Northern Plains, USA. *Ecology* 82:620–636.

Clarke, B. B., J. F. White, Jr., R. H. Hurley, M. S. Torres, S. Sun, and D. R. Huff. 2006. Endophyte-mediated suppression of dollar spot disease in fine fescues. *Plant Disease* 90:994–998.

Clay, K. 1984. The effect of the fungus *Atkinsonella hypoxylon* (Clavicipitaceae) on the reproductive system and demography of the grass *Danthonia spicata*. *New Phytologist* 98:165–175.

Clay, K. 1986. Induced vivipary in the sedge *Cyperus virens* and the transmission of the fungus *Balansia cyperi* (Clavicipitaceae). *Canadian Journal of Botany* 64:2984–2988.

Clay, K. 1987a. The effect of fungi on the interaction between host plants and their herbivores. *Canadian Journal of Plant Pathology* 9:380–388.

Clay, K. 1987b. Effects of fungal endophytes on the seed and seedling biology of *Lolium perenne* and *Festuca arundinacea*. *Oecologia* 73:358–362.

Clay, K. 1988a. Clavicipitaceous fungal endophytes of grasses: coevolution and the change from parasitism to mutualism. In *Coevolution of fungi with plants and animals*, eds. D. L. Hawksworth and K. Pirozynski, pp. 79–105. New York: Academic Press.

Clay, K. 1988b. Fungal endophytes of grasses: a defensive mutualism between plants and fungi. *Ecology* 69:10–16.

Clay, K. 1989. Clavicipitaceous endophytes of grasses: their potential as biocontrol agents. *Mycological Research* 92:1–12.

Clay, K. 1990a. Comparative demography of three graminoids infected by systemic, clavicipitaceous fungi. *Ecology* 71:558–570.

Clay, K. 1990b. Fungal endophytes of grasses. *Annual Review of Ecology and Systematics* 21:275–297.

Clay, K. 1990c. The impact of parasitic and mutualistic fungi on competitive interactions among plants. In *Perspectives on plant competition*, eds. J. B. Grace and D. Tilman, pp. 391–412. San Diego: Academic Press.

Clay, K. 1991a. Endophytes as antagonists of plant pests. In *Microbial ecology of leaves*, eds. J. H. Andrews and S. S. Hirano, pp. 331–357. Berlin: Springer-Verlag.

Clay, K. 1991b. Parasitic castration of plants by fungi. *Trends in Ecology and Evolution* 6:162–166.

Clay, K. 1993. The ecology and evolution of endophytes. *Agriculture, Ecosystems and Environment* 44:39–64.

Clay, K. 1994a. Hereditary symbiosis in the grass genus *Danthonia*. *New Phytologist* 126:223–231.

Clay, K. 1994b. The potential role of endophytes in ecosystems. In *Biotechnology of endophytic fungi of grasses*, eds. C. W. Bacon and J. F. White, Jr., pp. 73–86. Boca Raton, FL: CRC Press.

Clay, K. 1998. Fungal endophyte infection and the population dynamics of grasses. In *Population biology of grasses*, ed. G. P. Cheplick, pp. 255–285. Cambridge: Cambridge University Press.

Clay, K. 2001. Symbiosis and the regulation of communities. *American Zoologist* 41:810–824.

Clay, K., and V. K. Brown. 1997. Infection of *Holcus lanatus* and *H. mollis* by *Epichloë* in experimental grasslands. *Oikos* 79:363–370.

Clay, K., and G. P. Cheplick. 1989. Effect of ergot alkaloids from fungal endophyte-infected grasses on fall armyworm (*Spodoptera frugiperda*). *Journal of Chemical Ecology* 15:169–182.

Clay, K., G. P. Cheplick, and S. Marks. 1989. Impact of the fungus *Balansia henningsiana* on *Panicum agrostoides*: frequency of infection, plant growth and reproduction, and resistance to pests. *Oecologia* 80:374–380.

Clay, K., T. N. Hardy, and A. M. Hammond, Jr. 1985. Fungal endophytes of grasses and their effects on an insect herbivore. *Oecologia* 66:1–5.

Clay, K., and J. Holah. 1999. Fungal endophyte symbiosis and plant diversity in successional fields. *Science* 285:1742–1744.

Clay, K., J. Holah, and J. A. Rudgers. 2005. Herbivores cause a rapid increase in hereditary symbiosis and alter plant community composition. *Proceedings of the National Academy of Sciences USA* 102:12465–12470.

Clay, K., and P. X. Kover. 1996. The Red Queen hypothesis and plant/pathogen interactions. *Annual Review of Phytopathology* 34:29–50.

Clay, K., and A. Leuchtmann. 1989. Infection of woodland grasses by fungal endophytes. *Mycologia* 81:805–811.

Clay, K., S. Marks, and G. P. Cheplick. 1993. Effects of insect herbivory and fungal endophyte infection on competitive interactions among grasses. *Ecology* 74:1767–1777.

Clay, K., and C. Schardl. 2002. Evolutionary origins and ecological consequences of endophyte symbiosis with grasses. *American Naturalist* 160:S99–S127.

Clement, S. L. 2001. Overview of *Neotyphodium* incidence in seed bank collections and plants in managed and unmanaged habitats. In *Proceedings of the 4th International Neotyphodium/Grass Interactions Symposium*, Universität-Gesamthochschule, eds. V. H. Paul and P. D. Dapprich, Paderborn, Soest, Germany, pp. 113–122.

Clement, S. L., L. R. Elberson, N. A. Bosque-Pérez, and D. J. Schotzko. 2005. Detrimental and neutral effects of wild barley—*Neotyphodium* fungal associations on insect survival. *Entomologia Experimentalis et Applicata* 114:119–125.

Clement, S. L., L. R. Elberson, N. N. Youssef, C. M. Davitt, and R. P. Doss. 2001. Incidence and diversity of *Neotyphodium* fungal endophytes in tall fescue from Morocco, Tunisia, and Sardinia. *Crop Science* 41:570–576.

Clement, S. L., W. J. Kaiser, and H. Eichenseer. 1994. *Acremonium* endophytes in germplasms of major grasses and their utilization for insect resistance. In *Biotechnology of endophytic fungi of grasses*, eds. C. W. Bacon and J. F. White, Jr., pp. 185–199. Boca Raton, FL: CRC Press.

Clement, S. L., D. G. Lester, A. D. Wilson, and K. S. Pike. 1992. Behavior and performance of *Diuarphis noxia* (Homoptera: Aphididae) on fungal endophyte-infected and uninfected perennial ryegrass. *Journal of Economic Entomology* 85:583–588.

Clement, S. L., A. D. Wilson, D. G. Lester, and C. M. Davitt. 1997. Fungal endophytes of wild barley and their effects on *Diuraphis noxia* population development. *Entomologia Experimentalis et Applicata* 82:275–281.

Clement, S. L., N. N. Youssef, G. W. Bruehl, W. J. Kaiser, L. R. Elberson, and V. Bradley. 2004. Effects of different storage temperatures on grass seed germination and *Neotyphodium* survival. In *Proceedings of the 5th International Symposium on Neotyphodium/Grass Interactions*, eds. R. Kallenbach, C. Rosenkraus, Jr., and T. R. Lock, Fayetteville, AR, Paper 511.

Coley, A. B., H. A. Fribourg, M. R. Pelton, and K. D. Gwinn. 1995. Effects of tall fescue endophyte infestation on relative abundances of small mammals. *Journal of Environmental Quality* 24:472–475.

Collins, J. P. 2003. What can we learn from community genetics? *Ecology* 84:574–577.

Conover, M. R. 2003. Impact of the consumption of endophyte-infected perennial ryegrass by meadow voles. *Agriculture, Ecosystems and Environment* 97:199–203.

Conover, M. R., and T. A. Messmer. 1996. Feeding preferences and changes in mass of Canada geese grazing endophyte-infected tall fescue. *Condor* 98:859–862.

Crous, P. W., O. Petrini, G. F. Marais, Z. A. Pretorius, and F. Rehder. 1995. Occurrence of fungal endophytes in cultivars of *Triticum aestivum* in South Africa. *Mycoscience* 36:105–111.

Crutsinger, G. M., M. D. Collins, J. A. Fordyce, Z. Gompert, C. C. Nice, and N. J. Sanders. 2006. Plant genotypic diversity predicts community structure and governs an ecosystem process. *Science* 313:966–968.

Cullen, B. R., D. F. Chapman, and P. E. Quigley. 2006. Comparative defoliation tolerance of temperate perennial grasses. *Grass and Forage Science* 61:405–412.

Cunningham, A. A., P. Daszak, and J. P. Rodriguez. 2003. Pathogen pollution: defining a parasitological threat to biodiversity conservation. *Journal of Parasitology* 89:S78–S83.

Cunningham, P. J., J. Z. Foot, and K. F. M. Reed. 1993. Perennial ryegrass (*Lolium perenne*) endophyte (*Acremonium lolii*) relationships: the Australian experience. *Agriculture, Ecosystems and Environment* 44:157–168.

Darbyshire, S. J. 2007. *Schedonorus* P. Beauv. In *Magnoliophyta: Commelinidae (in part): Poaceae, part 1. Flora of North America north of Mexico,* vol. 24, eds. M. E. Barkworth, K. M. Capels, S. Long, L. K. Anderton, and M. B. Piep, pp. 445–448. New York: Oxford University Press.

Daszak, P., A. A. Cunningham, and A. D. Hyatt. 2000. Emerging infectious diseases of wildlife-threats to biodiversity and human health. *Science* 287:443–449.

Davidson, A. W., and D. A. Potter. 1995. Response of plant-feeding, predatory, and soil inhabiting invertebrates to *Acremonium* endophyte and nitrogen fertilization in tall fescue turf. *Journal of Economic Entomology* 88:367–379.

de Battista, J. P., C. W. Bacon, R. Severson, R. D. Plattner, and J. H. Bouton. 1990. Indole acetic acid production by the fungal endophyte of tall fescue. *Agronomy Journal* 82:878–880.

de Jong, T., and P. Klinkhamer. 2005. *Evolutionary ecology of plant reproductive strategies.* Cambridge: Cambridge University Press.

de Sassi, C., C. B. Müller, and J. Krauss. 2006. Fungal plant endosymbionts alter life history and reproductive success of aphid predators. *Proceedings of the Royal Society of London, Series B – Biological Sciences* 273:1301–1306.

Dean, A. M. 1983. A simple model of mutualism. *American Naturalist* 121:409–417.

Del-Val, E. K., and M. J. Crawley. 2005. Are grazing increaser species better tolerators than decreasers? An experimental assessment of defoliation tolerance in eight British grassland species. *Journal of Ecology* 93:1005–1016.

Desprez-Loustau, M-L., C. Robin, M. Buée, R. Courtecuisse, J. Garbaye, F. Suffert, I. Sache, and D. M. Rizzo. 2007. The fungal dimension of biological invasions. *Trends in Ecology and Evolution* 22:472–480.

Devarajan, P. T., and T. S. Suryanarayanan. 2006. Evidence for the role of phytophagous insects in dispersal of non-grass fungal endophytes. *Fungal Diversity* 23:111–119.

Dicke, M., and L. E. M. Vet. 1999. Plant-carnivore interactions: evolutionary and ecological consequences. In *Herbivores: between plants and predators,* eds. H. Olff, V. K. Brown, and R. H. Drent, pp. 483–520. Cambridge: Cambridge University Press.

do Valle Ribeiro, M. A. M. 1993. Transmission and survival of *Acremonium* and the implications for grass breeding. *Agriculture, Ecosystems and Environment* 44:195–213.

Doebeli, M., and N. Knowlton. 1998. The evolution of interspecific mutualisms. *Proceedings of the National Academy of Sciences USA* 95:8676–8680.

Doss, R. P., and R. E. Welty. 1995. A polymerase chain reaction-based procedure for detection of *Acremonium coenophialum* in tall fescue. *Phytopathology* 85:913–917.

Douglas, A. E. 1994. *Symbiotic interactions.* Oxford: Oxford University Press.

Dudt, J. F., and D. J. Shure. 1994. The influence of light and nutrients on foliar phenolics and insect herbivory. *Ecology* 75:86–98.

Easton, H. S. 2007. Grasses and *Neotyphodium* endophytes: co-adaptation and adaptive breeding. *Euphytica* 154:295–306.

Easton, H. S., G. C. M. Latch, B. A. Tapper, and O. J.-P. Ball. 2002. Ryegrass host genetic control of concentrations of endophyte-derived alkaloids. *Crop Science* 42:51–57.

Eerens, J. P. J., J. G. H. White, and R. J. Lucas. 1993. The influence of the *Acremonium* endophyte on the leaf extension rate of moisture stressed ryegrass plants. In *Proceedings of the Second International Symposium on Acremonium/Grass*

Interactions, eds. D. E. Hume, G. C. M. Latch and H. S. Easton, Palmerston North, New Zealand, pp. 200–203.

Eerens, J. P. J., R. J. Lucas, H. S. Easton, and J. G. H. White. 1998a. Influence of the endophyte (*Neotyphodium lolii*) on morphology, physiology, and alkaloid synthesis of perennial ryegrass during high temperature and water stress. *New Zealand Journal of Agricultural Research* 41:219–226.

Eerens, J. P. J., R. J. Lucas, H. S. Easton, and J. G. H. White. 1998b. Influence of the ryegrass endophyte (*Neotyphodium lolii*) in a cool moist environment. I. Pasture production. *New Zealand Journal of Agricultural Research* 41:39–48.

Eerens, J. P. J., M. H. P. W. Visker, R. J. Lucas, H. S. Easton, and J. G. H. White. 1998c. Influence of the ryegrass endophyte (*Neotyphodium lolii*) in a cool moist environment. II. Sheep production. *New Zealand Journal of Agricultural Research* 41:191–199.

Eerens, J. P. J., M. H. P. W. Visker, R. J. Lucas, H. S. Easton, and J. G. H. White. 1998d. Influence of the ryegrass endophyte (*Neotyphodium lolii*) in a cool moist environment. IV. Plant parasitic nematodes. *New Zealand Journal of Agricultural Research* 41:209–217.

Ehrlich, P. R., and P. H. Raven. 1964. Butterflies and plants: a study in coevolution. *Evolution* 18:586–608.

Eichenseer, H., D. L. Dahlman, and L. P. Bush. 1991. Influence of endophyte infection, plant age and harvest interval on *Rhopalosiphum padi* survival and its relation to quality of N-formyl and N-acetyl loline in tall fescue. *Entomologia Experimentalis et Applicata* 60:29–38.

Elbersen, H. W., and C. P. West. 1996. Growth and water relations of field-grown tall fescue as influenced by drought and endophyte. *Grass and Forage Science* 51:333–342.

Elmi, A. A., and C. P. West. 1995. Endophyte infection effects on stomatal conductance, osmotic adjustment, and drought recovery of tall fescue. *New Phytologist* 131:61–67.

Elmqvist, T., and P. A. Cox. 1996. The evolution of vivipary in flowering plants. *Oikos* 77:3–9.

Elton, C. S. 1926. *Animal ecology.* Chicago: University of Chicago Press.

Ewald, P. W. 1987. Transmission modes and evolution of the parasitism-mutualism continuum. *Annals of the New York Academy of Sciences* 503:295–306.

Ewald, P. W. 1994. *Evolution of infectious disease.* Oxford: Oxford University Press.

Faeth, S. H. 1994. Induced plant responses: effects on parasitoids and other natural enemies of phytophagous insects. In *Parasitoid community ecology*, eds. B. A. Hawkins and W. S. Sheehan, pp. 245–260. Oxford: Oxford University Press.

Faeth, S. H. 2002. Are endophytic fungi defensive plant mutualists? *Oikos* 98:25–36.

Faeth, S. H. 2008. Asexual fungal symbionts alter reproductive allocation and herbivory over time in their native perennial grass hosts. *American Naturalist* (in review).

Faeth, S. H., and T. L. Bultman. 2002. Endophytic fungi and interactions among host plants, herbivores, and natural enemies. In *Multitrophic level interactions*, eds. T. Tscharntke and B. A. Hawkins, pp. 89–123. Cambridge: Cambridge University Press.

Faeth, S. H., L. P. Bush, and T. J. Sullivan. 2002a. Peramine alkaloid variation in *Neotyphodium*-infected Arizona fescue: effects of endophyte and host genotype and environment. *Journal of Chemical Ecology* 28:1511–1526.

Faeth, S. H., and W. F. Fagan. 2002. Fungal endophytes: common host plant symbionts but uncommon mutualists. *Integrative and Comparative Biology* 42:360–368.

Faeth, S. H., D. R. Gardner, C. J. Hayes, A. Jani, S. K. Wittlinger, and T. A. Jones. 2006. Temporal and spatial variation in alkaloid levels in *Achnatherum robustum*, a native grass infected with the endophyte *Neotyphodium*. *Journal of Chemical Ecology* 32:307–324.

Faeth, S. H., S. M. Haase, S. S. Sackett, T. J. Sullivan, R. K. Keithley, and C. E. Hamilton. 2002b. Does fire maintain symbiotic, fungal endophyte infections in native grasses? *Symbiosis* 32:211–228.

Faeth, S. H., K. P. Hadeler, and H. R. Thieme. 2007. An apparent paradox of horizontal and vertical disease transmission. *Journal of Biological Dynamics* 1:45–62.

Faeth, S. H., and C. E. Hamilton. 2006. Does an asexual endophyte symbiont alter life stage and long-term survival in a perennial host grass? *Microbial Ecology* 52:748–755.

Faeth, S. H., and K. E. Hammon. 1997. Fungal endophytes oak trees. I. Long-term patterns of abundance and associations with leafminers. *Ecology* 78:810–819.

Faeth, S. H., M. L. Helander, and K. T. Saikkonen. 2004. Asexual *Neotyphodium* endophytes in a native grass reduce competitive abilities. *Ecology Letters* 7:304–313.

Faeth, S. H., and K. Saikkonen. 2007. Variability is the nature of the endophyte-grass interaction. In *Proceedings of the 6th International Symposium on Fungal Endophytes of Grasses*, eds. A. J. Popay and E. R. Thom, pp. 37–48. Christchurch, New Zealand: New Zealand Grassland Association.

Faeth, S. H., and T. J. Sullivan. 2003. Mutualistic asexual endophytes in a native grass are usually parasitic. *American Naturalist* 161:310–325.

Faeth, S. H., and D. Wilson. 1996. Induced responses in trees: mediators of interactions between macro- and micro-herbivores? In *Multitrophic interactions in terrestrial systems*, eds. A. C. Gange and V. K. Brown, pp. 201–215. Oxford: Blackwell Scientific.

Fair, J., W. K. Lauenroth, and D. P. Coffin. 1999. Demography of *Bouteloua gracilis* in a mixed prairie: analysis of genets and individuals. *Journal of Ecology* 87:233–243.

Fales, S. L., A. S. Laidlaw, and M. G. Lambert. 1996. Cool-season grass ecosystems. In *Cool-season forage grasses*, eds. L. E. Moser, D. R. Buxton, and M. D. Casler, pp. 267–296. Madison, WI: American Society of Agronomy, Crop Science Society of America, and Soil Science Society of America.

Ferdy, J.-B., and B. Godelle. 2005. Diversification of transmission mode and the evolution of mutualism. *American Naturalist* 166:613–627.

Ferraro, D. O., and M. Oesterheld. 2002. Effect of defoliation on grass growth. A quantitative review. *Oikos* 98:125–133.

Fine, P. E. M. 1975. Vectors and vertical transmission: an epidemiologic perspective. *Annuals of the New York Academy of Science* 266:173–194.

Finkes L. K., A. B. Cady, J. C. Mulroy, K. Clay, and J. A. Rudgers. 2006. Plant-fungus mutualism affects spider composition in successional fields. *Ecology Letters* 9:344–353.

Flor, H. H. 1955. Host-parasite interaction in flax-rust—its genetics and other implications. *Phytopathology* 45:680–685.

Fornoni, J., J. Nunez-Farfan, P. L. Valverde, and M. D. Rausher. 2004. Evolution of mixed strategies of plant defense allocation against natural enemies. *Evolution* 58:1685–1695.

Fortier, G. M., N. Bard, M. Jansen, and K. Clay. 2000. Effects of tall fescue endo-phyte infection and population density on growth and reproduction in prairie voles. *Journal of Wildlife Management* 64:122–128.

Fortier, G. M., M. A. Osmon, M. Roach, and K. Clay. 2001. Are female voles food limited? Effects of endophyte-infected tall fescue on home range size in female prairie voles (*Microtus ochrogaster*). *American Midland Naturalist* 146:63–71.

Foster, K. R., and T. Wenseleers. 2006. A general model for the evolution of mutual-isms. *Journal of Evolutionary Biology* 19:1283–1293.

Fowler, N. L., and K. Clay. 1995. Environmental heterogeneity, fungal parasitism and the demography of the grass *Stipa leucotricha*. *Oecologia* 103:55–62.

Francis, S. M., and D. B. Baird. 1989. Increases in the proportion of endophyte-infected perennial ryegrass in overdrilled pastures. *New Zealand Journal of Agricultural Research* 32:437–440.

Frank, A. B., S. Bittman, and D. A. Johnson. 1996. Water relations of cool-season grasses. In *Cool-season forage grasses*, eds. L. E. Moser, D. R. Buxton, and M. D. Casler, pp. 127–164. Madison, WI: American Society of Agronomy, Crop Science Society of America, and Soil Science Society of America.

Frank, S. A. 1993. Coevolutionary genetics of plants and pathogens. *Evolutionary Ecology* 7:45–75.

Frank, S. A. 1994. Genetics of mutualisms: the evolution of altruism between species. *Journal of Theoretical Biology* 170:393–400.

Franzluebbers, A. J., N. Nazih, J. A. Stuedemann, J. J. Fuhrmann, H. H. Schomberg, and P. G. Hartel. 1999. Soil carbon and nitrogen pools under low and high endo-phyte infected tall fescue. *Soil Science Society of America Journal* 63:1687–1694.

Freeland, W. J., and D. H. Janzen. 1974. Strategies in herbivory by mammals: the role of plant secondary compounds. *American Naturalist* 108:269–289.

Freeman, E. M. 1904. The seed fungus of *Lolium temulentum* L., the darnel. *Philosophical Transactions of the Royal Society of London [Biol.]* 196:1–27.

Freeman, S., and R. J. Rodriguez. 1993. Genetic conversion of a fungal plant patho-gen to a nonpathogenic, endophytic mutualist. *Science* 260:75–78.

Fritz, R. S. and E. L. Simms, eds. 1992. *Plant resistance to herbivores and pathogens: ecology, evolution, and genetics*. Chicago: University of Chicago Press.

Futuyma, D. J., and M. Slatkin, eds. 1983. *Coevolution*. Sunderland, MA: Sinauer Associates.

Gallagher, R. T., A. D. Hawkes, P. S. Steyn, and R Vleggaar. 1984. Tremorgenic neu-rotoxins from perennial ryegrass causing ryegrass staggers disorder of livestock—structure elucidation of lolitrem-B. *Journal of the Chemical Society, Chemical Communications* 9:614–616.

Gandon, S. 1998. Local adaptation and host-parasite interactions. *Trends in Ecology and Evolution* 13:214–216.

Gandon, S. 2002. Local adaptation and the geometry of host-parasite coevolution. *Ecology Letters* 5:246–256.

Gandon, S., Y. Capowiez, Y. Dubois, Y. Michalakis, and I. Olivieri. 1996. Local adap-tation and gene-for-gene coevolution in a metapopulation model. *Proceedings of the Royal Society of London, Series B – Biological Sciences* 263:1003–1009.

Gentile, A., M. S. Rossi, D. Cabral, K. D. Craven, and C. L. Schardl. 2005. Origin, divergence, and phylogeny of *Epichloë* endophytes of native Argentine grasses. *Molecular Phylogenetics and Evolution* 35:196–208.

Gibson, D. J. 2002. *Methods in comparative plant population ecology*. Oxford: Oxford University Press.

Giddings, G. D., N. R. S Hamilton, and M. D. Hayward. 1997. The release of genetically modified grasses. Part 1: pollen dispersal to traps in *Lolium perenne*. *Theoretical and Applied Genetics* 94:1000–1006.

Gilbert, G. S. 2002. Evolutionary ecology of plant diseases in natural ecosystems. *Annual Review of Phytopathology* 40:13–43.

Gilchrist, M. A., D. L. Sulsky, and A. Pringle. 2006. Identifying fitness and optimal life-history strategies for an asexual filamentous fungus. *Evolution* 60:970–979.

Godt, M. J. W., and J. L. Hamrick. 1998. Allozyme diversity in the grasses. In *Population biology of grasses*, ed. G. P. Cheplick, pp. 11–29. Cambridge: Cambridge University Press.

Goldson, S. L., J. R. Proffitt, L. R. Fletcher, and D. B. Baird. 2000. Multitrophic interaction between the ryegrass *Lolium perenne*, its endophyte *Neotyphodium lolii*, the weevil pest *Listronotus bonariensis*, and its parasitoid *Microctonus hyperodae*. *New Zealand Journal of Agricultural Research* 43:227–233.

Gomulkiewicz, R., S. L. Nuismer, and J. N. Thompson. 2003. Coevolution in variable mutualisms. *American Naturalist* 162:S80–S93.

Gonthier, D. J., T. J. Sullivan, K. L. Brown, B. Wurtzel, R. Lawal, K. VandenOever, Z. Buchan, T. L. Bultman. 2008. Stroma-forming endophyte *Epichloë glyceriae* provides wound-inducible herbivore resistance to its grass host. *Oikos* 117:629–633.

Gotelli, N. J., and G. L. Entsminger. 2007. EcoSim: null models software for ecology. Version 7.0. Acquired Intelligence Inc. & Kesey-Bear, Jericho, VT 05465. http://garyentsminger.com/ecosim.htm.

Grewal, S. K., P. S. Grewal, and R. Gaugler. 1995. Endophytes of fescue grasses enhance susceptibility of *Popillia japonica* larvae to an entomophagous nematode. *Entomologia Experimentalis et Applicata* 74:219–224.

Grime, J. P. 1979. *Plant strategies and vegetation processes*. New York: John Wiley & Sons.

Groppe, K., and T. Boller. 1997. PCR assay based on a microsatellite-containing locus for detection and quantification of *Epichloë* endophytes in grass tissue. *Applied and Environmental Microbiology* 63:1543–1550.

Groppe, K., I. Sanders, A. Wiemken, and T. Boller. 1995. A microsatellite marker for studying the ecology and diversity of fungal endophytes (*Epichloë* spp.) in grasses. *Applied and Environmental Microbiology* 61:3943–3949.

Groppe, K., T. Steinger, I. Sanders, B. Schmid, A. Wiemken, and T. Boller. 1999. Interaction between the endophytic fungus *Epichloë bromicola* and the grass *Bromus erectus*: effects of endophyte infection, fungal concentration and environment on grass growth and flowering. *Molecular Ecology* 8:1827–1835.

Groppe, K., T. Steinger, B. Schmid, B. Baur, and T. Boller. 2001. Effects of habitat fragmentation on choke disease (*Epichloë bromicola*) in the grass *Bromus erectus*. *Journal of Ecology* 89:247–255.

Gundel, P. E., P. H. Maseda, M. M. Vila-Aiub, C. M. Ghersa, and R. Benech-Arnold. 2006. Effects of *Neotyphodium* fungi on *Lolium multiflorum* seed germination in relation to water availability. *Annals of Botany* 97:571–577.

Gundel, P. E., W. B. Batista, M. Texeira, M. A. Martinez-Ghersa, M. Omacini, and C. M. Ghersa. 2008. *Neotyphodium* endophyte infection frequency in annual grass populations: relative importance of mutualism and transmission efficiency. *Proceedings of the Royal Society B* 275:897–905.

Guillaumin, J.-J., M. Frain, N. Picheon, and C. Ravel. 2001. Survey of fungal endophytes in wild grass species of the Auvergne region (central France). In *Proceedings of the Fourth International Neotyphodium/Grass Interactions*

Symposium, Fachbereich Agrarwirtschaft, eds. V. H. Paul and P. D. Dapprich, Soest, Germany, pp. 85–92.

Guo, B. Z., J. W. Hendrix, Z.-Q. An, and R. S. Ferriss. 1992. Role of *Acremonium* endophyte of fescue on inhibition of colonization and reproduction of mycorrhizal fungi. *Mycologia* 84:882–885.

Gwinn, K. D., and Gavin, A. M. 1992. Relationship between endophyte infection level of tall fescue seed lots and *Rhizoctonia zeae* seedling disease. *Plant Disease* 76:911–914.

Gyllenberg, M., D. Preoteasa, and K. Saikkonen. 2002. Vertically transmitted symbionts in structured host metapopulations. *Bulletin of Mathematical Biology* 64:959–978.

Hahn, H., W. Huth, W. Schöberlein, and W. Diepenbrock. 2003. Detection of endophytic fungi in *Festuca* spp. by means of tissue print immunoassay. *Plant Breeding* 122:217–222.

Hamilton, C. E. 2002. Maintenance of systemic *Neotyphodium* infections in Arizona fescue: a test of three hypotheses. Master's thesis, Arizona State University, Tempe, AZ.

Hamilton, C. E., and S. H. Faeth. 2005. Asexual *Neotyphodium* endophytes in Arizona fescue: a test of the seed germination and pathogen resistance hypothesis. *Symbiosis* 38:69–85.

Hamilton, C. E., and S. H. Faeth. 2007. *Neotyphodium* infection and hybridisation as a function of environmental variation. In *Proceedings of the 6th International Symposium on Fungal Endophytes of Grasses*, eds. A. J. Popay and E. R. Thom, pp. 215–217. Christchurch, New Zealand: New Zealand Grassland Association.

Hammon, K. E., and S. H. Faeth. 1992. Ecology of plant-herbivore communities: a fungal component? *Natural Toxins* 1:197–208.

Hance, H. F. 1876. On a mongolian grass producing intoxication in cattle. *Journal of Botany* 14:210–212.

Hardy, T., K. Clay, and A. M. Hammond, Jr. 1985. Fall armyworm (Lepidoptera: Noctuidae): a laboratory bioassay and larval performance study for the fungal endophyte of perennial ryegrass. *Journal of Economic Entomology* 78:571–575.

Hardy, T., K. Clay, and A. M. Hammond, Jr. 1986. Leaf age and related factors affecting endophyte-mediated resistance to fall armyworm (Lepidoptera: Noctuidae) in tall fescue. *Environmental Entomology* 15:1083–1089.

Harper, J. L. 1977. *Population biology of plants*. San Diego: Academic Press.

Härri, S. A. 2007. Effects of endophytes on multitrophic interactions. PhD dissertation, University of Zurich, Zurich, Switzerland.

Hawkes, C. V., J. Belnap, C. D'Antonio, and M. K. Firestone. 2006. Arbuscular mycorrhizal assemblages in native plant roots change in the presence of exotic grasses. *Plant and Soil* 281:367–379.

Hawkes, C. V., and J. J. Sullivan. 2001. The impact of herbivory on plants in different resource conditions—a meta-analysis. *Ecology* 82:2045–2058.

Hawkes, C.V., I. F. Wren, D. J. Herman, and M. K. Firestone. 2005. Plant invasion alters nitrogen cycling by modifying the soil nitrifying community. *Ecology Letters* 8:976–985.

Heckman, D. S., D. H. Gleiser, B. R. Eidell, R. L. Stauffer, N. L. Kardos, and S. B. Hedges. 2001. Molecular evidence for the early colonization of land by fungi and plants. *Science* 293: 1129–1133.

Heemsbergen, D. A., M. P. Berg, M. Loreau, J. R. van Hal, J. H. Faber, and H. A. Verhoef. 2004. Biodiversity effects on soil processes explained by interspecific functional dissimilarity. *Science* 306:1019–1020.

Herms, D. A., and W. J. Mattson. 1992. The dilemma of plants: to grow or defend. *Quarterly Review of Biology* 67:283–335.

Hesse, U., H. Hahn, K. Andreeva, K. Förster, K. Warnstorff, W. Schöberlein, and W. Diepenbrock. 2004. Investigations on the influence of *Neotyphodium* endophytes on plant growth and seed yield of *Lolium perenne* genotypes. *Crop Science* 44:1689–1695.

Hesse, U., W. Schöberlein, L. Wittenmayer, K. Förster, K. Warnstorff, W. Diepenbrock, and W. Merbach. 2003. Effects of *Neotyphodium* endophytes on growth, reproduction and drought-stress tolerance of three *Lolium perenne* L. genotypes. *Grass and Forage Science* 58:407–415.

Hesse, U., W. Schöberlein, L. Wittenmayer, K. Förster, K. Warnstorff, W. Diepenbrock, and W. Merbach. 2005. Influence of water supply and endophyte infection (*Neotyphodium* spp.) on vegetative and reproductive growth of two *Lolium perenne* L. genotypes. *European Journal of Agronomy* 22:45–54.

Hiatt, E. E., III, N. S. Hill, J. H. Bouton, and J. A. Stuedemann. 1999. Tall fescue endophyte detection: commercial immunoblot test kit compared with microscopic analysis. *Crop Science* 39:796–799.

Hignight, K. W., G. A. Muilenburg, and A. J. P. van Wijk. 1993. A clearing technique for detecting the fungal endophyte *Acremonium* sp. in grasses. *Biotechnic & Histochemistry* 68:87–90.

Hill, N. S., D. P. Belesky, and W. C. Stringer. 1991. Competitiveness of tall fescue as influenced by *Acremonium coenophialum*. *Crop Science* 31:185–190.

Hill, N. S., J. G. Pachon, and C. W. Bacon. 1996. *Acremonium coenophialum*-mediated short-and long-term drought acclimation in tall fescue. *Crop Science* 36:665–672.

Hill, N. S., D. P. Belesky, and W. C. Stringer. 1998. Encroachment of endophyte-infected on endophyte-free tall fescue. *Annals of Botany* 81:483–488.

Hochwender, C. G., R. J. Marquis, and K. A. Stowe. 2000. The potential for and constraints on the evolution of compensatory ability in *Asclepias syriaca*. *Oecologia* 122:361–370.

Holland, B. R., C. L. Schardl, and J. Schmid. 2007. Endophyte survival without sex. In *Proceedings of the 6th International Symposium on Fungal Endophytes of Grasses*, eds. A. J. Popay and E. R. Thom, pp. 151–154. Christchurch, New Zealand: New Zealand Grassland Association.

Holland, J. N., D. L. DeAngelis, and J. L. Bronstein. 2002. Population dynamics and mutualism: functional responses of benefits and costs. *American Naturalist* 159:231–244.

Holsinger, K. E. 2000. Demography and extinction in small populations. In *Genetics, demography and viability of fragmented populations*, eds. A. G. Young and G. M. Clarke, pp. 55–74. Cambridge: Cambridge University Press.

Hoveland, C. S. 1993. Importance and economic significance of the *Acremonium* endophytes to performance of animals and the grass plant. *Agriculture, Ecosystems, and Environment* 44: 3–12.

Hoveland, C. S., J. H. Bouton, and R. G. Durham. 1999. Fungal endophyte effects on production of legumes in association with tall fescue. *Agronomy Journal* 91:897–902.

Hoveland, C. S., R. G. Durham, and J. H. Bouton. 1997. Tall fescue response to clipping and competition with no-till seeded alfalfa as affected by fungal endophyte. *Agronomy Journal* 89:119–125.

Huigens, M. E., H. F. Luck, R. H. G. Klaassen, M. F. P. M. Maas, M. J. T. N. Timmerman, and R. Stouthamer. 2000. Infectious parthenogenesis. *Nature* 405:178–179.

Huitu, O., M. Helander, P. Lehtonen, and K. Saikkonen. 2008. Consumption of grass endophytes alters the ultraviolet spectrum of vole urine. *Oecologia* 156:333–340.

Humphrey, L. D., and D. A. Pyke. 1998. Demographic and growth responses of a guerrilla and a phalanx perennial grass in competitive mixtures. *Journal of Ecology* 86:854–865.

Hunt, M. G., and J. A. Newman. 2005. Reduced herbivore resistance from a novel grass-endophyte association. *Journal of Applied Ecology* 42:762–769.

Hunt, M. G., S. Rasmussen, P. C. D. Newton, A. J. Parsons, and J. A. Newman. 2005. Near-term impacts of elevated CO_2, nitrogen, and fungal endophyte-infection on *Lolium perenne* L. growth, chemical composition and alkaloid production. *Plant, Cell and Environment* 28:1345–1354.

Isaac, S. 1992. *Fungal-plant interactions*. London: Chapman & Hall.

Jani, A. J. 2005. Association of *Neotyphodium* endophytes and alkaloids with community structure in native grasses. Master's thesis, Arizona State University, Tempe, AZ.

Janzen, D. H. 1966. Coevolution of mutualism between ants and acacias in Central America. *Evolution* 20:249–275.

Jeffries, M. J., and J. H. Lawton. 1984. Enemy free space and the structure of ecological communities. *Biological Journal of the Linnean Society* 23:269–286.

Johnson, M. C., R. C. Anderson, R. J. Kryscio, and M. R. Siegel. 1983. Sampling procedures for determining endophyte in tall fescue seed lots by ELISA. *Phytopathology* 73:1406–1409.

Johnson, N. C., J. H. Graham, and F. A. Smith. 1997. Functioning of mycorrhizal associations along the mutualism-parasitism continuum. *New Phytologist* 135:575–585.

Johnson, N. C., J. Wolf, M. A. Reyes, A. Panter, G. W. Koch, and A. Redman. 2005. Species of plants and associated arbuscular mycorrhizal fungi mediate mycorrhizal responses to CO_2 enrichment. *Global Change Biology* 11:1156–1166.

Johnson-Cicalese, J., M. E. Secks, C. K. Lam, W. A. Meyer, J. A. Murphy, and F. C. Belanger. 2000. Cross species inoculation of Chewings and strong creeping red fescues with fungal endophytes. *Crop Science* 40:1485–1489.

Jones, T. A., M. H. Ralphs, D. R. Gardner, and N. J. Chatterton. 2000. Cattle prefer endophyte-free robust needlegrass. *Journal of Range Management* 53:427–431.

Jump, A. S., and J. Peñuelas. 2005. Running to stand still: adaptation and the response of plants to rapid climate change. *Ecology Letters* 8:1010–1020.

Kaltz, O., S. Gandon, Y. Michalakis, and J. A. Shykoff. 1999. Local maladaptation in the anther-smut fungus *Microbotryum violaceum* to its host plant *Silene latifolia*: evidence from a cross-inoculation experiment. *Evolution* 53:395–407.

Kaltz, O., and J. A. Shykoff. 1998. Local adaptation in host-parasite systems. *Heredity* 81:361–370.

Karban, R. 1997. Neighborhood affects a plant's risk of herbivory and subsequent success. *Ecological Entomology* 22:433–439.

Karban, R., and I. T. Baldwin. 1997. *Induced responses to herbivory*. Chicago: University of Chicago Press.

Karban, R., A. K. Brody, and W. C. Schnathorst. 1989. Crowding and a plant's ability to defend itself against herbivores and diseases. *American Naturalist* 134:749–760.

Kareiva, P. M., J. G. Kingsolver, and R. B. Huey, eds. 1993. *Biotic interactions and global change*. Sunderland, MA: Sinauer Associates.

Keeler, K. H. 1985. Cost:benefit models of mutualism. In *The biology of mutualism*, eds. D. H. Boucher, pp. 100–127. New York: Oxford University Press.

Kelley, S. E., and K. Clay. 1987. Interspecific competitive interactions and the maintenance of genotypic variation within two perennial grasses. *Evolution* 41:92–103.

Keogh, R. G., and T. Lawrence. 1987. Influence of *Acremonium lolii* presence on emergence and growth of ryegrass seedlings. *New Zealand Journal of Agricultural Research* 30:507–510.

Kessler, A., and I. T. Baldwin. 2001. Defensive function of herbivore-induced plant volatile emissions in nature. *Science* 291:2141–2144.

Kimmons, C. A., Gwinn, K. D., and E. C. Bernard. 1990. Nematode reproduction on endophyte-infected and endophyte-free tall fescue. *Plant Disease* 74:757–761.

Kniskern, J. M., and M. D. Rausher. 2006. Environmental variation mediates the deleterious effects of *Coleosporium ipomoeae* on *Ipomoea purpurea*. *Ecology* 87:675–685.

Knoch, T. R., S. H. Faeth, and D. Arnott. 1993. Endophytic fungi alter foraging by two species of desert seed-harvesting ants. *Oecologia* 95:470–475.

Koch, A. M., D. Croll, and I. R. Sanders. 2006. Genetic variability in a population of arbuscular mycorrhizal fungi causes variation in plant growth. *Ecology Letters* 9:103–110.

Koga, H., M. J. Christensen, and R. J. Bennett. 1993. Incompatibility of some grass-*Acremonium* endophyte associations. *Mycological Research* 97:1237–1244.

Kogel, K.-H., P. Franken, and R. Hückelhoven. 2006. Endophyte or parasite—what decides? *Current Opinion in Plant Biology* 9:358–363.

Koh, S., and D. S. Hik. 2007. Herbivory mediates grass-endophyte relationships. *Ecology* 88:2752–2757.

Koh, S., M. Vicari, J. P. Ball, T. Rakocevic, S. Zaher, D. S. Hik, and D. R. Bazely. 2006. Rapid detection of fungal endophytes in grasses for large-scale studies. *Functional Ecology* 20: 736–742.

Körner, C. 1991. Some often overlooked plant characteristics as determinants of plant growth: a reconsideration. *Functional Ecology* 5:162–173.

Koshino, H., S. Togiya, T. Yoshihara, and S. Sakamura. 1987. Four fungitoxic C-18 hydroxy unsaturated fatty acids from stromata of *Epichloë typhina*. *Tetrahedron Letters* 28:73–76.

Kover, P. X. 2000. Effects of parasitic castration on plant resource allocation. *Oecologia* 123:48–56.

Kover, P. X., and K. Clay. 1998. Trade-off between virulence and vertical transmission and the maintenance of a virulent plant pathogen. *American Naturalist* 152:165–175.

Kover, P. X., T. E. Dolan, and K. Clay. 1997. Potential versus actual contribution of vertical transmission to pathogen fitness. *Proceedings of the Royal Society of London, Series B – Biological Sciences* 264:903–909.

Krauss J., S. A. Härri, L. Bush, R. Husi, L. Bigler, S. A. Power, and C. B. Müller. 2007. Effects of fertiliser, fungal endophytes and plant cultivar on the performance of insect herbivores and their natural enemies. *Functional Ecology* 21:107–116.

Krings, M., T. N. Taylor, H. Hass, H. Kerp, N. Dotzler, and E. J. Hermsen. 2007. Fungal endophytes in a 400-million-yr-old land plant: infection pathways, spatial distribution, and host responses. *New Phytologist* 174:648–657.

Kula, A. A. R., D. C. Hartnett, and G. W. T. Wilson. 2005. Effects of mycorrhizal symbiosis on tallgrass prairie plant-herbivore interactions. *Ecology Letters* 8:61–69.

Kuldau, G. A., H. F. Tsai, and C. L. Schardl. 1999. Genome sizes of *Epichloë* and anamorphic hybrids. *Mycologia* 91:776–782.

Kunkel, B. A., and P. S. Grewal. 2003. Endophyte infection in perennial ryegrass reduces susceptibility of *Agrotis ipsilon* to an entomopathogenic nematode. *Entomologia Experimentalis et Applicata* 107:95–104.

Kunkel, B. A., P. S. Grewal, and M. F. Quiley. 2004 A mechanism of acquired resistance against an entomopathogenic nematode by *Agrotis ipsilon* feeding on perennial ryegrass harboring a fungal endophyte. *Biological Control* 29:100–108.

Lane, G. A., B. A. Tapper, E. Davies, D. E. Hume, G. C. M. Latch, D. J. Barker, H. S. Easton, and M. P. Rolston. 1997. Effect of growth conditions on alkaloid concentrations in perennial ryegrass naturally infected with endophytes. In *Neotyphodium/grass interactions*, eds. C. W. Bacon and A. C. Hill, pp. 179–182. New York: Plenum Press.

Latch, G. C. M. 1993. Physiological interactions of endophytic fungi and their hosts. Biotic stress tolerance imparted to grasses by endophytes. *Agriculture, Ecosystems and Environment* 44:143–156.

Latch, G. C. M., and M. J. Christensen. 1982. Ryegrass endophyte, incidence, and control. *New Zealand Journal of Agricultural Research* 25:443–448.

Latch, G. C. M., and M. J. Christensen. 1985. Artificial infections of grasses with endophytes. *Annals of Applied Biology* 107:17–24.

Latch, G. C. M., W. F. Hunt, and D. R. Musgrave. 1985. Endophytic fungi affect growth of perennial ryegrass. *New Zealand Journal of Agricultural Research* 28:165–168.

Latch, G. C. M., L. R. Potter, and B. R. Tyler. 1987. Incidence of endophytes in seeds from collections of *Lolium* and *Festuca* species. *Annals of Applied Biology* 111:59–64.

Law, R. 1985. Evolution in a mutualistic environment. In *The biology of mutualism*, ed. D. H. Boucher, pp. 145–170. New York: Oxford University Press.

Lehtonen, P., M. Helander, and K. Saikkonen. 2005a. Are endophyte-mediated effects on herbivores conditional on soil nutrients? *Oecologia* 142:38–45.

Lehtonen, P., M. Helander, S. A. Siddiqui, K. Lehto, and K. Saikkonen. 2006. Endophytic fungi decrease plant virus infections in meadow ryegrass (*Lolium pratense*). *Biology Letters* 2:620–623.

Lehtonen, P., M. Helander, M. Wink, F. Sporer, and K. Saikkonen. 2005b. Transfer of endophyte-origin defensive alkaloids from a grass to a hemiparasitic plant. *Ecology Letters* 8:1256–1263.

Lemons, A., K. Clay, and J. A. Rudgers. 2005. Connecting plant-microbial interactions above and belowground: a fungal endophyte affects decomposition. *Oecologia* 145:595–604.

Lennartsson, T., J. Tuomi, and P. Nilsson. 1997. Evidence for an evolutionary history of overcompensation in the grassland biennial *Gentianella campestris* (Gentianaceae). *American Naturalist* 149:1147–1155.

LeRoy, C. J., T. G. Whitham, P. Keim, and J. C. Marks. 2006. Plant genes link forests and streams. *Ecology* 87:255–261.

Leuchtmann, A. 1992. Systematics, distribution, and host specificity of grass endophytes. *Natural Toxins* 1:150–162.

Leuchtmann, A. 2003. Taxonomy and diversity of *Epichloë* endophytes. In *Clavicipitalean fungi: evolutionary biology, chemistry, biocontrol, and cultural impacts*, eds. J. F. White, Jr., C. W. Bacon, N. L. Hywel-Jones, and J. W. Spatafora, pp. 169–194. New York: Marcel Dekker.

Leuchtmann, A., and K. Clay. 1988a. *Atkinsonella hypoxylon* and *Balansia cyperi*, epiphytic members of the Balansiae. *Mycologia* 80:192–199.

Leuchtmann, A., and K. Clay. 1988b. Experimental infection of host grasses and sedges with *Atkinsonella hypoxylon* and *Balansia cyperi* (Balansiae, Clavicipitaceae). *Mycologia* 80:291–297.

Leuchtmann, A., and K. Clay. 1989a. Experimental evidence for genetic variation in compatibility between the fungus *Atkinsonella hypoxylon* and its three host grasses. *Evolution* 43:825–834.

Leuchtmann, A., and K. Clay. 1989b. Isozyme variation in the fungus *Atkinsonella hypoxylon* within and among populations of its host grasses. *Canadian Journal of Botany* 67:2600–2607.

Leuchtmann, A., and K. Clay. 1990. Isozyme variation in the *Acremonium/Epichloë* fungal endophyte complex. *Phytopathology* 80:1133–1139.

Leuchtmann, A., and K. Clay. 1993. Nonreciprocal compatibility between *Epichloë typhina* and four host grasses. *Mycologia* 85:157–163.

Leuchtmann, A., and K. Clay. 1996. Isozyme evidence for host races of the fungus *Atkinsonella hypoxylon* (Clavicipitaceae) infecting the *Danthonia* (Poaceae) complex in the southern Appalachians. *American Journal of Botany* 83:1144–1152.

Leuchtmann, A., and K. Clay. 1997. The population biology of grass endophytes. In *The Mycota. V. Plant Relationships. Part B*, eds. G. C. Carroll and P. Tudzynski, pp. 185–204. Berlin: Springer-Verlag.

Leuchtmann, A., D. Schmidt, and L. P. Bush. 2000. Different levels of protective alkaloids in grasses with stroma-forming and seed-transmitted *Epichloë/Neotyphodium* endophytes. *Journal of Chemical Ecology* 26:1025–1036.

Levine, J. M., and C. M. D'Antonio. 1999. Elton revisited: a review of evidence linking diversity and invasibility. *Oikos* 87:15–26.

Lewis, D. H. 1985. Symbiosis and mutualism: crisp concepts and soggy semantics. In *The biology of mutualism*, ed. D. H. Boucher, pp. 29–39. New York: Oxford University Press.

Lewis, G. C. 1992. Effect of ryegrass endophyte in mixed swards of perennial ryegrass and white clover under two levels of irrigation and pesticide treatment. *Grass and Forage Science* 47:302–305.

Lewis, G. C. 2004. Effects of biotic and abiotic stress on the growth of three genotypes of *Lolium perenne* with and without infection by the fungal endophyte *Neotyphodium lolii*. *Annals of Applied Biology* 144:53–63.

Lewis, G. C., A. K. Bakken, J. H. MacDuff, and N. Raistrick. 1996. Effect of infection by the endophytic fungus *Acremonium lolii* on growth and nitrogen uptake by perennial ryegrass (*Lolium perenne*) in flowing solution culture. *Annals of Applied Biology* 129:451–460.

Lewis, G. C., and R. O. Clement. 1986. A survey of ryegrass endophyte in the U.K. and its apparent ineffectuality on a seedling pest. *Journal of Agricultural Sciences Cambridge* 107:633–638.

Lewis, G. C., C. Ravel, W. Naffaa, C. Astier, and G. Charmet. 1997. Occurrence of *Acremonium* endophytes in wild populations of *Lolium* spp. in European countries and a relationship between level of infection and climate in France. *Annals of Applied Biology* 130:227–238.

Lewis, J. 1973. Longevity of crop and weed seeds: survival after 20 years in soil. *Weed Research* 13:179–191.

Leyronas, C., B. Mériaux, and G. Raynal. 2006. Chemical control of *Neotyphodium* spp. endophytes in perennial ryegrass and tall fescue seeds. *Crop Science* 46:98–104.

Li, B., X. Zheng, and S. Sun. 1997a. A survey of endophytic fungi in some native forage grasses of Northwestern China. In *Neotyphodium/grass interactions*, eds. C. W. Bacon and N. S. Hill, pp. 69–71. New York: Plenum Press.

Li, C., Z. Nan, J. Gao, and P. Tian. 1997b. Detection and distribution of *Neotyphodium-Achnatherum inebrians* association in China. In *Proceedings of the 5th International Symposium on Neotyphodium/Grass Interactions*, eds. R. Kallenbach, C. Rosenkraus, Jr., and T. R. Lock, Fayetteville, AR, pp. 210–212.

Li, C. J., Z. B. Nan, V. H. Paul, P. D. Dapprich, and Y. Liu. 2004. *Neotyphodium* species symbiotic with drunken horse grass (*Achnatherum inebrians*) in China. *Mycotaxon* 90: 141–147.

Linhart, Y. B., K. Keefover-Ring, K. A. Mooney, B. Breland, and J. D. Thompson. 2005. A chemical polymorphism in a multitrophic setting: thyme monoterpene composition and food web structure. *American Naturalist* 166:517–529.

Lipsitch, M., A. Nowak, D. Ebert, and R. May. 1995. The population dynamics of vertically and horizontally transmitted parasites. *Proceedings of the Royal Society of London, Series B – Biological Sciences* 260:321–327.

Lipsitch, M., S. Sillerand, and A. Nowak. 1996. The evolution of virulence in pathogens with vertical and horizontal transmission. *Evolution* 50:1729–1741.

Logendra, S., and M. D. Richardson. 1997. Ergosterol as an indicator of endophyte biomass in grass tissue. In *Neotyphodium/grass interactions*, eds. C. W. Bacon and N. S. Hill, pp. 267–270. New York: Plenum Press.

Lopez, J. E., S. H. Faeth, and M. Miller. 1995. The effect of endophytic fungi on herbivory by red-legged grasshoppers (Orthoptera: Acrididae) on Arizona fescue. *Environmental Entomology* 24:1576–1580.

Loreau, M., S. Naeem, P. Inchausti, J. Bengtsson, J. P. Grime, A. Hector, D. U. Hooper, M. A. Huston, D. Raffaelli, B. Schmid, D. Tilman, and D. A. Wardle. 2001. Ecological-biodiversity and ecosystem functioning: current knowledge and future challenges. *Science* 294:804–808.

Lyons, P. C., J. J. Evans, and C. W. Bacon. 1990. Effects of the fungal endophyte *Acremonium coenophialum* on nitrogen accumulation and metabolism in tall fescue. *Plant Physiology* 92:726–732.

Lyons, P. C., R. D. Plattner, and C. W. Bacon. 1986. Occurrence of peptide and clavine ergot alkaloids in tall fescue grass. *Science* 232:487–489.

Mack, K. M. L., and J. A. Rudgers. 2008. Balancing multiple mutualists: asymmetric interactions among plants, arbuscular mycorrhizal fungi, and fungal endophytes. *Oikos* 117:310–320.

Mack, M. C., and C. M. D'Antonio. 2003. The effects of exotic grasses on litter decomposition in a Hawaiian woodland: the importance of indirect effects. *Ecosystems* 6:723–738.

Maclean, B., C. Matthew, G. C. M. Latch, and D. J. Barker. 1993. The effect of endophyte on drought resistance in tall fescue. In *Proceedings of the Second International Symposium on Acremonium/Grass Interactions*, eds. D. E. Hume, G. C. M. Latch, and H. S. Easton, Palmerston North, New Zealand, pp. 165–169.

Malinowski, D. P., G. A. Alloush, and D. P. Belesky. 2000. Leaf endophyte *Neotyphodium coenophialum* modifies mineral uptake in tall fescue. *Plant and Soil* 227:115–126.

Malinowski, D. P., and D. P. Belesky. 2000. Adaptations of endophyte-infected cool-season grasses to environmental stresses: mechanisms of drought and mineral stress tolerance. *Crop Science* 40:923–940.

Malinowski, D. P., and D. P. Belesky. 2006. Ecological importance of *Neotyphodium* spp. grass endophytes in agroecosystems. *Grassland Science* 52:1–14.

Malinowski, D. P., D. P. Belesky, and J. M. Fedders. 1999a. Endophyte infection may affect the competitive ability of tall fescue grown with red clover. *Journal of Agronomy and Crop Science* 183:91–101.

Malinowski, D. P., D. P. Belesky, N. S. Hill, V. C. Baligar, and J. M. Fedders. 1998. Influence of phosphorus on the growth and ergot alkaloid content of *Neotyphodium coenophialum*-infected tall fescue (*Festuca arundinacea* Schreb.). *Plant and Soil* 198:53–61.

Malinowski, D. P., D. P. Belesky, and G. C. Lewis. 2005. Abiotic stresses in endophytic grasses. In *Neotyphodium in cool-season grasses*, eds. C. A. Roberts, C. P. West, and D. E. Spiers, pp. 187–199. Ames, IA: Blackwell Publications.

Malinowski, D. P., D. K. Brauer, and D. P. Belesky. 1999b. The endophyte *Neotyphodium coenophialum* affects root morphology of tall fescue grown under phosphorus deficiency. *Journal of Agronomy and Crop Science* 183:53–60.

Malinowski, D. P., A. Leuchtmann, D. Schmidt, and J. Nösberger. 1997a. Growth and water status in meadow fescue is affected by *Neotyphodium* and *Phialophora* species endophytes. *Agronomy Journal* 89:673–678.

Malinowski, D. P., A. Leuchtmann, D. Schmidt, and J. Nösberger. 1997b. Symbiosis with *Neotyphodium uncinatum* endophyte may increase the competitive ability of meadow fescue. *Agronomy Journal* 89:833–839.

Margulis, L. 1991. Symbiogenesis and symbionticism. In *Symbiosis as a source of evolutionary innovation*, eds. L. Margulis and R. Fester, pp. 1–14. Cambridge, MA: MIT Press.

Marks S., and K. Clay. 1990. Effects of CO_2 enrichment, nutrient addition, and fungal endophyte-infection on the growth of two grasses. *Oecologia* 84:207–214.

Marks S., and K. Clay. 1996. Physiological responses of *Festuca arundinacea* to fungal endophyte infection. *New Phytologist* 133:727–733.

Marks S., and K. Clay. 2007. Low resource availability differentially affects the growth of host grasses infected by fungal endophytes. *International Journal of Plant Sciences* 168:1269–1277.

Marks, S., K. Clay, and G. P. Cheplick. 1991. Effects of fungal endophytes on interspecific and intraspecific competition in the grasses *Festuca arundinacea* and *Lolium perenne*. *Journal of Applied Ecology* 28:194–204.

Marks, S., and D. E. Lincoln. 1996. Antiherbivore defense mutualism under elevated carbon dioxide levels: a fungal endophyte and grass. *Environmental Entomology* 25:618–623.

Márquez, L. M., R. S. Redman, R. J. Rodriquez, and M. J. Roossinck. 2007. A virus in a fungus in a plant: three-way symbiosis required for thermal tolerance. *Science* 315:513–515.

Márquez, S. S., G. F. Bills, and I. Zabalgogeazcoa. 2007. The endophytic community of *Dactylis glomerata*. In *Proceedings of the 6th International Symposium on Fungal Endophytes of Grasses*, eds. A. J. Popay and E. R. Thom, pp. 69–73. Christchurch, New Zealand: New Zealand Grassland Association.

Marsh, C. D., and A. B. Clawson. 1929. Sleepy grass (*Stipa vaseyi*) as a stock-poisoning plant. Technical Bulletin 114. Washington, DC: U.S. Department of Agriculture.

Marshall, D., B. Tunali, and L. R. Nelson. 1999. Occurrence of fungal endophytes in species of wild *Triticum*. *Crop Science* 39:1507–1512.

Maschinski, J., and T. G. Whitham. 1989. The continuum of plant responses to herbivory: the influence of plant association, nutrient availability, and timing. *American Naturalist* 134:1–9.

Massad, E. 1987. Transmission rates and the evolution of pathogenicity. *Evolution* 41:1127–1130.

Matthews, J. W., and K. Clay. 2001. Influence of fungal endophyte infection on plant-soil feedback and community interactions. *Ecology* 82:500–509.

Mayer, P. M., S. J. Tunnell, D. M. Engle, E. E. Jorgensen, and P. Nunn. 2005. Invasive grass alters litter decomposition by influencing macrodetritivores. *Ecosystems* 8:200–209.

McCormick, M. K., K. L. Gross, and R. A. Smith. 2001. *Danthonia spicata* (Poaceae) and *Atkinsonella hypoxylon* (Balansiae): environmental dependence of a symbiosis. *American Journal of Botany* 88:903–909.

McKone, M. J. 1989. Intraspecific variation in pollen yield in bromegrass (Poaceae; *Bromus*). *American Journal of Botany* 76:231–237.

McKone, M. J., C. P. Lund, and J. M. O'Brien. 1998. Reproductive biology of two dominant prairie grasses (*Andropogon gerardii* and *Sorghastrum nutans*; Poaceae): male-biased sex allocation in wind-pollinated plants? *American Journal of Botany* 85:776–783.

McNaughton, S. J. 1984. Grazing lawns—animals in herds, plant form, and coevolution. *American Naturalist* 124:863–886.

Meijer, G., and A. Leuchtmann. 1999. Multistrain infections of the grass *Brachypodium sylvaticum* by its fungal endophyte *Epichloë sylvatica*. *New Phytologist* 141:355–368.

Meijer, G., and A. Leuchtmann. 2000. The effects of genetic and environmental factors on disease expression (stroma formation) and plant growth in *Brachypodium sylvaticum* infected by *Epichloë sylvatica*. Oikos 91:446–458.

Meijer, G., and A. Leuchtmann. 2001. Fungal genotype controls mutualism and sex in *Brachypodium sylvaticum* infected by *Epichloë sylvatica*. *Acta Biologica Hungarica* 52:249–263.

Meister B., J. Krauss, S. A. Harri, M. V. Schneider, and C. B. Müller. 2006. Fungal endosymbionts affect aphid population size by reduction of adult life span and fecundity. *Basic and Applied Ecology* 7:244–252.

Merilä, J., and M. Björklund. 2004. Phenotypic integration as a constraint and adaptation. In *Phenotypic integration: studying the ecology and evolution of complex phenotypes*, eds. M. Pigliucci and K. Preston, pp. 107–129. Oxford: Oxford University Press.

Miles, C. O., M. E. di Menna, S. W. L. Jacobs, I. Garthwaite, G. A. Lane, R. A. Prestidge, S. L. Marshall, H. W. Wilkinson, C. L. Schardl, O. J.-P. Ball, and G. C. M. Latch. 1998. Endophytic fungi in indigenous Australasian grasses associated with toxicity to livestock. *Applied and Environmental Microbiology* 64:601–606.

Miles, C. O., G. A. Lane, M. E. Di Menna, I. Garthwaite, E. L. Piper, O. J.-P. Ball, G. C. M. Latch, J. M. Allen, M. B. Hunt, L. P. Bush, F. K. Min, I. Fletcher, and P. S. Harris. 1996. High levels of ergonovine and lysergic acid amide in toxic *Achnatherum inebrians* accompany infection by an *Acremonium*-like endophytic fungus. *Journal of Agricultural and Food Chemistry* 44:1285–1290.

Mills, N. J., and W. M. Getz. 1996. Modelling the biological control of insect pests: a review of host-parasitoid models. *Ecological Modelling* 92:121–143.

Mirlohi, A., M. R. Sabzalian, B. Sharifnabi, and M. K. Nekoui. 2006. Widespread occurrence of *Neotyphodium*-like endophyte in populations of *Bromus tomentellus* Boiss. in Iran. *FEMS Microbiology Letters* 256:126–131.

Mitchell, C. E., and A. G. Power. 2003. Release of invasive plants from fungal and viral pathogens. *Nature* 421:625–627.

Monnet, F., N. Vaillant, A. Hitmi, and H. Sallanon. 2005. Photosynthetic activity of *Lolium perenne* as a function of endophyte status and zinc nutrition. *Functional Plant Biology* 32:131–139.

Moon, C. D., K. D. Craven, A. Leuchtmann, S. L. Clement, and C. L. Schardl. 2004. Prevalence of interspecific hybrids amongst asexual fungal endophytes of grasses. *Molecular Ecology* 13:1455–1467.

Moon, C. D., C. O. Miles, U. Järlfors, and C. L. Schardl. 2002. The evolutionary origins of three new *Neotyphodium* endophyte species from grasses indigenous to the Southern Hemisphere. *Mycologia* 94:694–711.

Moon, C. D., B. Scott, C. L. Schardl, and M. J. Christensen. 2000. The evolutionary origins of *Epichloë* endophytes from annual ryegrasses. *Mycologia* 92:1103–1118.

Moran, N. A. 2007. Symbiosis as an adaptive process and source of phenotypic complexity. *Proceedings of the National Academy of Sciences USA* 104:8627–8633.

Moran, N. A., and J. J. Wernegreen. 2000. Lifestyle evolution in symbiotic bacteria: insights from genomics. *Trends in Ecology and Evolution* 15:321–326.

Moranz, R., and L. P. Brower. 1998. Geographic and temporal variation of cardenolide-based chemical defenses of queen butterfly (*Danaus gilippus*) in northern Florida. *Journal of Chemical Ecology* 24:905–922.

Morin, P. J. 1999. *Community ecology.* Malden, MA: Blackwell Science.

Morin, P. J. 2003. Community ecology and the genetics of interacting species. *Ecology* 84:577–580.

Morse, L. J., T. A. Day, and S. H. Faeth. 2002. Effect of *Neotyphodium* endophyte infection on growth and leaf gas exchange of Arizona fescue under contrasting water availability regimes. *Environmental and Experimental Botany* 48:257–268.

Morse, L. J., S. H. Faeth, and T. A. Day. 2007. *Neotyphodium* interactions with a wild grass are driven mainly by endophyte haplotype. *Functional Ecology* 21:813–822.

Moy, M., F. Belanger, R. Duncan, A. Feedhoff, C. Leary, R. Sullivan, and J. F. White, Jr. 2000. Identification of epiphyllous nets on leaves of grasses infected by clavicipitaceous endophytes. *Symbiosis* 28:291–302.

Müller, C. B., and J. Krauss. 2005. Symbiosis between grasses and asexual fungal endophytes. *Current Opinion in Plant Biology* 8:450–456.

Müller, J. 2003. Artificial infection by endophytes affects growth and mycorrhizal colonisation of *Lolium perenne*. *Functional Plant Biology* 30:419–424.

Nan, Z. B., and Li, C. J. 2001. *Neotyphodium* in native grasses in China and observations on endophyte/host interactions. In *Proceedings of the Fourth International Neotyphodium/Grass Interactions Symposium*, Fachbereich Agrarwirtschaft, Soest, Germany, eds. V. H. Paul. and P. D. Dapprich, pp. 41–50.

Neil, K., R. T. Tiller, and S. H. Faeth. 2003. Germination success of big Sacaton and *Neotyphodium*-infected and uninfected Arizona fescue. *Journal of Range Management* 56:612–622.

Nelson, C. J. 1988. Genetic association between photosynthetic characteristics and yield: review of the evidence. *Plant Physiology and Biochemistry* 26:543–554.

Nelson, C. J. 1996. Physiology and developmental morphology. In *Cool-season forage grasses*, eds. L. E. Moser, D. R. Buxton, and M. D. Casler, pp. 87–125. Madison, WI: American Society of Agronomy, Crop Science Society of America, and Soil Science Society of America.

Neuhauser, C., D. A. Andow, G. E. Heimpel, G. May, R. G. Shaw, and S. Wagenius. 2003. Community genetics: expanding the synthesis of ecology and genetics. *Ecology* 84:545–558.

Newman, J. A., M. L. Abner, R. G. Dado, D. J. Gibson, A. Brookings, and A. J. Parsons. 2003. Effects of elevated CO_2, nitrogen and fungal endophyte-infection on tall fescue: growth, photosynthesis, chemical composition and digestibility. *Global Change Biology* 9:425–437.

Newsham, K. K., G. C. Lewis, P. D. Greenslade, and A. R. McLeod. 1998. *Neotyphodium lolii*, a fungal leaf endophyte, reduces fertility of *Lolium perenne* exposed to elevated UV-B radiation. *Annals of Botany* 81:397–403.

Noss, R. F., E. T. La Roe, and J. M. Scott. 1995. Endangered ecosystems of the United States: a preliminary assessment of loss and degradation. Biological Report 28. Washington, DC: USGS National Biological Resources Division.

Novas, M. V., A. Gentile, and D. Cabral. 2003. Comparative study of growth parameters on diaspores and seedlings between populations of *Bromus setifolius* from Patagonia, differing in *Neotyphodium* endophyte infection. *Flora* 198:421–426.

Novas, M. V., D. Cabral, and A. M. Godeas. 2005. Interaction between grass endophytes and mycorrhizae in *Bromus setifolius* from Patagonia, Argentina. *Symbiosis* 40:23–30.

Nuismer, S. L., J. N. Thompson, and R. Gomulkiewicz. 1999. Gene flow and geographically structured coevolution. *Proceedings of the Royal Society of London, Series B – Biological Sciences* 266:605–609.

Nuismer, S. L., J. N. Thompson, and R. Gomulkiewicz. 2000. Coevolutionary clines across selection mosaics. *Evolution* 54:1102–1115.

Nurminiemi, M., J. Tufto, N. O. Nilsson, and O. A. Rognli. 1998. Spatial models of pollen dispersal in the forage grass meadow fescue. *Evolutionary Ecology* 12:487–502.

Obeso, J. R. 2002. The costs of reproduction in plants. *New Phytologist* 155:321–348.

Ohnmeiss, T. E., and I. T. Baldwin.1994. The allometry of nitrogen allocation to growth and an inducible defense under nitrogen-limited growth. *Ecology* 75:995–1002.

Olejniczak, P., and M. Lembicz. 2007. Age-specific response of the grass *Puccinellia distans* to the presence of a fungal endophyte. *Oecologia* 152:485–494.

Olff, H., and M. E. Ritchie. 1998. Effects of herbivores on grassland plant diversity. *Trends in Ecology and Evolution* 13:261–265.

Omacini, M., E. J. Chaneton, and C. M. Ghersa. 2005. A hierarchical framework for understanding the ecosystem consequences of endophyte-grass symbiosis. In *Neotyphodium in cool-season grasses*, eds. C. A. Roberts, C. P. West, and D. E. Spiers, pp. 141–162. Ames, IA: Blackwell Publications.

Omacini, M., E. J. Chaneton, C. M. Ghersa, and C. B. Müller. 2001. Symbiotic fungal endophytes control insect host-parasite interaction webs. *Nature* 409:78–81.

Omacini, M., E. J. Chaneton, C. M. Ghersa, and P. Otero. 2004. Do foliar endophytes affect grass litter decomposition? A microcosm approach using *Lolium multiflorum*. *Oikos* 104: 581–590.

Omacini, M., T. Eggers, M. Bonkowski, A. C. Gange, and T. H. Jones. 2006. Leaf endophytes affect mycorrhizal status and growth of co-infected and neighbouring plants. *Functional Ecology* 20:226–232.

Orr, S. P., J. A. Rudgers, and K. Clay. 2005. Invasive plants can inhibit native tree seedlings: testing potential allelopathic mechanisms. *Plant Ecology* 181:153–165.

Paige, K. N., and T. G. Whitham. 1992. Overcompensation in response to mammalian herbivory: the advantage of being eaten. *American Naturalist* 129:407–416.

Pan, J. J., and K. Clay. 2002. Infection by the systemic fungus *Epichloë glyceriae* and clonal growth of its host grass *Glyceria striata*. *Oikos* 98:37–46.

Pan, J. J., and K. Clay. 2003. Infection by the systemic fungus *Epichloë glyceriae* alters clonal growth of its grass host, *Glyceria striata*. *Proceedings of the Royal Society of London, Series B – Biological Sciences* 270:1585–1591.

Pan, J. J., and K. Clay. 2004. *Epichloë glyceriae* infection affects carbon translocation in the clonal grass *Glyceria striata*. *New Phytologist* 164:467–475.

Paracer, S., and V. Ahmadjian. 2000. *Symbiosis: an introduction to biological associations*. New York: Oxford University Press.

Parker, M. A. 1995. Plant fitness variation caused by different mutualist genotypes. *Ecology* 76:1525–1535.

Pavlu, V., M. Hejcman, L. Pavlu, J. Gaisler, P. Hejcmanova-Nezerkova, and L. Meneses. 2006. Changes in plant densities in a mesic species-rich grassland after imposing different grazing management treatments. *Grass and Forage Science* 61:42–51.

Pellmyr, O., J. Leebens-Mack, and C. J. Huth. 1996. Non-mutualistic yucca moths and their evolutionary consequences. *Nature* 380:155–156.

Peters, E. J., and A. H. B. M. Zam. 1981. Allelopathic effects of tall fescue genotypes. *Agronomy Journal* 73:56–58.

Petrini, O., T. N. Sieber, L. Toti, and O. Viret. 1992. Ecology, metabolite production, and substrate utilization in endophytic fungi. *Natural Toxins* 1:185–196.

Petroski, R. J., R. G. Powell, and K. Clay. 1992. Alkaloids of *Stipa robusta* (sleepygrass) infected with an *Acremonium* endophyte. *Natural Toxins* 1:84–88.

Pfender, W. F., and S. C. Alderman. 1999. Geographical distribution and incidence of orchardgrass choke, caused by *Epichloë typhina*, in Oregon. *Plant Disease* 83:754–758.

Piano, E., F. B. Bertoli, M. Romani, A. Tava, L. Riccioni, M. Valvassori, A. M. Carroni, and L. Pecetti. 2005. Specificity of host-endophyte association in tall fescue populations from Sardinia, Italy. *Crop Science* 45:1456–1463.

Pigliucci, M. 2004. Studying the plasticity of phenotypic integration in a model organism. In *Phenotypic integration: studying the ecology and evolution of complex phenotypes*, eds. M. Pigliucci and K. Preston, pp. 155–175. Oxford: Oxford University Press.

Pimentel, D. 1988. Herbivore population feeding pressure on plant hosts: feedback evolution and host conservation. *Oikos* 53:289–302.

Popay, A. J. 1997. Tiller mortality in mixtures of endophyte-free ryegrass and ryegrass infected with two different endophytes. In *Neotyphodium/grass interactions*, eds. C. W. Bacon and N. S. Hill, pp. 191–193. New York: Plenum Press.

Popay, A. J., and S. A. Bonos. 2005. Biotic responses in endophytic grasses. In *Neotyphodium in cool-season grasses*, eds. C. A. Roberts, C. P. West, and D. E. Spiers, pp. 163–185. Ames, IA: Blackwell Publishing.

Popay, A. J., and D. D. Rowan. 1994. Endophytic fungi as mediators of plant-insect interactions. In *Insect-plant interactions*, vol. V, ed. E. A. Bernays, pp. 83–103. Boca Raton, FL: CRC Press.

Popay, A. J., and E. R. Thom, eds. 2007. *Proceedings of the 6th International Symposium on Fungal Endophytes of Grasses*. Christchurch, New Zealand: New Zealand Grassland Association.

Popay, A. J., and R. T. Wyatt. 1995. Resistance to Argentine stem weevil in perennial ryegrass infected with endophytes producing different alkaloids. *Proceedings of the 48th New Zealand Plant Protection Conference*, pp. 229–236.

Post, D. M. 2002. The long and short of food-chain length. *Trends in Ecology and Evolution* 17:269–277.

Poulsen, M., and J. J. Boomsa. 2005. Mutualistic fungi control crop diversity in fungus-growing ants. *Science* 307:741–744.

Powell, R. G., and R. J. Petroski. 1992. Alkaloid toxins in endophyte-infected grasses. *Natural Toxins* 1: 163–170.

Powell, R. G., M. R. Tepaske, R. D. Plattner, J. F. White, and S. L. Clement. 1994. Isolation of resveratrol from *Festuca versuta* and evidence for the widespread occurrence of this stilbene in the Poaceae. *Phytochemistry* 35:335–338.

Power, M. E. 1992. Top-down and bottom-up forces in food webs: do plants have primacy? *Ecology* 73:733–746.

Prestidge, R. A. 1993. Causes and control of perennial ryegrass staggers in New Zealand. *Agriculture, Ecosystems and Environment* 44:283–300.

Prestidge, R. A., and R. T. Gallagher. 1988. Endophyte fungus confers resistance to ryegrass: Argentine stem weevil larval studies. *Ecological Entomology* 13:429–435.

Prestidge, R. A., R. P. Pottinger, and G. M. Barker. 1982. An association of *Lolium* endophyte with ryegrass resistance to Argentine stem weevil. In *Proceedings of the 35th New Zealand Weed and Pest Control Conference*, pp. 119–122.

Preston, K. A., and D. D. Ackerly. 2004. The evolution of allometry in modular organisms. In *Phenotypic integration: studying the ecology and evolution of complex phenotypes*, eds. M. Pigliucci and K. Preston, pp. 80–106. Oxford: Oxford University Press.

Price, P. W. 1984. *Insect ecology*, 2nd ed. New York: John Wiley & Sons.

Priestley, D. A. 1986. *Seed aging: implications for seed storage and persistence in the soil*. Ithaca, NY: Cornell University Press.

Pringle, A., and J. W. Taylor. 2002. The fitness of filamentous fungi. *Trends in Microbiology* 10:474–481.

Quinn, J. F., and J. R. Karr. 1993. Habitat fragmentation and global change. In *Biotic interactions and global change*, eds. P. M. Kareiva, J. G. Kingsolver, and R. B. Huey, pp. 451–463. Sunderland, MA: Sinauer Associates.

Rahman, M. H., and S. Saiga. 2005. Endophytic fungi (*Neotyphodium coenophialum*) affect the growth and mineral uptake, transport and efficiency ratios in tall fescue (*Festuca arundinacea*). *Plant and Soil* 272:163–171.

Rambo, J. L., and S. H. Faeth. 1999. Effect of vertebrate grazing on plant and insect community structure. *Conservation Biology* 13:1047–1054.

Rao, S., S. C. Alderman, J. Takeysu, and B. Matson. 2005. The *Botanophila-Epichloë* association in cultivated *Festuca* in Oregon: evidence of simple fungivory. *Entomologia Experimentalis et Applicata* 115:427–433.

Rao, S., and D. Baumann. 2004. The interaction of a *Botanophila* fly species with an exotic *Epichloë* fungus in a cultivated grass: fungivore or mutualist? *Entomologica Experimentalis et Applicata* 112:99–105.

Rasmussen, S., A. J. Parsons, S. Bassett, M. J. Christensen, D. E. Hume, L. J. Johnson, R. D. Johnson, W. R. Simpson, C. Stacke, C. R. Voisey, H. Xue, and J. A. Newman. 2007. High nitrogen supply and carbohydrate content reduce fungal endophyte and alkaloid concentration in *Lolium perenne*. *New Phytologist* 173:787–797.

Rasmussen, S., A. J. Parsons, K. Fraser, H. Xue, and J. A. Newman. 2008. Metabolic profiles of *Lolium perenne* are differentially affected by nitrogen supply, carbohydrate content, and fungal endophyte infection. *Plant Physiology* 146:1440–1453.

Ravel, C., G. Charmet, and F. Balfourier. 1995. Influence of the fungal endophyte *Acremonium lolii* on agronomic traits of perennial ryegrass in France. *Grass and Forage Science* 50:75–80.

Ravel, C., F. Balfourier, and J. J. Guillaumin. 1999. Enhancement of yield and persistence of perennial ryegrass inoculated with one endophyte isolate in France. *Agronomie* 19:635–644.

Ravel, C., C. Courty, A. Coudret, and G. Charmet. 1997a. Beneficial effects of *Neotyphodium lolii* on the growth and the water status in perennial ryegrass cultivated under nitrogen deficiency or drought stress. *Agronomie* 17:173–181.

Ravel, C., Y. Michalakis, and G. Charmet. 1997b. The effect of imperfect transmission on the frequency of mutualistic seed-borne endophytes in natural populations of grasses. *Oikos* 80:18–24.

Read, J. C., and B. J. Camp. 1986. The effect of the fungal endophyte *Acremonium coenophialum* in tall fescue on animal performance, toxicity, and stand maintenance. *Agronomy Journal* 78:848–850.

Redlin, S. C., and L. M. Carris, eds. 1996. *Endophytic fungi in grasses and woody plants.* St. Paul, MN: American Phytopathological Society Press.

Redman, R. S., D. D. Dunigan, and R. J. Rodriguez. 2001. Fungal symbiosis from mutualism to parasitism: who controls the outcome, host or invader? *New Phytologist* 151:705–716.

Redman, R. S., K. B. Sheehan, R. G. Stout, R. J. Rodriquez, and J. M. Henson. 2002. Thermotolerance generated by plant/fungal symbiosis. *Science* 298:1581–1581.

Ren, A. Z., Y. B. Gao, and F. Zhou. 2007. Response of *Neotyphodium lolii*-infected perennial ryegrass to phosphorus deficiency. *Plant, Soil and Environment* 53:113–119.

Renne, I. G., B. J. Rios, J. S. Fehmi, and B. F. Tracy. 2004. Low allelopathic potential of an invasive forage grass on native grassland plants: a cause for encouragement? *Basic and Applied Ecology* 5:261–269.

Reynolds, H. L., A. E. Hartley, K. M. Vogelsang, J. D. Bever, and P. A. Schultz. 2005. Arbuscular mycorrhizal fungi do not enhance nitrogen acquisition and growth of old-field perennials under low nitrogen supply in glasshouse culture. *New Phytologist* 167:869–880.

Rice, J. S., B. W. Pinkerton, W. C. Stringer, and D. J. Undersander. 1990. Seed production in tall fescue as affected by fungal endophyte. *Crop Science* 30:1303–1305.

Richardson, D. M., N. Allsopp, C. M. D'Antonio, S. J. Milton, and M. Rejmanek. 2000. Plant invasions—the role of mutualisms. *Biological Reviews* 75:65–93.

Richardson, M. D., G. W. Chapman, Jr., C. S. Hoveland, and C. W. Bacon. 1992. Sugar alcohols in endophyte-infected tall fescue. *Crop Science* 32:1060–1061.

Richardson, M. D., C. S. Hoveland, and C. W. Bacon. 1993. Photosynthesis and stomatal conductance of symbiotic and nonsymbiotic tall fescue. *Crop Science* 33:145–149.

Richmond, D. S., J. Cardina, and P. S. Grewal. 2006. Influence of grass species and endophyte infection on weed populations during establishment of low-maintenance lawns. *Agriculture, Ecosystems and Environment* 115:27–33.

Richmond, D. S., P. S. Grewal, and J. Cardina. 2003. Competition between *Lolium perenne* and *Digitaria sanguinalis*: ecological consequences for harbouring an endosymbiotic fungus. *Journal of Vegetation Science* 14:835–840.

Richmond, D. S., P. S. Grewal, and J. Cardina. 2004a. Influence of Japanese beetle (*Popillia japonica*) larvae and fungal endophytes on competition between turfgrasses and dandelions. *Crop Science* 44:600–606.

Richmond, D. S., B. A. Kunkel, N. Somesekar, and P. S. Grewal. 2004b. Top-down and bottom-up regulation of herbivores: *Spodoptera frugiperda* turns tables on endophyte-mediated plant defence and virulence of an entomopathogenic nematode. *Ecologica Entomology* 29:353–360.

Roberts, C. A., C. P. West, and D. E. Spiers, eds. 2005. *Neotyphodium* in cool-season grasses. Ames, IA: Blackwell Publishers.

Rodriguez, R. J., R. S. Redman, and J. M. Henson. 2004. The role of fungal symbionts in the adaptation of plants to high stress environments. *Mitigation and Adaptation Strategies for Global Change* 9:261–272.

Rodriguez, R. J., R. S. Redman, and J. M. Henson. 2005. Symbiotic lifestyle expression by fungal endophytes and the adaptation of plants to stress: unraveling the complexities of intimacy. In *The fungal community: its organization and role in the ecosystem*, 3rd ed., eds. J. Dighton, J. F. White, and P. Oudemans, pp. 683–695. Boca Raton, FL: CRC Press.

Rodriguez-Saona, C., and J. S. Thaler. 2005. Herbivore-induced responses and patch heterogeneity affect abundance of arthropods on plants. *Ecological Entomology* 30:156–163.

Rolston, M. P., M. D. Hare, K. K. Moore, and M. J. Christensen. 1986. Viability of *Lolium* endophyte fungus in seed stored at different moisture contents and temperatures. *New Zealand Journal of Experimental Agriculture* 14:297–300.

Root, R. B. 1973. Organisation of a plant-arthropod association in simple and diverse habitats: the fauna of collards (*Brassica oleracea*). *Ecological Monographs* 43:95–124.

Rowan, D. D., J. J. Dymock, and M. A. Brimble. 1990. Effect of fungal metabolite peramine and analogs on feeding and development of Argentine stem weevil (*Listronotus bonariensis*). *Journal of Chemical Ecology* 16:1683–1695.

Rowan, D. D., and G. C. M Latch. 1994. Utilization of endophyte-infected perennial ryegrass for increased insect resistance. In *Biotechnology of endophytic fungi of grasses*, eds. C. W. Bacon and J. F. White, Jr., pp. 169–183. Boca Raton, FL: CRC Press.

Roylance, J. T., N. S. Hill, and C. S. Agee. 1994. Ergovaline and peramine production in endophyte-infected tall fescue: independent regulation and effects of plant and endophyte genotype. *Journal of Chemical Ecology* 20:2171–2183.

Rudgers, J. A., and K. Clay. 2005. Fungal endophytes in terrestrial communities and ecosystems. In *The fungal community: its organization and role in the ecosystem*, 3rd ed., eds. J. Dighton, J. F. White, and P. Oudemans, pp. 423–442. Boca Raton, FL: CRC Press.

Rudgers, J. A., J. Holah, S. P. Orr, and K. Clay. 2007. Forest succession suppressed by an introduced plant-fungal symbiosis. *Ecology* 88:18–25.

Rudgers, J. A., J. M. Koslow, and K. Clay. 2004. Endophytic fungi alter relationships between diversity and ecosystem properties. *Ecology Letters* 7:42–51.

Rudgers, J. A., W. B. Mattingly, and J. M. Koslow. 2005. Mutualistic fungus promotes plant invasion into diverse communities. *Oecologia* 144: 463–471.

Ryley, M. J. 2003. *Nigrocornus scleroticus*, a common Old World Balansioid fungus. In *Clavicipitalean fungi: evolutionary biology, chemistry, biocontrol, and cultural impacts*, eds. J. F. White, Jr., C. W. Bacon, N. L. Hywel-Jones, and J. W. Spatafora, pp. 247–271. New York: Marcel Dekker.

Sabo, J. L., J. L. Bastow, and M. E. Power. 2002. Length-mass relationships for adult aquatic and terrestrial invertebrates in a California watershed. *Journal of the North American Benthological Society* 21:336–343.

Saccheri, I., and I. Hanski. 2006. Natural selection and population dynamics. *Trends in Ecology and Evolution* 21:341–347.

Sachs, J. L., and E. L. Simms. 2006. Pathways to mutualism breakdown. *Trends in Ecology and Evolution* 21:585–592.

Saikkonen, K. 2000. Kentucky 31, far from home. *Science* 287:1887.

Saikkonen, K., J. Ahlholm, M. Helander, S. Lehtimäki, and O. Niemeläinen. 2000. Endophytic fungi in wild and cultivated grasses in Finland. *Ecography* 23:360–366.

Saikkonen, K., S. H. Faeth, M. Helander, and T. J. Sullivan. 1998. Fungal endophytes: a continuum of interactions with host plants. *Annual Review of Ecology and Systematics* 29: 319–343.

Saikkonen, K., M. Helander, S. H. Faeth, F. Schulthess, and D. Wilson. 1999. Endophyte-grass-herbivore interactions: the case of *Neotyphodium* endophytes in Arizona fescue populations. *Oecologia* 121: 411–420.

Saikkonen, K., D. Ion, and M. Gyllenberg. 2002. The persistence of vertically transmitted fungi in grass metapopulations. *Proceedings of the Royal Society of London, Series B – Biological Sciences* 269:1397–1403.

Saikkonen, K., P. Lehtonen, M. Helander, J. Koricheva, and S. H. Faeth. 2006. Model systems in ecology: dissecting the endophyte-grass literature. *Trends in Plant Science* 11:428–433.

Saikkonen, K., P. Wäli, M. Helander, and S. H. Faeth. 2004. Evolution of endophyte-plant symbioses. *Trends in Plant Science* 9:275–280.

Salvaudon, L., V. Heraudet, and J. A. Shykoff. 2005. Parasite-host fitness trade-offs change with parasite identity: genotype-specific interactions in a plant-pathogen system. *Evolution* 59:2518–2524.

Samson, F., and F. Knopf. 1994. Prairie conservation in North America. *BioScience* 44:418–421.

Schardl, C. L. 1996. *Epichloë* species: fungal symbionts of grasses. *Annual Review of Phytopathology* 34:109–130.

Schardl, C. L. 2001. *Epichloë festucae* and related mutualistic symbionts of grasses. *Fungal Genetics and Biology* 33:69–82.

Schardl, C. L., and K. Clay. 1997. Evolution of mutualistic endophytes from plant pathogens. In *The Mycota. V. Plant Relationships. Part B*, eds. G. C. Carroll and P. Tudzynski, pp. 221–238. Berlin: Springer-Verlag.

Schardl, C. L., and K. D. Craven. 2003. Interspecific hybridization in plant-associated fungi and oomycetes: a review. *Molecular Ecology* 12:2861–2873.

Schardl, C. L., R. B. Grossman, P. Nagabhyru, J. R. Faulkner, and U. P. Mallik. 2007. Loline alkaloids: currencies of mutualism. *Phytochemistry* 68:980–996.

Schardl, C. L., and A. Leuchtmann. 2005. The *Epichloë* endophytes of grasses and the symbiotic continuum. In *The fungal community: its organization and role in the ecosystem*, 3rd ed., eds. J. Dighton, J. F. White, Jr., and P. Oudemans, pp. 475–503. Boca Raton, FL: CRC Press.

Schardl, C. L., A. Leuchtmann, K.-R. Chung, D. Penny, and M. R. Siegel. 1997. Coevolution by common descent of fungal symbionts (*Epichloë* spp.) and grass hosts. *Molecular Biology and Evolution* 14:133–143.

Schardl, C. L., A. Leuchtmann, and M. J. Spiering. 2004. Symbioses of grasses with seedborne fungal endophytes. *Annual Reviews in Plant Biology* 55: 315–340.

Schardl, C. L., A. Leuchtmann, H.-F. Tsai, M. A. Collett, D. M. Watt, and D. B. Scott. 1994. Origin of a fungal symbiont of perennial ryegrass by interspecific hybridization of a mutualist with the ryegrass choke pathogen, *Epichloë typhina*. *Genetics* 136:1301–1317.

Schardl, C. L., and C. D. Moon. 2003. Processes of species evolution in *Epichloë/Neotyphodium* endophytes of grasses. In *Clavicipitalean fungi: evolutionary biology, chemistry, biocontrol, and cultural impacts*, eds. J. F. White, Jr., C. W.

Bacon, N. L. Hywel-Jones, and J. W. Spatafora, pp. 273–310. New York: Marcel Dekker.

Schardl, C. L., and D. G. Panaccione. 2005. Biosynthesis of ergot and loline alkaloids. In *Neotyphodium in cool-season grasses*, eds. C. A. Roberts, C. P. West, and D. E. Spiers, pp. 75–92. Ames, IA: Blackwell Publications.

Schardl, C. L., D. G. Panaccione, and T. P. Tudzynski. 2006. Ergot alkaloids—biology and molecular biology. In *The alkaloids: chemistry and biology*, vol. 63, ed. G. A. Cordell, pp. 45–86. New York: Academic Press.

Schardl, C. L., and H. H. Wilkinson. 2000. Hybridization and cospeciation hypotheses for the evolution of grass endophytes. In *Microbial endophytes*, eds. C. W. Bacon and J. F. White, Jr., pp. 63–83. New York: Marcel Dekker.

Schlichting, C. D., and M. Pigliucci. 1998. *Phenotypic evolution: a reaction norm perspective*. Sunderland, MA: Sinauer Associates.

Schmidt, D. D., C. Voelkel, M. Hartl, S. Schmidt, and I. T. Baldwin. 2005. Specificity in ecological interactions. Attack from the same lepidopteran herbivore results in species-specific transcriptional responses in two solanaceous host plants. *Plant Physiology* 138:1763–1773.

Schoener, T. W. 1989. Food webs from the small to the large. *Ecology* 70:1559–1589.

Schulthess, F. M., and S. H. Faeth. 1998. Distribution, abundances, and associations of the endophytic fungal community of Arizona fescue (*Festuca arizonica*). *Mycologia* 90:569–578.

Schulz, B., and C. Boyle. 2005. The endophytic continuum. *Mycological Research* 109:661–686.

Scott, B. 2001. *Epichloë* endophytes: fungal symbionts of grasses. *Current Opinion in Microbiology* 4:393–398.

Selosse, M.-A., and F. Le Tacon. 1998. The land flora: a phototroph-fungus partnership? *Trends in Ecology and Evolution* 13: 15–20.

Selosse, M.-A., and C. L. Schardl. 2007. Fungal endophytes of grasses: hybrids rescued by vertical transmission? An evolutionary perspective. *New Phytologist* 173:452–458.

Shipley, B. 2000. *Cause and correlation in biology*. Cambridge: Cambridge University Press.

Siegel, M. R., and L. P Bush. 1996. Defensive chemicals in grass-fungal endophyte associations. *Recent Advances in Phytochemistry* 30:81–119.

Siegel, M. R., and L. P. Bush. 1997. Toxin production in grass/endophyte association. In *The Mycota. V. Plant relationships, Part A*, eds. G. C. Carroll and P. Tudzynski, pp. 185–208. Berlin: Springer-Verlag.

Siegel, M. R., M. C. Johnson, D. R. Varney, W. C. Nesmith, R. C. Buckner, L. P. Bush, P. B. Burrus, T. A. Jones, and J. A. Boling. 1984a. A fungal endophyte in tall fescue: incidence and dissemination. *Phytopathology* 74:932–937.

Siegel, M. R. and G. C. Latch. 1991. Expression of anti-fungal activity in agar culture by isolates of grass endophytes. *Mycologia* 83:529–537.

Siegel, M. R., G. C. M. Latch, L. P. Bush, F. F. Fannin, D. D. Rowan, B. A. Tapper, C. W. Bacon, and M. C. Johnson. 1990. Fungal endophyte-infected grasses: alkaloid accumulation and aphid response. *Journal of Chemical Ecology* 16:3301–3315.

Siegel, M. R., G. C. M. Latch, and M. C. Johnson. 1985. *Acremonium* fungal endophytes of tall fescue and perennial ryegrass: significance and control. *Plant Disease* 69:179–183.

Siegel, M. R., G. C. M. Latch, and M. C. Johnson. 1987. Fungal endophytes of grasses. *Annual Review of Phytopathology* 25:293–315.

Siegel, M. R., D. R. Varney, M. C. Johnson, W. C. Nesmith, R. C. Buckner, L. P. Bush, P. B. Burris II, and J. R. Hardison. 1984b. A fungal endophyte of tall fescue: evaluation of control methods. *Phytopathology* 74:937–941.

Silvertown, J., and Charlesworth, D. 2001. *Introduction to plant population biology*, 4th edition. Oxford: Blackwell Science.

Sleper, D. A., and C. P. West. 1996. Tall fescue. In *Cool-season forage grasses*, eds. L. E. Moser, D. R. Buxton, and M. D. Casler, pp. 471–502. Madison, WI: American Society of Agronomy, Crop Science Society of America, and Soil Science Society of America.

Solomon, E. P., L. R. Berg, and D. W. Martin. 2006. *Biology*, 7th ed. Belmont, CA: Brooks/Cole-Thomson Learning.

Spiering, M. J., D. H. Greer, and J. Schmid. 2006a. Effects of the fungal endophyte, *Neotyphodium lolii*, on net photosynthesis and growth rates of perennial ryegrass (*Lolium perenne*) are independent of *in planta* endophyte concentration. *Annals of Botany* 98:379–387.

Spiering, M. J., G. A. Lane, M. J. Christensen, and J. Schmid. 2005. Distribution of the fungal endophyte *Neotyphodium lolii* is not a major determinant of the distribution of fungal alkaloids in *Lolium perenne* plants. *Phytochemistry* 66:195–202.

Spiering, M. J., C. D. Moon, H. H. Wilkinson, and C. L. Schardl. 2006b. Gene clusters for insecticidal loline alkaloids in the grass-endophytic fungus *Neotyphodium uncinatum*. *Genetics* 169:1403–1414.

Spyreas, G., D. J. Gibson, and M. Basinger. 2001. Endophyte infection levels of native and naturalized fescues in Illinois and England. *Journal of the Torrey Botanical Society* 128:25–34.

Stachowicz, J. J. 2001. Mutualism, facilitation, and the structure of ecological communities. *Bioscience* 51:235–246.

Stearns, S. C. 1992. *The evolution of life histories*. Oxford: Oxford University Press.

Steiner, U., M. A. Ahimsa-Mueller, A. Markert, S. Kucht, J. Gross, N. Kauf, M. Kuzma, M. Zych, M. Lamshoeft, M. Furmanowa, V. Knoop, C. Drewke, and E. Leistner. 2006. Molecular characterization of a seed transmitted clavicipitaceous fungus occurring on dicotyledonous plants (Convolvulaceae). *Planta* 224:533–544.

Stevens, D. R., and M. J. Hickey. 1990. Effects of endophytic ryegrass on the production of ryegrass/white clover pastures. In *Proceedings of the International Symposium on Acremonium/Grass Interactions*, eds. S. S. Quisenberry and R. E. Joost, pp. 58–61. Baton Rouge: Louisiana Agricultural Experiment Station.

Stohlgren, T. J., D. T. Barnett, and J. T. Kartesz. 2003. The rich get richer: patterns of plant invasions in the United States. *Frontiers in Ecology and the Environment* 1:11–14.

Stone, J. K., C. W. Bacon, and J. F. White, Jr. 2000. An overview of endophytic microbes: endophytism defined. In *Microbial endophytes*, eds. C. W. Bacon and J. F. White, Jr., pp. 3–29. New York: Marcel Dekker.

Stovall, M. E., and K. Clay. 1988. The effect of the fungus, *Balansia cyperi* Edg., on growth and reproduction of purple nutsedge, *Cyperus rotundus* L. *New Phytologist* 109:351–359.

Stovall, M. E., and K. Clay. 1991. Fungitoxic effects of *Balansia cyperi* (Clavicipitaceae). *Mycologia* 83:288–295.

Strauss, S. Y., and A. A. Agrawal. 1999. The ecology and evolution of plant tolerance to herbivory. *Trends in Ecology and Evolution* 14:179–185.

Strauss, S. Y., and R. E. Irwin. 2004. Ecological and evolutionary consequences of multispecies plant-animal interactions. *Annual Review of Ecology, Evolution and Systematics* 35:435–466.

Strong, D. R., Jr., J. H. Lawton, and R. Southwood. 1984. *Insects on plants. Community patterns and mechanisms.* Oxford: Blackwell Science.

Sugawara, K., H. Ohkubo, M. Yamashita, and Y. Mikoshiba. 2004. Flowers for *Neotyphodium* endophytes detection: a new observation method using flowers of host grasses. *Mycoscience* 45:222–226.

Sullivan, T. J. and S. H. Faeth. 2004. Gene flow in the endophyte *Neotyphodium* and implications for coevolution with *Festuca arizonica. Molecular Ecology* 13:649–656.

Sullivan, T. J. and S. H. Faeth. 2008. Local adaptation in *Festuca arizonica* by hybrid and nonhybrid *Neotyphodium* endophytes. *Microbial Ecology* 55:697–704.

Sullivan, T. J., J. Rodstrom, J. Vandop, J. Librizzi, C. Graham, C. L. Schardl, and T. L. Bultman. 2007. Symbiont-mediated changes in *Lolium arundinaceum* inducible defenses: evidence from changes in gene expression and leaf composition. *New Phytologist* 176:673–679.

Sutherland, B. L., and J. H. Hoglund. 1990. Effect of ryegrass containing the endophyte *Acremonium lolii* on associated white clover. In *Proceedings of the International Symposium on Acremonium/Grass Interactions,* eds. S. S. Quisenberry and R. E. Joost, pp. 67–71. Baton Rouge: Louisiana Agricultural Experiment Station.

Sutherland, B. L., D. E. Hume, and B. A. Tapper. 1999. Allelopathic effects of endophyte-infected perennial ryegrass extracts on white clover seedlings. *New Zealand Journal of Agricultural Research* 42:19–26.

Tadych, M., and J. F. White, Jr. 2007. Ecology of epiphyllous stages of endophytes and implications for horizontal dissemination. In *Proceedings of the 6th International Symposium on Fungal Endophytes of Grasses,* eds. A. J. Popay and E. R. Thom, pp. 157–161. Christchurch, New Zealand: New Zealand Grassland Association.

Tahvanainen, J. O., and R. B. Root. 1972. Influence of vegetational diversity on population ecology of a specialized herbivore, *Phyllotreta cruciferae* (Coleoptera-Chrysomelidae). *Oecologia* 10:321–346.

Tan, Y. Y., M. J. Spiering, V. Scott, G. A. Lane, M. J. Christensen, and J. Schmid. 2001. In planta regulation of extension of an endophytic fungus and maintenance of high metabolic rates in its mycelium in the absence of apical extension. *Applied and Environmental Microbiology* 67: 5377–5383.

Tanaka, A., M. J. Christensen, D. Takemoto, P. Park, and B. Scott. 2006. Reactive oxygen species play a role in regulating a fungus-perennial ryegrass mutualistic interaction. *Plant Cell* 18:1052–1066.

Taylor, P. D., T. Day, D. Nagy, G. Wild, J.-B. Andre, and A. Gardner. 2006. The evolutionary consequences of plasticity in host-pathogen interactions. *Theoretical Population Biology* 69:323–331.

Thompson, F. N., and J. A. Stuedemann. 1993. Pathophysiology of fescue toxicosis. *Agriculture, Ecosystem and Environment* 44:263–281.

Thompson, J. N. 1994. *The coevolutionary process.* Chicago: University of Chicago Press.

Thompson, J. N. 1997. Evaluating the dynamics of coevolution among geographically structured populations. *Ecology* 78:1619–1623.

Thompson, J. N. 1999. The raw material for coevolution. *Oikos* 84:5–16.

Thompson, J. N. 2005. *The geographic mosaic of coevolution.* Chicago: University of Chicago Press.

Tibbets, T. M., and S. H. Faeth. 1999. *Neotyphodium* endophytes in grasses: deterrents or promoters of herbivory by leaf-cutting ants? *Oecologia* 118:297–305.

Tiffin, P., and M. D. Rauscher. 1999. Genetic constraints and selection acting on tolerance to herbivory in the common morning glory *Ipomoea purpurea. American Naturalist* 145:700–716.

Tilman, D., J. Knops, D. Wedin, P. Reich, M. Ritchie, and E. Siemann. 1997. The influence of functional diversity and composition on ecosystem processes. *Science* 277:1300–1302.

Timper, P., R. N. Gates, and J. H. Bouton. 2005. Response of *Pratylenchus* spp. in tall fescue infected with different strains of the fungal endophyte *Neotyphodium coenophialum. Nematology* 7:105–110.

Tintjer, T., A. Leuchtmann, and K. Clay. 2008. Variation in horizontal and vertical transmission of the endophyte *Epichloë elymi* infecting the grass *Elymus hystrix. New Phytologist* 179:236–246.

Tintjer, T., and J. A. Rudgers. 2006. Grass-herbivore interactions altered by strains of a native endophyte. *New Phytologist* 170:513–521.

Tomimatsu, H., and M. Ohara. 2006. Evolution of hierarchical floral resource allocation associated with mating system in an animal-pollinated hermaphroditic herb, *Trillium camschatcense* (Trilliaceae). *American Journal of Botany* 93:134–141.

Tracy, B. F., and I. J. Renne. 2005. Reinfestation of endophtye-infected tall fescue in renovated endophyte-free pastures under rotational stocking. *Journal of Agronomy* 97:1473–1477.

Tredway, L. P., J. F. White, Jr., B. S. Gaut, P. V. Reddy, M. D. Richardson, and B. B. Clarke. 1999. Phylogenetic relationships within and between *Epichloë* and *Neotyphodium* endophytes as estimated by AFLP markers and rDNA sequences. *Mycological Research* 103:1593–1603.

Trevathan, L. E. 1996. Performance of endophyte-free and endophyte infected tall fescue seedlings in soil infected with *Cochiobolus sativus. Canadian Journal of Plant Pathology* 18: 415–418.

Trumble, J. T., D. M. Kolodny-Hirsch, and I. P. Ting. 1993. Plant compensation for arthropod herbivory. *Annual Review of Entomology* 38:93–120.

Tsai, H.-F., J.-S. Liu, C. Staben, M. J. Christensen, G. C. M. Latch, M. R. Siegel, and C. L. Schardl. 1994. Evolutionary diversification of fungal endophytes of tall fescue grass by hybridization with *Epichloë* species. *Proceedings of the National Academy of Sciences USA* 91:2542–2546.

Turelli, M., D. W. Schemske, and P. Bierzychudek. 2001. Stable two-allele polymorphisms maintained by fluctuating fitnesses and seed banks: protecting the blues in *Linanthus parryae. Evolution* 55:1283–1298.

Turkington, R., and L. A. Mehrhoff. 1990. The role of competition in structuring pasture communities. In *Perspectives on plant competition*, eds. J. B. Grace and D. Tilman, pp. 307–340. San Diego: Academic Press.

Unterseher, M., A. Reiher, K. Finstermeier, P. Otto, and W. Morawetz. 2007. Species richness and distribution patterns of leaf-inhabiting endophytic fungi in a temperate forest canopy. *Mycological Progress* 6:201–212.

Valdez Barillas, J. R., M. W. Paschke, M. H. Ralphs, and R. D. Child. 2007. White locoweed toxicity is facilitated by a fungal endophyte and nitrogen-fixing bacteria. *Ecology* 88:1850–1856.

Vanderkoornhuyse, P., K. P. Ridgway, I. J. Watson, A. H. Fitter, and J. P. W. Young. 2003. Co-existing grass species have distinctive arbuscular mycorrhizal communities. *Molecular Ecology* 12:3085–3095.

van Heeswijck, R., and G. McDonald. 1992. *Acremonium* endophytes in perennial ryegrass and other pasture grasses in Australia and New Zealand. *Australian Journal of Agricultural Research* 43:1683–1709.

van Horn, R., and K. Clay. 1995. Mitochondrial DNA variation in the fungus *Atkinsonella hypoxylon* infecting sympatric *Danthonia* grasses. *Evolution* 49:360–371.

van Valen, L. 1973. A new evolutionary law. *Evolutionary Theory* 1:1–30.

van Zijll de Jong, E., K. M. Guthridge, G. C. Spangenberg, and J. W. Foster. 2003. Development and characterization of EST-derived simple sequence repeat (SSR) markers for pasture grass endophytes. *Genome* 46:277–290.

van Zijll de Jong, E., K. F. Smith, G. C. Spangenberg, and J. W. Foster. 2005. Molecular genetic marker-based analysis of the grass-endophyte symbiosis. In *Neotyphodium in cool-season grasses*, eds. C. A. Roberts, C. P. West, and D. E. Spiers, pp. 123–140. Ames, IA: Blackwell Publishing.

Vicari, M., and D. R. Bazely. 1993. Do grasses fight back? The case for anti-herbivore defenses. *Trends in Ecology and Evolution* 8:137–141.

Vicari, M., P. E. Hatcher, and P. G. Ayres. 2002. Combined effect of foliar and mycorrhizal endophytes on an insect herbivore. *Ecology* 83:2452–2464.

Vila-Aiub, M. M., P. E. Gundel, and C. M. Ghersa. 2005. Fungal endophyte infection changes growth attributes in *Lolium multiflorum* Lam. *Austral Ecology* 30:49–57.

Vinton, M. A., E. S. Kathol, K. P. Vogel, and A. A. Hopkins. 2001. Endophytic fungi in Canada wild rye in natural grasslands. *Journal of Range Management* 54:390–395.

Vujanovic, V., and J. Brisson. 2002. A comparative study of endophytic mycobiota in leaves of *Acer saccharum* in eastern North America. *Mycological Progress* 1:147–154.

Wagenius, S. 2006. Scale dependence of reproductive failure in fragmented *Echinacea* populations. *Ecology* 87:931–941.

Wäli, P. R., J. U. Ahlholm, M. Helander, and K. Saikkonen. 2007. Occurrence and genetic structure of the systemic grass endophyte *Epichloë festucae* in fine fescue populations. *Microbial Ecology* 53:20–29.

Wäli, P. R., M. Helander, O. Nissinen, and K. Saikkonen. 2006. Susceptibility of endophyte-infected grasses to winter pathogens (snow molds). *Canadian Journal of Botany* 84:1043–1051.

Warren, J. M., A. F. Raybould, T. Ball, A. J. Gray, and M. D. Hayward. 1998. Genetic structure in the perennial grasses *Lolium perenne* and *Agrostis curtisii*. *Heredity* 81:556–562.

Watson, R. N., R. A. Prestidge, and O. J.-P. Ball. 1993. Suppression of white clover by ryegrass infected with *Acremonium* endophyte. In *Proceedings of the Second International Symposium on Acremonium/Grass Interactions*, eds. D. E. Hume, G. C. M. Latch, and H. S. Easton, Palmerston North, New Zealand, pp. 218–221.

Wei, Y. K., Y. B. Gao, H. Xu, D. Su, X. Zhang, Y. H. Wang, F. Lin, L. Chen, L. Y. Nie, and A. Z. Ren. 2006. Occurrence of endophytes in grasses native to northern China. *Grass and Forage Science* 61:422–429.

Welty, R. E., and M. D. Azevedo. 1985. Survival of endophyte hyphae in seeds of tall fescue stored one year. *Phytopathology* 77:893–900.

Welty, R. E., M. D. Azevedo, and K. L. Cook. 1986a. Detecting viable *Acremonium* endophytes in leaf sheaths and meristems of tall fescue and perennial ryegrass. *Plant Disease* 70:431–435.

Welty, R. E., M. D. Azevedo, and T. M. Cooper. 1987. Influence of moisture content, temperature, and length of storage on seed germination and survival of endophytic fungi in seeds of tall fescue and perennial ryegrass. *Phytopathology* 77: 893–900.

Welty, R. E., G. M. Milbrath, D. Faulkenberry, M. D. Azevedo, L. Meek, and K. Hall. 1986b. Endophyte detection in tall fescue seed by staining and ELISA. *Seed Science and Technology* 14:105–116.

Wennström, A. 1996. The distribution of *Epichloë typhina* in natural plant populations of the host plant *Calamagrostis purpurea. Ecography* 19:377–381.

Werren, J. H. 1997. Biology of *Wolbachia. Annual Review of Entomology* 423:587–609.

Werren, J. H. and S. L. O'Neill. 1997. The evolution of heritable symbionts. In *Influential passengers: inherited microorganisms and arthropod reproduction*, eds. S. L. O'Neill, J. H. Werren, and A. A. Hoffmann, pp. 1–41. Oxford: Oxford University Press.

West, C. P. 1994. Physiology and drought tolerance of endophyte-infected grasses. In *Biotechnology of endophytic fungi of grasses*, eds. C. W. Bacon and J. F. White, Jr., pp. 87–99. Boca Raton, FL: CRC Press.

West, C. P. 2007. Plant influence on endophyte expression. In *Proceedings of the 6th International Symposium on Fungal Endophytes of Grasses*, eds. A. J. Popay and E. R. Thom, pp. 117–121. Christchurch, New Zealand: New Zealand Grassland Association.

West, C. P., H. W. Elberson, A. A. Elmi, and G. W. Buck. 1995. *Acremonium* effects on tall fescue growth: parasite or stimulant? In *Proceedings of the 50th Southern Pasture and Forage Crop Improvement Conference*, pp. 102–111.

West, C. P., E. Izekor, K. E. Turner, and A. A. Elmi. 1993. Endophyte effects on growth and persistence of tall fescue along a water-supply gradient. *Agronomy Journal* 85:264–270.

West, C. P., F. Volaire, and F. Lelievre. 2007. Tiller survival after drought of 'Grasslands Flecha' tall fescue as influenced by endophyte. In *Proceedings of the 6th International Symposium on Fungal Endophytes of Grasses*, eds. A. J. Popay and E. R. Thom, pp. 267–269. Christchurch, New Zealand: New Zealand Grassland Association.

West-Eberhard, M. J. 2003. *Developmental plasticity and evolution*. Oxford: Oxford University Press.

Westoby, M. 1978. What are the biological bases of varied diets? *American Naturalist* 112:627–631.

White, J. F., Jr. 1987. Widespread distribution of endophytes in the Poaceae. *Plant Disease* 71:340–342.

White, J. F., Jr. 1988. Endophyte-host associations in forage grasses. XI. A proposal concerning origin and evolution. *Mycologia* 80:442–446.

White, J. F., Jr. 1997. Systematics of the graminicolous Clavicipitaceae: applications of morphological and molecular approaches. In *Neotyphodium/grass interactions*, eds. C. W. Bacon and N. S. Hill, pp. 27–39. New York: Plenum Press.

White, J. F., Jr., and G. T. Cole. 1985. Endophyte associations in forage grasses. III. In vitro inhibition by *Acremonium coenophialum. Mycologia* 77:487–489.

White, J. F., Jr., T. I. Martin, and D. Cabral. 1996. Endophyte-host associations in grasses. 22. Conidia formation by *Acremonium* endophytes on the phylloplanes of *Agrostis hiemalis* and *Poa rigidifolia. Mycologia* 88:174–178.

White, J. F., Jr., G. Morgan-Jones, and A. C. Morrow. 1993. Taxonomy, life cycle, reproduction and detection of *Acremonium* endophytes. *Agriculture, Ecosystems and Environment* 44:13–37.

White, J. F., Jr., R. F. Sullivan, G. A. Balady, T. J. Gianfagna, Q. Yue, W. A. Meyer, and D. Cabral. 2001a. A fungal endosymbiont of the grass *Bromus setifolius:* distribution in some Andean populations, identification, and examination of beneficial properties. *Symbiosis* 31:241–257.

White, J. F. Jr., R. Sullivan, and M. Moy. 2001b. An overview of the biology and systematics of *Neotyphodium* endophytes. In *Proceedings of the 4th International Neotyphodium/Grass Interactions Symposium*, Fachbereich Agrarwirtschaft, eds. P. D. Dapprich and V. H. Paul, Soest, Germany, pp. 17–30.

White, R. H., M. C. Engelke, S. J. Morton, J. M. Johnson-Cicalese, and B. A. Ruemmele. 1992. *Acremonium* endophyte effects on tall fescue drought tolerance. *Crop Science* 32:1392–1396.

Whitham, T. G., W. P. Young, G. D. Martinsen, C. A. Gehring, J. A. Schweitzer, S. M. Shuster, G. M. Wimp, D. G. Fischer, J. K. Bailey, R. L. Lindroth, S. Woolbright, and C. R. Kuske. 2003. Community and ecosystem genetics: a consequence of the extended phenotype. *Ecology* 84:559–573.

Wiggins, N. L., C. McArthur, N. D. Davies, and S. McLean. 2006. Spatial scales of the patchiness of plant poisons: a critical influence on foraging efficiency. *Ecology* 87:2236–2243.

Wilkinson, H. H., and C. L. Schardl. 1997. The evolution of mutualism in grass-endophyte associations. In *Neotyphodium/grass interactions*, eds. C. W. Bacon and N. S. Hill, pp. 13–25. New York: Plenum Press.

Wilkinson, H. H., M. R. Siegel, J. D. Blankenship, A. C. Mallory, L. P. Bush, and C. L. Schardl. 2000. Contribution of fungal loline alkaloids to protection from aphids in an endophyte-grass mutualism. *Molecular Plant-Microbe Interactions* 13:1027–1033.

Wille, P. A., R. A. Aeschbacher, and T. Boller. 1999. Distribution of fungal endophyte genotypes in doubly infected host grasses. *Plant Journal* 18:349–358.

Williams, M. J., P. A. Backman, E. M. Clark, and J. F. White. 1984. Seed treatments for control of the tall fescue endophyte *Acremonium coenophialum*. *Plant Disease* 68:49–52.

Wilson, D. 1993. Fungal endophytes: out of sight but should not be out of mind. *Oikos* 68:379–384.

Wilson, D. 1995. Endophyte—the evolution of a term, and clarification of its use and definition. *Oikos* 73:274–276.

Wilson, A. D. 1996. Resources and testing of endophyte-infected germplasm in national grass repository collections. In *Endophytic fungi in grasses and woody plants*, eds. S. C. Redlin and L. M. Carris, pp. 179–195. St. Paul, MN: American Pathological Society Press.

Wilson, A. D., S. L. Clement, and W. J. Kaiser. 1991a. Endophytic fungi in a *Hordeum* germplasm collection. *Plant Genetics Resources Newsletter* 87:1–4.

Wilson, A. D., S. L. Clement, and W. J. Kaiser. 1991b. Survey and detection of endophytic fungi in *Lolium* germplasm by direct staining and aphid assays. *Plant Disease* 75:169–173.

Wise, M. J., and W. G. Abrahamson. 2007. Effects of resource availability on tolerance of herbivory: a review and assessment of three opposing models. *American Naturalist* 169:443–454.

Wiyakrutta, S., N. Sriubolmas, W. Panphut, N. Thongon, K. Danwisetkanjana, N. Ruangrungsi, and V. Meevootisom. 2004. Endophytic fungi with antimicrobial, anti-cancer and anti-malarial activities isolated from Thai medicinal plants. *World Journal of Microbiology and Biotechnology* 20:265–272.

Wolin, C. L. 1985. The population dynamics of mutualistic systems. In *The biology of mutualism*, eds. D. H. Boucher, pp. 248–269. Oxford: Oxford University Press.

Wolock-Madej, C. W., and K. Clay. 1991. Avian seed preference and weight loss experiments: the effect of fungal endophyte-infected seeds. *Oecologia* 88:296–302.

Woolhouse, M. E. J., D. T. Haydon, and R. Antia. 2005. Emerging pathogens: the epidemiology and evolution of species jumps. *Trends in Ecology and Evolution* 20:238–244.

Yamamura, N. 1993. Vertical transmission and evolution of mutualism from parasitism. *Theoretical Population Biology* 44:95–109.

Yamamura, N., M. Higashi, N. Behera, and J. Y. Wakano. 2004. Evolution of mutualism through spatial effects. *Journal of Theoretical Biology* 226:421–428.

Yates, S. G., J. C. Fenster, and R. J. Bartlett. 1989. Assay of tall fescue seeds extracts, fractions, and alkaloids using the large milkweed bug. *Journal of Agriculture and Food Chemistry* 37: 354–357.

Young, C. A., M. K. Bryant, M. J. Christensen, B. A. Tapper, G. T. Bryan, and B. Scott. 2005. Molecular cloning and genetic analysis of a symbiosis-expressed gene cluster for lolitrem biosynthesis from a mutualistic endophyte for perennial ryegrass. *Molecular Genetics and Genomics* 274:13–29.

Yue, Q., J. Johnson-Cicalese, T. J. Gianfagna, and W. A. Meyer. 2000a. Alkaloid production and chinch bug resistance in endophyte-inoculated Chewings and strong creeping red fescues. *Journal of Chemical Ecology* 26:279–292.

Yue, Q., C. J. Miller, J. F. White, Jr., and M. D. Richardson. 2000b. Isolation and characterization of fungal inhibitors from *Epichloë festucae*. *Journal of Agriculture and Food Chemistry* 48:4687–4692.

Zabalgogeazcoa, I., E. P. Benito, A. P. Eslava, A. G. Ciudad, B. R. Vázquez de Aldana, and B. García Criado. 2001. Viruses in endophytes infecting natural populations of *Festuca rubra*. In *Fourth International Neotyphodium/Grass Interactions Symposium*, Fachbereich, Agrarwirtschaft, Soest, Germany, eds. V. H. Paul and P. D. Dapprich, pp. 109–112.

Zabalgogeazcoa, I., A. García Cuidad, B. R. Vázquez de Aldana, and B. García Criado. 2006a. Effects of the infection by the fungal endophyte *Epichloë festucae* in the growth and nutrient content of *Festuca rubra*. *European Journal of Agronomy* 24:374–384.

Zabalgogeazcoa, I., M. Romo, E. Keck, B. R. Vázquez de Aldana, A. García Ciudad, and B. García Criado. 2006b. The infection of *Festuca rubra* subsp. *pruinosa* by *Epichloë festucae*. *Grass and Forage Science* 61:71–76.

Zabalgogeazcoa, I., B. R. Vázquez de Aldana, A. Garcia Ciudad, and B. Garcia Criado. 1999. The infection of *Festuca rubra* by the fungal endophyte *Epichloe festucae* in Mediterranean permanent grasslands. *Grass and Forage Science* 54:91–95.

Zabalgogeazcoa, I., B. R. Vázquez de Aldana, A. García Cuidad, and B. García Criado. 2003. Fungal endophytes in grasses from semi-arid permanent grasslands of western Spain. *Grass and Forage Science* 58:94–97.

Zhang, N. X., V. Scott, T. H. Al-Samarrai, Y. Y. Tan, M. J. Spiering, L. K. McMillan, G. A. Lane, D. B. Scott, M. J. Christensen, and J. Schmid. 2006. Transformation of the ryegrass endophyte *Neotyphodium lolii* can alter its in planta mycelial morphology. *Mycological Research* 110:601–611.

Index